Globalisation and Fisheries

Proceedings of an OECD-FAO Workshop

OECD

FAO FIAT PANIS

ORGANISATION FOR ECONOMIC CO-OPERATION AND DEVELOPMENT

The OECD is a unique forum where the governments of 30 democracies work together to address the economic, social and environmental challenges of globalisation. The OECD is also at the forefront of efforts to understand and to help governments respond to new developments and concerns, such as corporate governance, the information economy and the challenges of an ageing population. The Organisation provides a setting where governments can compare policy experiences, seek answers to common problems, identify good practice and work to co-ordinate domestic and international policies.

The OECD member countries are: Australia, Austria, Belgium, Canada, the Czech Republic, Denmark, Finland, France, Germany, Greece, Hungary, Iceland, Ireland, Italy, Japan, Korea, Luxembourg, Mexico, the Netherlands, New Zealand, Norway, Poland, Portugal, the Slovak Republic, Spain, Sweden, Switzerland, Turkey, the United Kingdom and the United States. The Commission of the European Communities takes part in the work of the OECD.

OECD Publishing disseminates widely the results of the Organisation's statistics gathering and research on economic, social and environmental issues, as well as the conventions, guidelines and standards agreed by its members.

Photo Credits: © FAO, Micaud; Indonesia , © FAO, Dioses; Peru , © FAO, Faidutti; Thailand

Corrigenda to OECD publications may be found on line at: *www.oecd.org/publishing/corrigenda*.

Foreword

The fisheries sector depends critically on a natural renewable resource base that must be managed in a sustainable and responsible way. At the same time, the industry is operating in an increasingly globalised world with markets for goods, capital, labour and technology growing more interdependent. The process of globalisation is creating a number of opportunities and risks for the sector. It is important to understand how governments, international organisations, industry and civil society are meeting these challenges. It is also essential to ensure that government policy makers create an enabling environment for the sector while successfully managing risks, so that various elements of the fisheries value chain can reap the benefits from globalisation without endangering the resource base.

The Workshop

Convened jointly by the OECD's Committee for Fisheries and the United Nations FAO Fisheries and Aquaculture Department, the Workshop benefited from the financial support of Canada, Finland, Iceland, Korea, Norway, Chinese Taipei and the World Bank.

The Workshop was chaired by Lori Ridgeway, Director-General, Department of International Policy and Integration, DFO Canada, and was structured to ensure that a dialogue could take place between fisheries policy makers and representatives from business communities, civil society organisations and academia. The field of attendance included Ministers, keynote speakers, and experts in the fisheries field from both developed and developing countries. The Workshop also provided a forum to explore new business opportunities across a range of countries.

The two day Workshop consisted of a series of sessions related to: the harvesting and aquaculture sectors; fish processing; the retailing and distribution of fish; and linkages across the value chain. A final round table panel session summarised the issues. The Workshop benefited from the perspectives of major international businesses involved in the fishing industry, donor institutions and representatives from a number of OECD and non-OECD countries who were present at the Workshop.

Acknowledgements

The OECD Fisheries Policies Division and the FAO Fisheries and Aquaculture Department express their appreciation to all participants for contributing to the success of the Workshop on Opportunities and Challenges of Fisheries Globalisation. The Workshop and this publication were made possible through voluntary funding from Canada, Finland, Iceland, Korea, Norway, Chinese Taipei and the World Bank, which is gratefully acknowledged.

The workshop was jointly organised by the OECD Fisheries Policies Division and the FAO Fisheries and Aquaculture Department. The publication has been prepared and edited by Simon Gregg (OECD), Emily Andrews-Chouicha (OECD), Yvonne Davidsson (FAO), William Emerson (FAO), Lahsen Ababouch (FAO), Ingrid Kelling (OECD) and Carl-Christian Schmidt (OECD).

Table of Contents

List of Acronyms

Acronym	Definition
ACIAR	Australian Centre for International Agricultural Research
ACP	African, Caribbean and Pacific States
ADB	Asian Development Bank
APEC	Asia-Pacific Economic Cooperation
ASEAN	Association of South East Asian Nations
BAAC	Bank for Agriculture and Agricultural Cooperatives, Indonesia
BCM	Central Bank of Mauritania
BKK	Badan Kredit Kecamatan
BMP	Better Management Practices
BRI	Bank Rakyat Indonesia
BRI-UD	Unit Desa (BRI-UD) or Village Bank
CCSBT	Commission for the Conservation of Southern Bluefin Tuna
CECAF	Fishery Committee for the Eastern Central Atlantic
CCRD	Credit Committee for Rural Development
CFP	Common Fisheries Policy
CIB	Consolidated Investment Budget
CNROP	National Centre for Ocean Research and Fisheries
CTH	Change in Tariff Heading
DC	Developing country
DFID	United Kingdom – Department for International Development
DSPCM	Delegation for Fisheries Surveillance and Inspections at Sea
DWF	Distant Waters Fishing Fleet
DWFN	Distant Waters Fishing Nations
EBA	Everything but Arms Initiative
EC	European Community
ECOWAS	Economic Community of West African States
EDF	European Development Fund
EEC	European Economic Community
EEZ	Exclusive Economic Zone
ENEMP	National Maritime Training College
EPA	Economic Partnership Agreement
EU	European Union
FAO	Food and Agriculture Organization of the United Nations
GATT	General Agreement on Tariffs and Trade
GCFM	General Fisheries Council for the Mediterranean
GDP	Gross domestic product
GRT	Gross registered tonnage
GSP	Generalised system of preferences
GT	Gross tonnage
HACCP	Hazard analysis critical control point
HIPC	Heavily indebted poor countries

IATTC	Inter American Tropical Tuna Commission
ICCAT	International Commission for the Conservation of Atlantic Tuna
ICNAF	International Commission for the Northwest Atlantic Fisheries
IDRC	International Development Research Centre, Canada
IMF	International Monetary Fund
IMROP	Mauritanian Institute for Ocean Research and Fisheries
IPOA	FAO International Plans of Action
IRM	Islamic Republic of Mauritania
ITQ	Individual transferable quota
IWC	International Whaling Commission
LDC	Least developed country
MAED	Ministry of Economic Affairs and Fisheries
MDG	Millennium Development Goals
MEEZ	Mauritanian Exclusive Economic Zone
MFN	Most favoured nation
MFP	Mauritanian Fisheries Policies
MPEDA	Marine Products Export Development Authority, India
MPEM	Ministry of Fisheries and the Maritime Economy
MSMEs	Micro, Small and Medium Enterprises
MSY	Maximum sustainable yield
NABARD	National Bank for Agriculture and Rural Development
NACA	Network for Aquaculture Centres in Asia Pacific
NACSA	National Centre for Sustainable Aquaculture
NASCO	North Atlantic Salmon Conservation Organisation
NBC	National Bank of Cambodia
NFF	National Federation of Fisheries
NGO	Non-governmental organisation
NSO	National Statistics Office
OECD	Organisation for Economic Co-operation and Development
PNBA	Banc d'Arguin National Park
PND	Diawling National Park
PRSP	Poverty Reduction Strategy Paper
QUEDANCOR	Quedan and Rural Credit Corporation
RDB	Rural Development Bank
RD&E	Research, Development, and Extension
RFMO	Regional Fisheries Management Organisation
RoO	Rules of Origin
SCM	Subsidies and Countervailing Measures Agreement of the WTO
SEAFO	South-East Atlantic Fisheries Organisation
SHG	Self-Help Groups (Self Help Group-Bank Linkage Program)
SIFFS	South Indian Federation of Fishermen Societies
SIOFA	South Indian Ocean Fisheries Organisation
SIPPO	Swiss Import Promotion Programme
SMCP	Mauritanian Fishery Product Marketing Company
SPS	WTO Agreement on Sanitary and Phytosanitary Measures
TAC	Total allowable catch
TBT	WTO Agreement on Technical Barriers to Trade
UICN	World Conservation Union
UNCLOS	United Nations Convention on the Law of the Sea
UNFSA	United Nations Fish Stocks Agreement

VA	Value-added
VASEP	National Fish & Seafood Association
WB	World Bank
WCPFC	Western and Central Pacific Ocean Fisheries Commission
WTO	World Trade Organization

Chair's Summary Report

Lori Ridgeway, Chair of the Workshop
Director-General, Department of International Policy and Integration, DFO Canada

Introduction

"Globalisation" is a key context underlying most current policy discussions in fisheries including both capture and aquaculture fisheries. Globalisation is a concept that implies a "system" of complex linkages among international participants, including states -- a system that needs to work effectively and responsibly. The challenges, opportunities and expectations accompanying increased globalisation are at the root of some of the most heated policy discussions taking place in fisheries.

Opinions differ on whether globalisation is a positive or negative factor in the world economy and community, where these opinions vary according to whether one is facing perceived opportunities or risks from it. The overall goal for the global community should be to understand the policy, governance and other changes needed so that the benefits of globalisation can be maximized, and risks minimised or managed.

Against this background, the OECD's Committee for Fisheries, as part of its programme of work, is undertaking a large project on globalisation and fisheries[1], to better understand what is needed to make the global "system" work better, and ensure a wider sharing of its benefits.

As part of this work programme, the Committee for Fisheries and its Secretariat, and the FAO Fisheries and Aquaculture Department jointly hosted a Workshop on the Challenges and Opportunities of Fisheries Globalisation, which took place in Paris on 16-17 April 2007. The Workshop brought together 130 policy makers and industry representatives (*e.g.*, producers/harvesters, processors, buyers, retailers), from both developed and developing countries to share views on the topic of globalisation and fisheries

The principal aims of the Workshop were to inform policy makers of participating countries to better understand the developments and challenges of globalisation in the fisheries sector from the range of perspectives along the vale chain (see below), and to understand, from various points of view, the nature of policy issues raised and possible solutions. The challenge for policy makers is to understand, gauge and provide appropriate policy responses to the risks inherent in globalisation in fisheries

1. "Globalisation" is also a broader cross-cutting topic at the OECD, intended to be addressed by a number of its policy committees.

(*e.g.* additional pressures of global demand on the wild fish stocks which in many cases are already fully- or over-exploited; reliance on growing aquaculture production and its ability to absorb global demand; issues in meeting high sanitary and phytosanitary standards in relation to seafood consumption in developed markets; overcoming continuing barriers to trade in fish and fish products, services and investment; the increasing role of diverse players (such as consumers) in setting "standards" for production based on their tolerance for health, safety, quality and ethical risks in fisheries; satisfying the need for low commercial risks on the part of business in an inherently risky industry; the need to share value added among existing and new participants in fisheries along the value chain) while, at the same time, ensure that states and their industries/communities can respond and benefit from the emerging and changing opportunities of a global seafood market.

As globalisation can give rise to a complex set of relationships, a simplifying paradigm has been adopted by the Committee to ensure a balanced and comprehensive approach to the issue. The "value chain", figuratively and simplistically[2] illustrated in the figure below shows the stages at which "value" is added to a product as it is transformed to meet a market need. Value-added increases as one moves from left to right in this chain. The fishing value chain starts on the left, with both harvesting and aquaculture "production", and together these products move through to processing. The product is then consumed locally, or distributed or traded into markets (sometimes back into aquaculture production as feed), while some products go to retailers and then to consumers. It is illustrative of a framework that can be used to examine the impacts of globalisation in a complete way.

→ **Value** →

Harvesting
↕
Production

→ **Processing** → **Distribution and trade** → **Retailers** → **Consumers**

The challenge is to understand the impacts of globalisation at each level of activity, understand the linkages up and down the chain from producers to consumers and vice versa; and understand the policy, institutional practices and governance needs at each stage in the value chain in order to maximize benefits and minimize risks of globalisation.

The agenda of the Workshop was organised with such a framework in mind, in order to best organise information and feedback from each community of interest, with a series of sessions that addressed various aspects and issues of globalisation in each element of the chain.

The following describes: *1)* key messages that flow from the discussion at the Workshop; *2)* a detailed account of the discussions on each element of the value chain that took place at the Workshop; and *3)* an overview of the presentations by ministers and invited experts.

2. The illustration shown in this diagram is very simplified. For instance it does not include the tertiary sector -- or related services -- spawned by higher-value activities at points throughout the chain.

Key Messages from the Workshop

Regarding the **benefits of globalisation**, while the workshop did not shy away from issues and challenges, it was markedly positive about globalisation, in general. Globalisation was viewed as also *encouraging* and *enabling* responsible production, as the global community finds ways to work together to link demand and supply and overcome barriers of a fragmented global system of production and consumption. Globalisation will provide opportunities for a wide set of players, providing that certain conditions -- most notably responsibility and sustainability -- are met. It will be the responsibility of all -- including both producing states and markets[3] -- to ensure that benefits accrue in a sustainable, resilient and inclusive global system that generates higher returns for all.

In a world where resources and economic returns are both being squeezed, **risk management** is increasingly characterizing each stage of the value chain, whether that be sustainability, health and safety, commercial and shareholder risk, or reputational risk (for retailers in developed states). Along the value chain, the accompanying increasing **relative concentration**, and hence power, within, and of, downstream components in the value chain has changed as globalisation has progressed. Not only do highest margins accrue as value added increases, reducing returns for producers and harvesters of commodities in developed and developing states alike, the potential mismatch between the increasing role in globalised fisheries for small-scale producers, in particular in developing countries, and increasing buyer concentration in developed countries creates particularly important policy challenges. This concerns both wild fisheries and aquaculture and is likely to be a growing problems as new sources for raw material are increasingly being produced in developing countries.

Sustainable and responsible fisheries management and production (aquaculture) is a *sine qua non* for all value chain elements in the fisheries sector.

- *Production*: not to undermine future production and market potential (resource and environmental sustainability, reputation);
- *Processors*: need security of supply, quality;
- *Investors and financial intermediation*: un-sustainability and poor practices are "bad bets" financially;
- *Exporters*: face standards (health and safety sustainability, labelling requirements);
- *Retailers/food service industry*: to protect their market "brand" and ensure sustainable high quality supply sources that meet market needs;
- *Consumers*: a sustainability "ethic" is increasing, as are concerns over food safety.

Pressure has been growing on policy makers and managers over the past decade from various stakeholder communities to ensure that fisheries are sustainable and well-managed (including the environmental impacts of their activities), and this pressure is expected to continue, if not increase.

3. By not providing a market haven for irresponsible product and by not setting unfair standards that interfere unduly with the trading system

For the **harvesting sector**, in an increasingly interdependent world, policies and norms should be adopted globally, to avoid a "race to the bottom". In a globalised world, in the absence of appropriate fisheries management policies and implementation, fishing effort will move to fisheries that are characterised by open access conditions, at a time when standards demanded by society increasingly value sustainable practices. It is also important that fishing activities are carried out within a framework that can generate flexibility and economic efficiency, while safeguarding social and cultural values that underpins community resilience. Concurrently, capture fisheries will face increased competition from aquaculture -- and both of them from alternative food sources -- which may put pressure on harvesting/production practices and on fishers' living standards. Hence, for policy makers, the key concern is to ensure that the fisheries management system is aligned with prevailing resource and market realities, so that the fishing sector is attractive for fishers, investors and consumers/society.

Aquaculture has a promising future in both developing and developed countries. In developing countries aquaculture provides particular opportunities in relation to capture fisheries, but is often characterised by small-scale family holdings that supply the domestic market and produce fish and crustaceans for exports. Meeting the diverse needs of small-scale producers is a priority for developing countries (*e.g.,* finance, capacity-building and technology transfer). Meeting the diverse requirements of the export market is often a challenge for small-scale producers. Collective action, including through cooperatives and "clustering", can however enhance their capacity to participate in the global value chain. For aquaculture in developed countries, the key challenge for policy makers is to ensure that aquaculture can fulfil its potential, given relatively higher cost structures, competition for oceans space, and in some cases environmental reputational issues. Hence, to keep the momentum to meet future global demand (and protein needs) and avoid the negative environmental effects associated with aquaculture, policy makers in all states need to articulate and develop aquaculture strategies and action plans to ensure sustainable production, market acceptance and the ability for products to enter into trade.

Hygiene and sanitary standards in the fisheries sector irrespective of various perceptions of "fairness" were outlined as generally "non-negotiable" in developed markets, where tolerance for health and safety risk is low and likely falling. Despite challenges for developing states, the workshop (which included many developing country representatives) took a positive view overall of the merit of health and safety standards, given the realities of demand in major markets that place a high premium on safety and security of food products generally. It was noted that reforms being undertaken to meet standards and associated capacity building, are enabling products to enter more markets. There are interesting examples from Africa and Asia about what can and is being done to enable products to meet rigorous standards in developed country markets. Rather, the issues raised were more in their application, stability, predictability and transparency, not in their existence *per se*.

An issue for private sanitary and hygiene standards as well as eco-labels (see below) is the extent to which they potentially create barriers to trade that may be incompatible with WTO principles (as articulated in the Agreement on the Application of Sanitary and Phytosanitary (SPS) Measures and the Agreement on Technical Barriers to Trade (TBT)), recognizing that private standards are not subject to WTO disciplines. In this regard a major challenge for policy makers is to how to deal with private standards that may be trade distorting but which remain outside the purview of the WTO.

Harmonisation and equivalency of public standards is also a challenge that needs to be addressed. For developing countries, in particular, technical assistance is necessary to ensure that these countries can meet requirements set by developed markets and still remains reliable suppliers. In particular, small-scale producers face the risk of being excluded from the global value chain if they cannot comply with hygiene and sanitary standards. Providing technical assistance should reflect enlightened self-interest by OECD countries that increasingly rely on developing countries to supply their markets.

Retailers are also developing their own sanitary and hygiene standards, *inter alia* through buying specifications that help to protect the value of their "brand". Voluntary private standards (including the demand for certain eco-labels), on top of safety/quality standards, are becoming a requirement for doing business in many developed markets. This has raised concerns about market access for developing country producers, as private standards are often costly to comply with. While there is recognition that minimum standards are necessary to ensure food safety and quality, private standards can be higher than those set by governments. Hence an emerging policy issue is the impact on the trading system of a system of fragmented public and private standards, where the international dispute recourse only relates to government standards.

Demands for eco-labels and sustainability "campaigns", which aim to ensure that fish and fish products are sourced from sustainably managed fisheries by informing consumers of the sustainability of the fisheries from which products are sourced, have increased in recent years[4]. The proliferation of labelling schemes and campaigns with potentially incompatible criteria and requirements however, pose a challenge for producers and policy-makers alike, compounded by other standards and labelling requirements[5]. For producers who are seeking certification for their fisheries, this may entail substantial costs. For policy makers the key challenge is to ensure collectively that practical minimum criteria for both capture fisheries and aquaculture are in place, mechanisms are in place to ensure they can be met even by relatively information-poor but sustainable fisheries, and that action is taken as far as possible to ensure consumers are not bombarded with misleading information about what they are purchasing. The most important action governments can take is to ensure that fisheries are well managed, so that if markets demand it, then the relevant fisheries can meet certification requirements.

Retailers and other operators are starting to invest in partnerships along the value chain to ensure that their sourcing requirements are adequately met. However, globalisation is creating an increasingly complex environment that involves a larger number of interactions. For example, as OECD countries outsource processing activities, the supply chains become increasingly complex, reflecting the need for more sophisticated logistics and traceability schemes[6].

Many **developing countries** are expressing a feeling of frustration at not being able to fully reap the benefits of the globalisation process in terms of access to value-added

4. The Marine Stewardship Council (MSC) now boasts over 700 products from certified fisheries across markets in 26 countries.

5. Such as health claims, organics, Country of Origin and the like.

6. Integrated global traceability, which can handle health and safety, sustainability (and increasingly, legality) of fisheries products is an emerging policy and operational issue.

activities. Resource access, capacity to meet international management standards, perceptions of biases in access arrangements for some, trading rules, tariff escalation and tariff peaks, high standards and perceived non-tariff barriers (NTBs), the increasing advent of private standards, lack of resources, logistics and knowledge makes it often difficult to enter, and find a sure niche[7] in, the complex, information-demanding and "high tech" markets of developed countries. Aid agencies and donor institutions have an important role in this area. It is critical that developed countries devote attention and resources to ensuring a level playing field in the globalised fisheries system.

Investment opportunities in the fisheries sector, including the capture sector, are numerous. **Financial institutions**, however, caution that investment will increasingly only proceed when certain conditions (including good governance practices and sufficient country credit ratings) are in place. In this regard policy makers should note that vertically-integrated companies seem, based on current evidence, only to invest in the capture sector when it is characterised by well-established property rights and a good fisheries management system that can ensure sustainable fisheries. Otherwise, vertical integration generally starts at the level of processing.

Thus overall, while globalisation is a positive development as it creates new opportunities, there is concern about how the **benefits** from globalisation are being – and will be -- shared. Conditions for access to both resources and markets -- and the capacity to manage both of them well – will play an important role in shaping the globalisation process and how the benefits from globalisation are distributed.

Summary of the Discussion

To capture the richness of the discussions that took place at the Workshop the following synthesis of the discussions is organised around the value chain elements.

Capture fisheries

Workshop participants shared the view that the sustainable management of fisheries resources should be at the centre of government's policy concerns. While acknowledging the difficulties in achieving responsible and sustainable fisheries it was observed that more efforts in the area were needed globally. In this regard it was observed that the use of fisheries to pursue purely short-term social objectives (rather than ensuring sustainable fisheries that can provide longer-term resilience and durable social outcomes) has resulted in lack of coherence in policy making. This is equally important for developing countries where, often, long-term food security is a dominant need.

Seafood businesses are reliant on secure and stable supplies of fish. This is the basis for the economic viability of all the parts of the fisheries value chain. It is therefore crucially important that policy makers get management right. In this respect, secure, enforceable and long term "user rights" for fishers and/or their communities, as relevant, were seen as a primary way forward for fisheries managers. The views of the financial and other downstream sectors were in this respect clear: unless capture fisheries are run as a long-term viable business, it is unattractive for investment (whether financial or

7. The example of moving towards specialty markets like organic farmed products is interesting in this context, as a way of creating new marketing opportunities for developing countries' in areas where market premium also occur (unlike what has been experienced, to date, for eco-labelled products).

direct investment). The result could be lower investment levels and lack of technological or other investments in global fisheries.

International fisheries, *i.e.* fisheries on the high seas outside the EEZs, are a particularly important area of challenge. Some fisheries – including straddling stocks -- are under the management authority of regional fisheries management bodies, others are trans-boundary stocks shared between or among two or more countries, and still others remain under no management regime. The increasing demand for fish, coupled with the ease of transferring fishing vessels to alternative flags has led certain fishing operations to explore international stocks that are managed under real or *de facto*[8] open access conditions. Furthermore it is difficult to monitor and control vast areas of ocean space. With regard to international fisheries the workshop identified several policy implementation gaps that should be urgently addressed. These include: lack of implementation and enforcement of the conservation and management measures adopted by RFMOs and other international fisheries bodies; developing countries' view that high seas allocation criteria have been biased to existing players and prevent access by emerging players; and the need to strengthen flag-state control, in particular in states with little or no capacity to control the activity on the high seas, of vessels carrying their flags.

Another important issue for discussion was the role of international access agreements (alone or in combination with trade agreements) in providing both capacity-building and access to returns from unutilised national fisheries resources for some developing states (in Africa and also small island developing states). Discussions centered on the pre-conditions necessary to ensure that those developing states are able, as capacity is developed, to increasingly exploit higher value added activities -- and higher job and income growth associated with them.

Aquaculture

The aquaculture sector in developing and developed countries has witnessed spectacular production increases over the past two decades; and there is nothing to suggest that this will change. Production increases are driven by further technological developments, increased demand for seafood and the fact that aquaculture (compared to capture fisheries production) provides more stable supplies, something that food service companies and retailers in particular value.

It is useful to observe that there are basically two very different production systems in aquaculture in place and that these serve very different political, economic and social objectives, and therefore need to be addressed differently. Among OECD countries, the aquaculture sector is composed of high tech companies, high stocking densities of carnivorous species while in the developing countries, fish farming is often characterised by small scale, extensive family holdings that may even be an add-on to other income-generating activities. This difference makes it sometimes difficult for developing countries' fish farmers to respond to market requirements and developments. This may require different institutional set ups. Pooling or clustering of small scale farmers into larger cooperative ventures has been successful and may provide and answer to future developments in this segment.

With a view to sharing knowledge, the Network of Aquaculture Centres in Asia-Pacific (NACA) was highlighted as a particularly interesting venture, and it is in the

8. For instance, where collective management is weak or ineffective.

process of being replicated in the Americas and more recently, considered for Africa. NACA is an intergovernmental organisation that uses networks of expert contact points and capacity-building activities to promote rural development through improvement of incomes, production and foreign exchange earnings. NACA also seeks to diversify farm production.

Availability of finance in small-scale aquaculture can be an important impediment to further growth in developing countries and hence on the ability to benefit from globalisation. The emergence of micro finance institutions may assist, in some cases, to overcome the limited access to formal credit available to small-scale operators. Concurrently, carving out speciality markets (*e.g.* organic aquaculture, which may also come with prospects for capacity-building by some certification schemes) may provide for a new orientation and strategy to further growth potential. For example the Swiss Import Promotion Programme provides assistance to farmers in developing countries to convert into the more lucrative organic farming segment. This segment is foreseen to have substantial future growth, as consumers in developing countries are increasingly oriented towards health concerns and organic foods distinguishing themselves from general eco-labelled products by commanding market premium.

Many OECD countries are in the process of reassessing the role of aquaculture in their domestic fisheries, and the public policies to be applied in aquaculture to increase its ability to reach its potential. One example concerns the 2002 initiative by the EU Commission to propose a strategy for the sustainable development of European aquaculture. Aquaculture growth in the EU has been lagging behind (and stagnating since 2000) the growth in global aquaculture. As a result, the EU is analysing the causes for this wide difference to see if a better framework for future aquaculture development can be created. Other OECD countries are also examining their aquaculture policies with a similar view, given their possible disadvantages competitively overall in global aquaculture, and the struggles for aquaculture to find a strong niche in diversified developed economies, especially where it suffers reputational issues in relation to oceans space and environmental effects.

A key policy concern for aquaculture is the mixed "image" that the sector often faces as a safe and sustainable source of fisheries products. The use of medicines, the risk of contaminants, impacts of escapees, impacts of intensive aquaculture on marine ecosystems, fish welfare etc. in the production systems creates potential image and marketing problems for the aquaculture sector. An interesting problem is to identify the role of governments versus producers in terms of addressing such issues, and how public regulation finds its correct balance between constraining and enabling private operators.

Seafood processing

The Workshop demonstrated how seafood processing has undergone major changes over the past two decades as this part of the value chain has been in a favourable position to benefit from globalisation. The regulatory environment in the seafood processing value chain element is primarily concerned with trade, and seafood safety and hygiene. It is also in the processing sector that major consolidations have taken place; mergers and acquisitions are used to reduce risk and increase shareholder value. International consolidation in the value chain makes it easier to take advantage of cost differences, keeping an eye on market developments and more generally to ensure that costs are reduced and margins maximised in a sector where the price-cost squeeze is an increasing issue.

Processors of fish and fish products, in most developed countries, indicated clearly that they rely increasingly on imported raw material for their businesses as domestic fish stocks are fully or overexploited. Such processors are seeking to reduce the costs of the imported raw material, *inter alia*, through a reduction in tariff levels, which, for certain products and in certain markets, remain high. This will both help the processing industry and consumers. Such imports – including aquaculture products -- may help drive prices on domestic raw material down, and may have an adverse impact on the prices fishers receive for domestic raw material. In this regard it was argued as crucial that the domestic fleet operate under conditions that ensure efficiency and allow for them to "manage to market" (*e.g.*, quality, specialised products) and reap maximum returns, within the parameters of conservation constraints.

Unequivocally, major processors are now sourcing raw material globally, as local sources cannot support investments at the scales that allow efficient and profitable production. Examples are many, but the most vivid example is whole fish from the North America and Europe being flown to China for filleting and packaging and for re-export to the North American and EU markets This takes place as developed markets increasingly focus on higher value-added secondary and tertiary activities, similar to other industries facing globalisation and increased competition for lower costs primary producers. An interesting observation is that a similar shift will occur within developing countries. As labour costs grow and greater economic development takes hold, China is also concerned about a shift of processing to other low-cost countries such as Vietnam and India.

However, changes in seafood processing were shown as differing in the developing and developed countries. In developed countries seafood processing is focussed on developing high value-added products (*e.g.*, ready and/or portion-controlled, uniform-quality meals) that calls for sophisticated production equipment and methods, and hence access to capital. In developing countries, supported by a pool of less expensive labour, the existence of tariff escalation and stringent sanitary and hygiene controls in major import markets, processing is often focused on less sophisticated methods of transformation (*e.g.* filleting, canning). The further outsourcing of production to developing countries is inhibited in particular by sanitary and hygiene requirements (see above) that can be hard to meet other than in segmented production streams (in partnership with developed markets).

For developing countries' policy makers the key towards more value added production is to implement a strict sanitary and hygiene control system and to lobby for a reduction in tariff rates, in particular on high value-added products. In this respect, developed countries also play a crucial role in ensuring that the necessary capacity building and expertise is available in developing countries.

Retail and Distribution

Retail and distribution, including food services, provided the workshop with some of the more revealing emerging changes in the nature of the value chain and in the relative power of different components – especially large buyers and retailers in developed country markets. Accompanying this are a number of new issues relating to key players in the global distribution network, as described above.

Companies in this element of the value chain act as the proxy for the final consumer in global markets, and are -- especially in some markets as the UK and US -- acting upon emerging or expected consumer and/or ENGO (environmental non-governmental organisation) concerns to force changes further upstream in the value chain (production and harvesting).

To ensure strong and responsible market "branding", many large companies (buyers) in this value chain element are developing individual buying specifications, whether they relate to food safety and quality standards or sustainability standards. These standards may be higher than public standards for a variety of reason. In some cases, sustainability standards are being adopted because of a perceived lapse in public regulation which may endanger future sustainable supplies. In such cases, private standards can aid in the choice of specific longer-term suppliers to these large retail establishments, and thus reduce supply risk. In other cases, the private standards may be driven by the desire to meet current or future expected consumer ethics, to forestall activities and protests from ENGOs for retailing choices (especially in the UK), or some combination thereof.

Retailers were frank in saying that in some cases their actions precede consumer awareness, and form part of retailer marketing/branding that is then intended to become part of consumer values. This is why it is difficult, analytically, at this time to show price premium for eco-certified products (other than organic, for which there is a defined market segment that is willing to pay more for certified products). Save one restaurant chain, which is establishing supplier relationships that involve capacity building, retailers were also fairly frank is suggesting that they do not see it is part of their responsibility to do the technical consultations and capacity building as an adjunct to their core mission but rather, that is left to the public sector. This aggravates the need for public capacity building. Again, as described above, the international system of recourse regarding unjustified trade barriers in the WTO is built around public standards that exceed "reasonable" risk. It was noted in the workshop that it will be a test of the international trade governance to deal with the onset of private standards, especially when they are intended to replace lagging public standards, or where branding seeks higher-than- public standards.

The result of health and safety standards, organics, sustainability and other aspects of corporate social responsibility, as well as NGO and ENGO campaigns, has been a proliferation of standards and labels that are confusing to consumers and buyers alike. Buyers are enforcing their own rigour and simplification buy choosing selected standards they will respect, especially where international guidelines exist (e.g., MSC eco-label, which is the only full eco-label that conforms to the FAO multilateral eco-labelling guidelines for capture fisheries).

As seafood demand continues to grow, increasing demand is being satisfied from aquaculture sources in both developed and developing countries. This is likely to further change the market dynamics in the future and will set additional new standards on responsible[9] aquaculture. There is as yet no internationally accepted guideline on aquaculture sustainability, which will give an advantage to exiting schemes whether or not they will meet future multilaterally agreed guidelines (*e.g.*, Aquaculture Certification Council (ACC).

9. Broadly defined, including environmental responsibility.

PART 1

Opening of the Workshop and Keynote Addresses

Chapter 1

Welcome Address by Einar Gudfinnsson

Minister of Fisheries, Iceland

The Minister of Fisheries for Iceland, Mr. Einar Gudfinnsson, opened the workshop providing an overview of the importance of globalisation to Iceland and also highlighting the increasing consumer demand for fisheries sustainability and Iceland's experience on the use of market measures in fisheries management. The fishing industry is facing certain risks, such as declining fish stocks and the effects of overcapacity. Concurrently value systems across the globe are converging, a process that is furthered through the globalisation process; this has resulted in increased awareness among consumers of the problems faced by natural resources. Traditional controls to achieve sustainability in the industry are still warranted but can usefully be complemented by new approaches, including labelling.

In capture fisheries, rights-based systems and the use of transferable quotas can help align the incentives of stakeholders with those of fisheries policy makers and managers, for example, by devolving user rights and responsibilities to communities.

Aquaculture is increasingly an important source of seafood and has the potential to meet increasing consumer demand, particularly from developing countries. However, this sector must also be managed in a sustainable way.

Two trends of globalisation were highlighted: outsourcing along the fisheries value chain, and the rise of eco-labels and resulting consequences for developing countries. The Minister also noted that the benefits from globalisation will only accrue when countries do not concurrently raise new barriers to trade.

It's a great pleasure for me to join you at this Workshop on globalisation in the fisheries sector, and the opportunities and challenges it raises. I will discuss in general about globalisation and if the international and national governance structures are sufficiently strong to cope with the pressures from globalisation. Furthermore I will discuss the urgent need for a better management of shared stocks which will, when achieved, open multiple opportunities in the fisheries business. Thirdly the consumer perception of fisheries sustainability and finally the changes in the fisheries value chain as a result of globalisation.

Increased trade is a cornerstone of development and has the capacity to increase prosperity. Iceland has witnessed first-hand the positive economic effects of trade through globalisation. In the course of a few decades, the Icelandic population has come from poverty to enjoy one of the highest living standards in the world. We not only believe, we know that trade works and millions of people could be lifted out of poverty like the Icelandic nation in the past. The positive impacts of globalisation are, among other things, improved economic growth, which is a goal of all authorities of all countries. I have at least never heard any government say that it doesn't want a higher growth. With increased economic growth, education can be improved, poverty reduced, children can be fed and so on and so forth.

There are though different views on the pros and cons of globalisation. Some say globalisation can increase threats to sustainability of natural resources where effective management measures are not in place. Others point out that while the process of market opening may be beneficial overall for jobs and growth; it also brings out disruptive transformation, particularly for less qualified and more vulnerable workers and it can increase inequality. The process of increased globalisation will continue and there is a need to carefully study its impact on the fisheries sector and the economy at large. This is necessary in order to identify appropriate policies and measures, which can help the fisheries sector to adjust successfully to new global developments, mitigate negative effects and derive the maximum benefit. We cannot erect protective walls and resist change.

In order for trade to have the capacity to increase prosperity in the fisheries sector we have to have an appropriate fisheries management system in place. Increased trade has contributed to rising fish prices and improved market access, which in turn has increased profitability of fishing operations. If there isn't an adequate management system in place that provides fishers with the right set of incentives, then trade can result in fishers resorting to IUU fishing or damaging fishing practices. They could try to increasing their effort, which will cause or exacerbate overfishing which, in the long term, reduce yields, income, and welfare. Negative impacts such as overfishing and overcapitalization may also result from heavy subsidization in the absence of an effective fisheries management system. I strongly believe that we must try to limit harmful subsidies that contribute to overcapacity and overfishing. Hopefully we will see progress to that extent in the current Doha rounds at the WTO. Subsidies and other economic distortions, have served as obstructions to free trade in fisheries production. It therefore lays a burden on fish producing nation and definitely leads to overexploitation of fish stocks. One can therefore say with full confidence that isolationist policies are adversary to the idea of good and responsible fisheries management. Not least with reference to this fact I would like to urge all those who want to exploit new potential in fisheries to bear this in mind. A prerequisite to a progress and maximum yield in exploiting our natural marine resources is free trade for fishery products on a global scale.

About ninety percent of the global fish catch is within zones of national jurisdiction. The bulk of fisheries management problems occur, therefore, in these areas. Governments must be aware of the state of their fisheries to be able to implement effective policies aimed at preventing resource depletion and the wastage of fisheries inputs and to facilitate stock rehabilitation. Commercial fisheries need more effective management. All fisheries require management especially where it is lacking, wastage of fisheries inputs and fish resources is inevitable. It should neither be surprising nor necessarily derogative of national governance worldwide. Benefits from good governance are quite obvious in the developed countries and this has been the case for some time. This is because their capacity for governance is more evolved. Furthermore, countries have very much to gain from improvements of fisheries performance. Much of the global fishing capacity and stocks are already fully exploited. We must be mindful of that.

Consumer perception of fisheries sustainability is increasing. Societal awareness of fisheries resources and the impacts of fishing activities on the environment continue to increase. Many ethical concerns relate to the societies perceptions of what is good or bad and therefore to the question of what are our duties and obligations. Ethical issues in fisheries and aquaculture influence significantly the scope and approaches used in fisheries, by fishermen and by the broader society in relation to fisheries resources. The perception of what constitutes as an ethical behavior in relation to fisheries differ from one society to another. The value systems of societies are converging as a result of increased globalisation of public awareness of problems that natural resources face. Development of this kind is influenced by the prominence of global issues such as, the increasing role of the market and vastly expanded trade in fish and fish products, the use of new biotechnologies in fisheries and aquaculture, excessive pressure on fishery resources and many other factors. The ideal objective of developing an ethical framework for consumers and producers in the aquaculture and fisheries sector is to ensure that they, who engage in these activities, will contribute to food safety, tackle the issues of malnutrition and hunger in less developed countries. Despite efforts, most of all fishing areas in the world have fisheries that are, unfortunately, not utilized sustainably and are there for coping with major problems of over-fishing and poor economic performance, which are symptoms of overcapacity. Overcapacity and over-fishing are increasingly significant signs of conflict and social stress both over the allocation of these limited resources among different users and the access right to fisheries resources. The basic solution to the problems of fisheries governance will ultimately require setting up a rights based management system. Such a system will clarify, more fully, who gets what and will align the behavioral incentives of stakeholders. It also needs to be designed so that, even though catches are limited, the management system will create a positive incentive to capture the benefits of technological innovations and other value adding strategies and not to over-invest in capacity. Systems of transferable quotas and individual quantitative need to be effectively put in place, in situations where individuals compete among themselves, to reinforce their rights and responsibilities to communities. Similarly, where communities are the predominant social unit, effective community based system of transferable quotas will need to be designed to try to allocate utilization rights and responsibilities to communities. And finally, management systems will have to be able to distinctly address the issue of payment for the right or privilege to use the world's fisheries resources for their private gain.

Parallel to the increase of trade in fish and fishery products, the issue of fish trade and food security has become very important. In this respect eco-labeling has become a very important issue in fisheries. The Scandinavian countries have for the last years been working on issues regarding eco-labeling or another kind of documentation of responsible fishery management. This is done to meet increased requirements from large buyers of seafood products. To begin with the main objective was to create a wide international agreement on guidelines for eco-labeling schemes. These kinds of guidelines were endorsed in a meeting in Committee of Fisheries at FAO in March 2005. Labels on products convey product information to consumers. In addition to fulfilling necessary regulatory requirements in import markets, a number of labels have been developed to enable marketing people to target specific segments of consumers in their marketing efforts. Among the more recent labels are those that are claiming to be fair trade and environmentally sound practices. Some countries fear that their sovereign right to fishery management might be infringed by eco-labeling schemes. Though, eco-labels schemes may be attractive to consumers in some markets it is not clear if consumers are willing to pay more to cover the costs. If not, then the producers will have to pay the price. About 50 percent of fish exports come from developing countries and with that in mind eco-labeling will have particular consequences for producers in these countries. We must bear in mind that eco-labels can also have the effect of an additional tool in fisheries management, alongside with more traditional controls. A large proportion of consumers in many important markets for fish and fishery products are increasingly concerned with the sustainability of fish stocks. It is therefore clear that these consumers take a special interest in products with specific labels that certify the use of environmentally friendly production methods and the sustainability of the resource.

A question we need to ask our self is what affect has the globalisation had on the fisheries value chain for the last decades? The fisheries value chain is, as we know it, the process that raw material passes through from capture fisheries, i.e. aquaculture and wild harvesting, to the delivery of a final product to the consumer. It includes the processors and traders of the raw material from where it enters a factory until the produced product has been sold, the distributors who move the product from one location to the next, retailers who make buying decisions based on consumer preferences and finally the consumer buys the product. This chain has changed because of globalisation and outsourcing has become a keyword. Globalisation has in fact transformed the nature of international economic interaction. It is not entirely the internationalisation of free trade and financial flows between national economies. Countries are producing where they have a comparative advantage and are less dependant of being able to make everything themselves. A good example of this is the outsourcing of filleting to China instead of doing it domestically. Trade and transactions between countries have increased dramatically with improved transportation and communication linkages. In fact, the world's economies affect each other and penetrate each other economic domains because companies are becoming more and more multinational. They work across borders. The affect has been that outsourcing of activities has transformed and linked financial and labour markets globally. The impacts of globalisation are regional, national, local and international.

To sum up: One of the main challenges governments face today is to manage fisheries in a responsible way to ensure responsible and sustainable use of resources as well as economically efficiently. A sound management system is essential to ensure sustainable exploitation of fishing stocks and to fight back against overcapacity and over-fishing. Parallel to increased globalisation, the issue of fish trade and food security has become very important. There are both positive and negative affects resulting from globalisation. Positive affects are greater than negative affects and we cannot resist change and erect protective walls. Globalisation in the fishing and aquaculture sector is not showing any signs of slowing down. In fact, indicators point out the exact opposite.

Chapter 2

Welcome Address by Abraham Iyambo

Minister of Fisheries and Marine Resources, Namibia

Mr. Iyambo spoke of a new era of globalisation, characterised by increased communication, complexity, opportunities and challenges. In this new era developing countries wish to enter at all stages of the value chain and enjoy a level playing field in international trade. In the fishing sector, developing countries face the challenge of creating value addition in their own countries, being seen as equal partners, and obtaining the skills and training necessary to compete in international markets.

A major role of developing countries in globalisation has been to permit foreign access to their natural resources, predominantly by developed countries. The lack of opportunity given to developing countries to add value to seafood was highlighted. Hence a fair deal for fisheries products is needed to create trade justice, including through the establishment of international regulations that are more equitable.

The Minister also explained a loophole in the international law of the sea. The 1982 Law of the Sea Convention established Exclusive Economic Zones, with a proviso that coastal states' surplus resources may be explored by other states. LOS therefore allows access to developing country resources. As many developing states have limited harvesting capacities, experience or financial resources to monitor and control fishing activities, this may result in overexploitation. As independence came relatively late to Namibia, the Namibian government has been able to learn a great deal from the mistakes of others; that is why Namibia has no foreign access agreements.

Increasingly, developed countries depend on developing countries' fish resources and developing countries depend on developed countries for export markets; for example 25% of total exports from Namibia are seafood. Consequently, developed and developing countries should act as partners. However, this is not always the case. Present trade conditions imposed by large trading countries distort trade, rather than allowing developing countries to compete equally and earn a fair share of the profits. The existence of tariff escalation is a case in point. This keeps developing countries from establishing their own value addition as well as forgoing employment that such additional processing would create. A way forward would be to create partnerships that encourage value addition in development countries.

The organisers of this world auspicious occasion requested me to talk about "Globalisation in the Fisheries Sector: Challenges and Opportunities for Developing Countries". I thank His Excellency Ichiro Nomura, Assistant Director-General of Fisheries and Aquaculture Department of FAO for the invitation extended to be with you at this event jointly organised by FAO and OECD.

We truly live in a global village. We travel, communicate across borders and continents, as we have never done before. The availability and speed of communications and the complexity of our daily work has increased greatly. All of us have become global citizens and we are one human fate.

Globalisation has brought us infinite opportunities, advantages and agonies. We are now able to effectively interact with the rest of the world through the Internet. It was not so many years ago that we depended on inadequate telephone, telefax and telex services. The latest technology bears countless advantages, and challenges as well.

Globalisation has boosted international trade and made previously unreachable markets accessible for accessing knowledge and business exchanges. Seafood trade has grown enormously. The developing world is a pillar of this growth as its contribution has been phenomenal. Evidence of this is that in 1950, developing countries accounted for only 32% of the world seafood exports. By 2004, this skyrocketed to about 51% and the share is growing unabatedly as a stimulant to our economies.

It will not be an exaggeration to state that for developing countries, globalisation is not a new phenomenon. The wave of globalisation engulfed Africa from the 15th Century through the 18th, the 19th and now far into the 21st Century. In the past centuries, we were called Colonies of Europe. We are now called "Third world or developing countries". The main objective of colonisation was unhindered access to unprocessed natural resources and the pursuit of profit and domination. Indeed no value addition. A question should be posed. Are developing nations, still the target for gaining access to unprocessed or raw materials (fish) for industries in developed nations? In colonial times, this was done in a fairly crude and straightforward way. The armies of their countries of origin protected big trading companies. Wars, genocide and extermination took place. Indigenous populations refused to succumb to unfair deals. That is why the United Nations calls for fair play and compassion.

Today, however, the same injustice is still being perpetrated, although in a much more sophisticated way. Are we all getting a fair deal for our fisheries products? I argue that some developing nations are being exploited. History will judge us all and the contemporary generation has to answer. Are the subtle and intricate Agreements between trading blocs and sovereign Nations fair, beneficial or destructive? Lawyers and negotiators have substituted armies and soldiers. But, the end result is still very much the same: exploitation of the developing countries by strong and rich nations and companies. The record is sad. A myriad of complicated rules and regulations that are largely not in favour of developing nations have been introduced by some in order to control world trade opportunities.

When many in the developing countries achieved independence, they were quite optimistic about the future, - naturally. They thought that now that they have ownership over the resources, they would get a fair share of the cake. Unfortunately, they knew too little about how the world turns. A rude awakening indeed. That's the dilemma of poverty and disempowerment. With the adoption of the Law of the Sea in the 1980s, Coastal States were given the legal foundation. Developing countries declared 200-mile EEZ. But they did not have the skills, experience and financial resources to keep watch over their rich domain. According to Article 62 of the Law of the Sea, Coastal States were "required to determine their capacity to harvest the living resources of the Exclusive Economic Zone...." A very thin writing was added that says: Where the coastal State does not have the capacity to harvest the entire allowable catch, it shall give other States access to the surplus...". Thus, a dangerous loophole was created by this provision. It does not consider value addition. In other words, many developing countries "rented out" the resources to others, often at ridiculously low rental fees.

Namibia is one of the youngest independent countries represented at this gathering. We gained our independence only in March 1990. But it was not until 1994 that we gained control over our most important fishing port, Walvis Bay. Through protracted negotiations, the reintegration was achieved peacefully. Gaining our independence very late in the 20th century, and over a half Century later than former colonies such as India we had certain advantages. We were able to learn from the mistakes of others to confront the challenges and to use to our advantages the development in the globalised world. That is why we in Namibia have never accepted Access Agreements. That is why we placed priority on learning how to manage our resources and how to control the fishing activities within our EEZ. That is why both Namibian and foreign investors have invested heavily in Namibia. We trust them. That is why the Namibian fisheries sector accounts for around 25% of total exports for Namibia. Conservation and harvesting go hand in hand and are being managed prudently.

The world fisheries resources are in a sorry situation and unhealthy state. The developed world is becoming increasingly dependent on obtaining fish from the developing world. For the last 30 years, the developed world has been a net importer of seafood, while the developing world has been a net exporter. And the gaps are widening. Since the late 1980s, developed countries have experienced sustained decline in their fish landings. In 1950, developing countries accounted for only 25% of the global landings. In 2005, 55 years later, today they account for over 80% of the total landings.

The challenge is that we should think differently; drop the crisis of old thinking and perception. We should not apply concepts of an outdated worldview. May I borrow from a renowned Scientist who once said: "The world we have made as a result of the level of thinking we have done thus far creates problems that we cannot solve at the same level at which we created them... We shall require a substantial new manner of thinking if mankind is to survive". We need fairness in this globalised world. Good old days, where people exploited resources at whim are over. The developed world has to deal with an increasingly awakening group of developing sovereign States that have the political will to responsibly manage their fisheries.

The challenges that we face at this time of increasing globalisation are many, and they are defined not by us, but by the developed world. Some of the requirements introduced we agree with and are essential; for example the health and hygiene standards. There cannot be quality standard designed for products from developed countries and developing countries. We cannot compromise health of fish consumers. In Namibia, we have seafood processors who are, objectively judged, as world class. I apologize for saying this, but some of our processing plants are far better than most of those found in Europe, Asia or the USA. We have taken the issues of health, hygiene, and quality seriously. I know that the same is true for many other developing countries in Africa, Asia and Latin America.

What we find difficult to deal with are the burdensome conditions being imposed by the big trading blocs. What we find offending are strategies meant to distort trade and not to compete equally. There are other globalisation issues that affect us but that we know we can do nothing about, such as the changing structure of trade and the ever-growing power of the multinationals, especially the supermarket chains. However, there are ways of dealing with this. In. Namibia, some of our companies have found ways to deal with it and to profit from the advantages. The changing structure of the market represents a challenge to us. Fortunately, market orientation has proven to be a way in which we can exploit it to our advantage. I recently commissioned a study on "Value Addition in the Namibian Fisheries industry", and I am happy to report that this study revealed that, with the right kind of thinking, our companies could handle this challenge in cooperation with their customers.

A major development we have seen lately, and one, which is perhaps most often used as the shining example of globalisation in the fisheries industry, is the shipment of raw materials to areas or countries where labour and operating costs are low. While this has opened up new opportunities for developing countries, not least in employment creation, it has also created problems of a different kind.

Globalisation of the fishing sector has opened the industry up for criticism from various civic-minded groups. While we may differ with some of their positions, many of their concerns are legitimate that we should all be attending to. One such issue is what has been labelled "food miles" or "the C02 footprint". It is the flip side of globalisation. We spend a lot of energy on transporting fish over great distances to reach the market. Yes, that we do. We have to in order to satisfy demand in our markets. The concern is on the probable global misuse of transportation systems because of evil intentions or sole pursuit of profits. Let us discuss these issues.

Developing countries want to trade. But they also deserve faire share of technology and profits. Developing nations must be allowed to do value addition in their own countries, rather than just providing raw materials and export jobs. Under the present trade Agreements and Regulations, the poor and weak are penalised when they try to beneficiate and add value to the raw materials in their own countries for export. Stringent conditions are hamstringing us. I will give you a simple example. EU import duty on round frozen tuna from an ACP country is at present zero percent. EU import duty on tuna loins and on canned tuna is 24%. In other words, it is easier for our companies to export the round frozen raw material than to export the finished consumer product. Where is fairness and justice? We need the employment that processing would create. It is just logic that in partnership, processing facilities that produce for exports to the developed world are established.

I could give you other examples that are equally bad. I don't want to point a finger at any one in particular. But the rules and regulations written by bureaucrats of developed countries are definitely designed to protect them. As a friend of mine recently said: "King Leopold the 2nd could not have done it better".

I see great opportunities for developing nations in this era of globalisation. I see potential for providing the developed world with the fish it needs. I see opportunities for the developing world to develop their economies and prosper. The main ingredient necessary is fairness and partnership. Developed and developing countries should be equal partners in feeding the world. We should be given a fair chance to use this market to create value in our own countries, to contribute to national development, the eradication of poverty and prosperity for all. We need help to develop the skills that the developed world possesses.

Chapter 3

Globalisation Overview

Keynote Opening Speech

Alastair Macfarlane, General Manager,
New Zealand Seafood Industry Council

Alastair MacFarlane's Keynote Speech introduced a brief history of globalisation in the fishing industry and outlined the principal elements of globalisation in each element of the value chain.

The creation of EEZs and the exclusion of Distant Water Fishing Nations from many EEZs provided coastal states with opportunities to globalise and trade. By now fish is a widely traded commodity, with 60% of international trade coming from developing countries. Albeit more recently, there has also been substantial growth in trade in aquaculture species, with omnivorous species such as tilapia, likely to grow exponentially in the future. The Uruguay Round of the WTO propelled globalisation by providing rules for food safety and thus ensured a more secure and stable trading environment. These developments also facilitated the outsourcing of processing to third countries, particularly China, in accordance with internationally adopted food safety standards. The development and harmonisation of these health and safety standards have since played a significant role in globalisation in providing an enabling, secure and known trading environment.

An emerging issue is that of developing countries, which, while agreeing on the need for these standards, believe that the same standards often work as barriers to trade. At the consumer end of the value chain, socially responsible branding is becoming increasingly common, particularly in food products. Supermarkets wish to reduce their risk of bad publicity which can have significant detrimental effects on sales and reputation. Risk management and competitive positioning by retailers are drivers for independent verification of fisheries sustainability. While the retail sector is driving the creation of standards, it was observed that there is no consistency across the standards, of which there are already many in the fisheries market. It is also uncertain how the end consumers will respond to the plethora of standards. In addition products from certified fisheries are unlikely to fetch a price premium.

Executive Summary

Fish and fish products have been actively traded in recorded history for at least 2000 years and perhaps longer in traditional societies. Trade was enabled by technologies to preserve fish – initially through drying, salting, pickling and smoking and later by canning and refrigeration. In trading sense, the fish trade has been early example of globalisation.

Over the last century, the catch sector has been transformed by technology enabling more and more distant water fishing. In the last thirty years, and the advent of the UN Convention on the Law of the Sea (UNCLOS) enabling coastal States to declare Exclusive Economic Zones, the global catching sector has been further transformed away from distant water fleets supplying domestic markets in developed countries to coastal States supplying those same developed countries with fish through international trade. More than 50% of fish products traded internationally originate from developed countries.

An examination of the investment profiles of the thirty largest publicly listed companies – 10 each from Europe, The Americas and Asia – reveals that the overwhelming majority are processors of fish products and actively forward linked into international markets. But only a minority has any investment in the catching or aquaculture sectors. Investment in the catch sector is concentrated in relatively abundant pelagic fisheries and/or in catching sectors where there are defined, secure access rights. Investment in aquaculture is concentrated in salmon, with some invested in other aquaculture of other carnivorous species (*e.g.* sea bream). The chapter concludes that investment in primary fish and fish product production is conservative, risk-averse and concentrated on proven areas of profitability with secure access rights.

The chapter discusses the relatively recent internationalising of the processing sector and suggests that the Sanitary and Phytosanitary and Technical Barriers to Trade Agreements arising from the Uruguay Round have facilitated this phenomenon. Rather than being a barrier to trade, the Agreements have enabled controlling and authorities and consumers to have confidence in the safety of seafood processed and marketed under established brands.

A short description of business to business relationships identifies that moving from commodity trade based relationships to more integration between production, processing and marketing is based on trust leading to investment. The chapter asserts at several points that uncertainty over resource sustainability or access is unlikely to encourage globalising investment and will continue to trap primary producers in States that are unable to offer security into ongoing commodity trading relationships with international markets.

The first part the chapter concludes with a short examination of the influence of the social responsibility brand values of some supermarket companies in Northern Europe and North America on their sourcing policies.

The second part of the chapter discusses some external influences on the globalising fish and fish products sector. In regard to governance in the catching sector, the chapter discusses the reluctance of globalised businesses to invest in catching sectors without secure access rights or reasonable certainty of fish stock abundance. It suggests that the willingness of management organisations to compromise their decisions discourages investment and discourages globalisation of the catch sector.

In regard to aquaculture, the chapter notes the importance of security of access or tenure over water space as a necessity to secure investment. Opposition to the location of aquaculture on the grounds of defending amenity values will ensure that aquaculture relocates to places with a more welcoming investment environment.

The role of governments as technical regulators (as in the case of food safety) is discussed. The chapter suggests that international agreements to manage food safety by scientifically based risk assessment has facilitated globalisation of the processing sector – and assisted the outsourcing of processing from developed to developing countries. The enabling role of government in this context is contrasted with the command and control approach still predominantly adopted in fisheries management.

The role of NGOs in catalysing public perceptions is discussed. The chapter conjectures that NGOs have been successful where they have been able to neutralise opposition from the seafood sector. By so doing, Governments are absolved from having to compromise governance decisions where they may fear a backlash from the economic sector that may have to face economic consequences.

Following its conclusion, the chapter contains a Case Study of the New Zealand wild capture seafood industry as an example of sector that is undergoing vertical integration towards a globalised business sector.

Setting the Scene

Seafood – An International Business for Thousands of Years

Perhaps it was the inherent fragility of seafood – the fact that fish is inedible within a day or two after catch without preservation – that led people to find ways to preserve their catches. Traditional drying, salting, smoking and pickling are so successful that they can render fish products stable and safe to eat for months. Preserved fish products – amphorae of pickled fish sauce that was a staple of Ancient Rome – have been found in the hold of a wrecked Roman galleon in the Mediterranean dating back a couple of thousand years[1]. This is taken to be evidence of trade between Rome and its Empire, and could be seen as an early example of globalised trade in a fish product.

The well known book "Cod – a Biography of the Fish That Changed the World"[2] discusses the importance of dried salted cod, bacalao, to the Portuguese and Spanish empires. Over 500 years ago, Portuguese fishermen would sail to Newfoundland, catch cod in the hugely abundant north-east Atlantic fishery, dry it on shore and then sail back home at the end of the season with the preserved catch. Over the years they were joined by fishermen from other parts of Europe. The bacalao cod was then widely traded as a staple food ingredient, enabling Roman Catholic populations in southern Europe for example to fulfil their dietary obligation to eat fish on Fridays.

Until the advent of modern food processing in the 19th and 20th centuries, most meat for consumption was traded as livestock and generally only locally – with capital (breeding) stock being traded over greater distances, including internationally. Salted meat was a phenomenon of shipboard life, rather than a popular traded food item on land.

1 Washington Post, 13 November 2006

2 Kurlansky, M. (1998), *Cod – A Biography of the Fish That Changed the World*, Penguin Books, London

People did smoke and salt meat for long term consumption, but that was a home preserving activity like brewing and other preserving, rather than an item of trade. But preserved fish has been traded between coastal and inland communities for centuries. This is a pattern followed by many societies the world over, including traditional societies. And whereas there is a strong bias, especially in Western societies, towards protecting self-sufficiency in agriculture and resistance to trusting international trade as being insufficiently reliable to assure fundamental food security, there is a long history of ready acceptance of trade as the conduit for assuring access to fish and fish products.

Technological advances in the 19th and 20th centuries enabled fishing to take place further and further away from the coast, through the development of motorised fishing vessels, the development of the trawl fishing method and the adoption of refrigeration and freezing on board vessels. By the mid 20th century, the main fish consuming populations in Europe, North Asia and North Asia had their demands for fish satisfied by their fishing industries fishing more and more in the high seas beyond their territorial limits while international trade remained relatively undeveloped, beyond the traditional products traded over previous centuries. By the middle of the 20th century, concern at the impact of uncontrolled expansion of fishing beyond territorial limits, inability legally to reach international agreements to share fish resources and ensure sustainable harvest practices were among the catalysts for the negotiation of the UN Convention on the Law of the Sea. UNCLOS provided a legal basis for States to assert their national interests in the living marine resources within 200 nautical miles of their coastlines.

The products of fishing in the high seas take their rule of origin from the flag state of the vessel. Thus distant water fishing for landing and sale in the flag state of the vessel is domestic trade. Distant water fishing nations in the post World War 2 period up to the mid to late 1980s presented themselves as domestically "self-sufficient" in fish and fish products, although a large proportion of their "domestic catch" was taken hundreds or even thousands of kilometres from their territories. These same distant water fishing nations were also in that period the main suppliers of fish to the relatively small international trade in fish and fish products.

From the late 1970s to the mid 1980s coastal States encouraged the expansion of their hitherto coastal fisheries out into their new exclusive economic zones. As a consequence there was a dramatic reduction in the opportunities for distant water fishing and contraction of distant water fishing fleets. UNCLOS and the creation of exclusive economic zones in the last thirty years have been the most important catalysts for international trade in fish and fish products. As distant water fishers were excluded from fisheries now "nationalised" in coastal State EEZs, they had to face up the unavoidable reality that consumer demand had to be met increasingly by importing fish. Imports had to be transparently sought from newly developing coastal State fisheries expanding into their adjacent EEZs, previously exploited by their distant water fleets.

The FAO reported[3] that by 2004, 53 million tonnes of fish (live weight equivalent) was exported – representing about 50% of global production for food consumption. Developing countries account for about 52% by value of international trade, and the trade in fish has become a more important contributor of export earnings to developing country economies than trade in any other food commodity. Import markets for fish and fish products are dominated by the developed world, in particular the European Union, Japan,

3 FAO, (2007) *The State of World Fisheries and Aquaculture (SOFIA)*, FAO, Rome.

and the USA. Together they provide markets for 74% of internationally traded fish and fish products, including intra-EU member state trade.

Trade is a first step towards the greater integration of business relationships that can be defined as globalisation. Understanding the motivations of business towards greater integration (globalisation) can be enhanced by better understanding the trade inter-relationships within the broad relationships outlined above.

Figure 3.1[4] illustrates the dominance of the developed regions of the world in providing import markets for fish and fish products. However the role of intra-EU member state trade needs to be kept in mind. It accounts for about USD 12 billion annually of both imports and exports. The higher value of internal exports within the EU may be a reflection of the value added by processing within member states of imported raw material from outside the Community before onward sale within the Community.

4 FAO (2006) *Fishery Yearbook, Fishery Statistics, Commodities, Vol.99 2004*, Table I-1.1, E-1.1.

Figure 3.1. Average Annual Import Export Trade by Region 2002-04

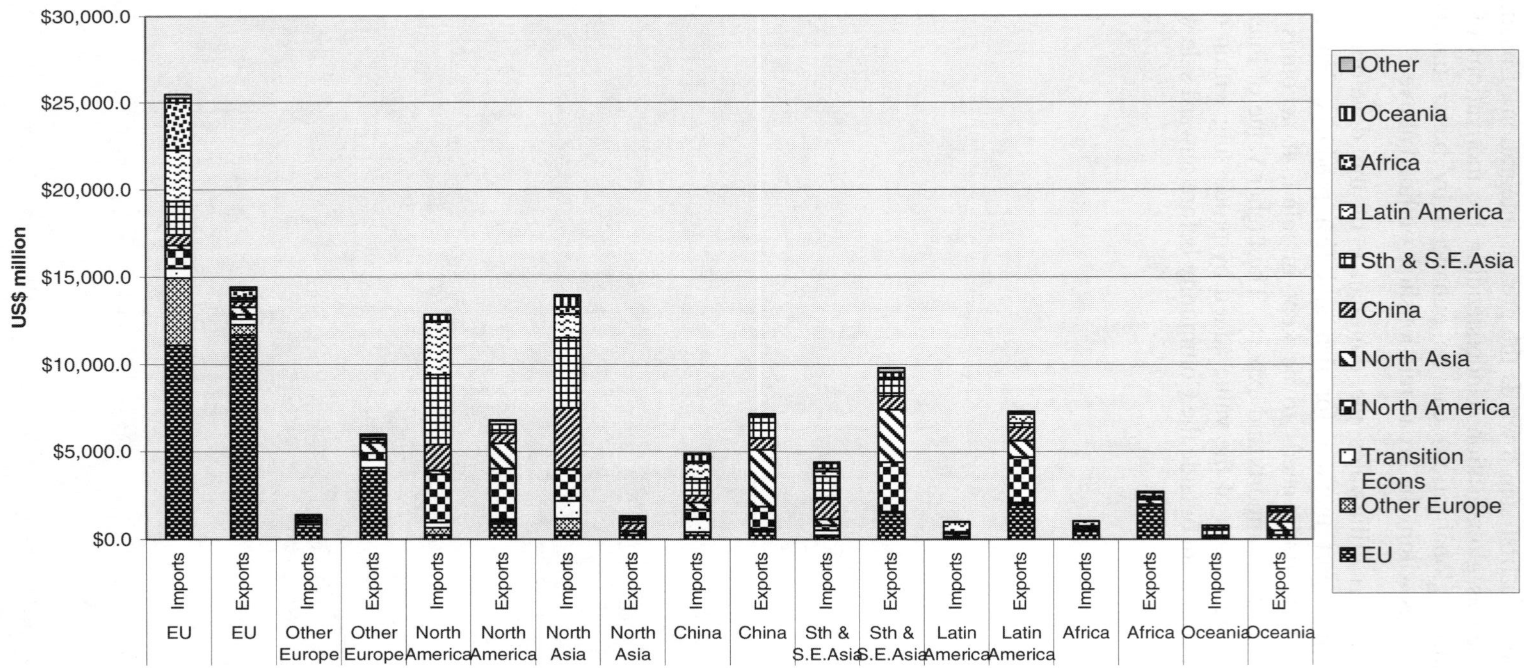

Source : FAO

When one takes away the impact of intra-European Union trade, the EU remains slightly ahead as the leading export market for seafood exporters, as illustrated in Figure 3.2. The EU, North Asia and North America account for 73% of seafood imports, not including intra-EU trade, with roughly equal shares of global imports, while other developed countries only account for a further 3%. Markets in South East Asia and China, account for much of the remainder of global imports. But a significant portion of imports are further processed in China and South East Asia and re-exported, mainly to markets in North Asia and North America.

Figure 3.2. Average Annual Global Import Market Shares of Fish and Fish Products 2002–04(excluding intra-EU)

(USD 56 billion c.i.f.)

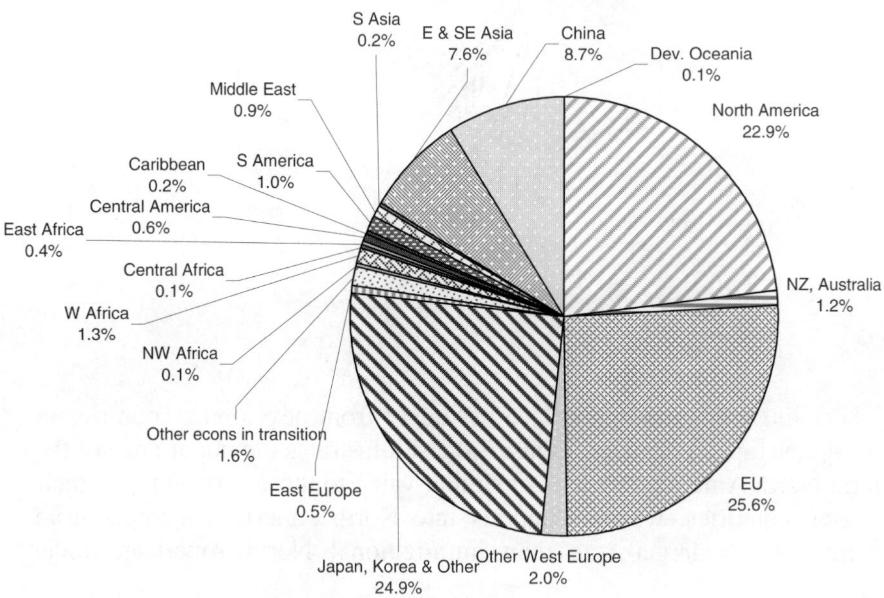

Source : FAO

Global export trade, net of intra-EU trade, is split between 37% undertaken by developed countries and 63% by developing countries and economies in transition, as illustrated in Figure 3.3. Exports from developing countries in Asia are on a par with the value of exports from developed countries.

Figure 3.3. Average Annual global Export Market Shares of Fish and Fish Products 2002-04 (Excl intra-EU)

(USD 47.4 billion f.o.b.)

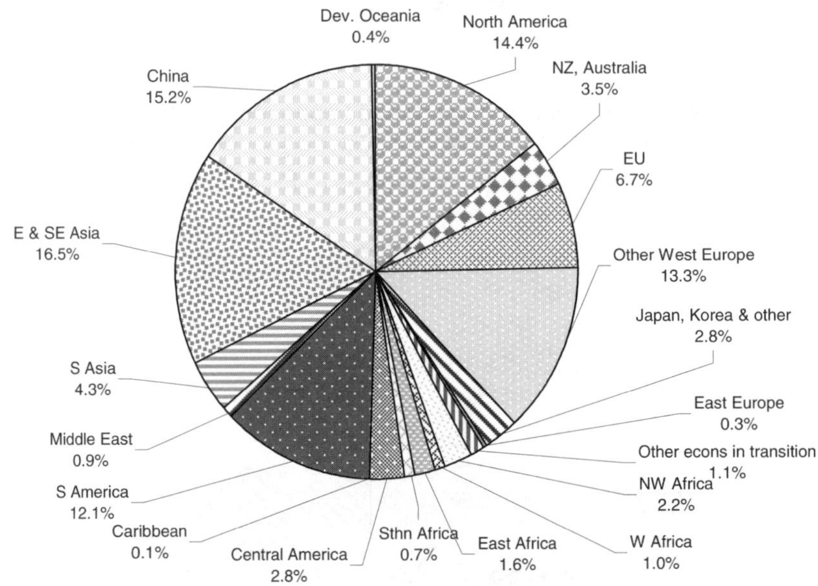

Source: FAO.

Figure 3.4 further illustrates the importance of supply from developing countries in meeting the needs of developed countries. Figure 3.4 also illustrates the importance of the intra-EU trade. Intra-North American trade is of similar value to those economies as their exports to developing countries, whereas imports into North American markets from developing countries play a larger role than intra-regional North American trade.

Figure 3.4. Average Annual Developed Regions' Fish Market Trade Flows 2002-04

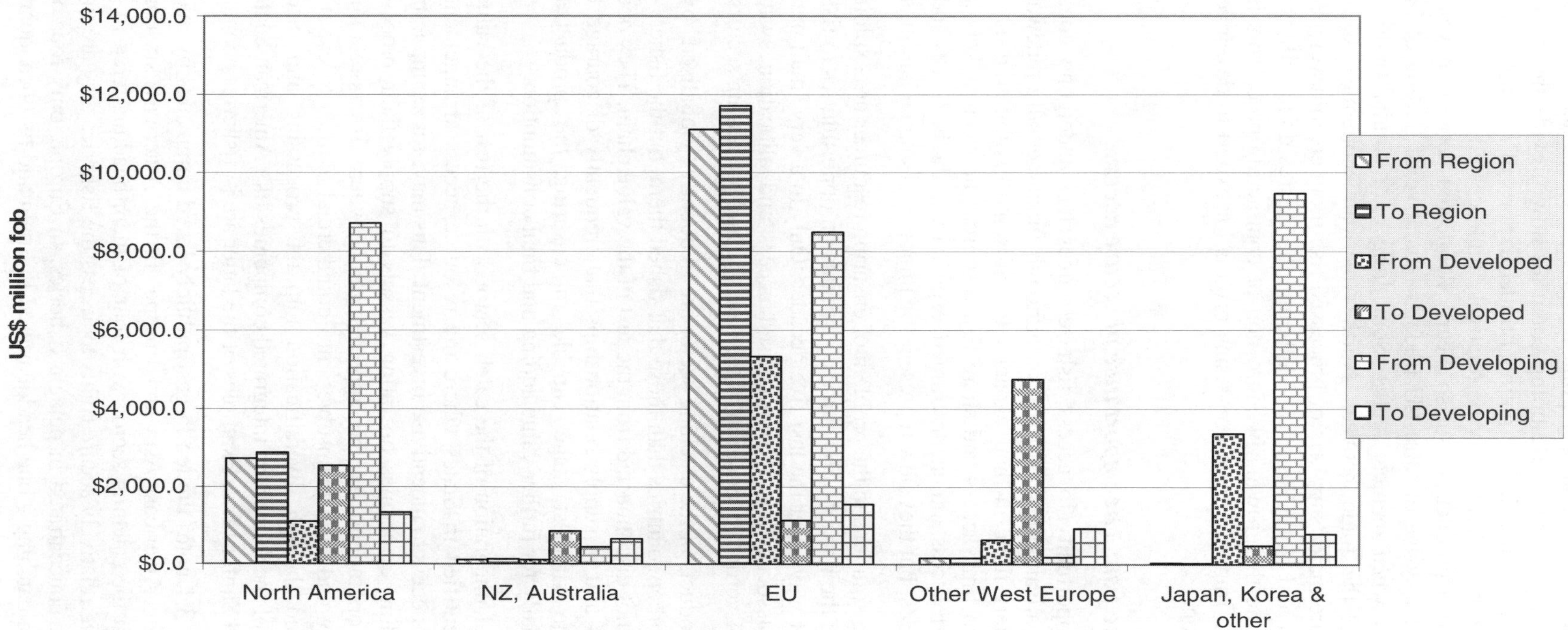

Source : FAO

These figures collectively point to some fundamentals of the international trade in fish and fish products. Developing countries now provide the majority of raw material fish and fish products in international trade. Exports of fish products from Africa and the Caribbean are predominantly of relatively unprocessed products. However the trade in processed fish products is not so clearly delineated between developing and developed countries. The value of intra-EU trade is indicative of trading of further processed fish products between member states. Something of a similar trade pattern may also be included in the trade between North American economies. China and Thailand in particular are major suppliers of processed fish products to developed country markets in North Asia and North America – although the figures do not illustrate this well. Nor do they illustrate the growing phenomenon of contracting processors in Asia and China to process fish imported from non-Asian origins for export to developed countries for final consumption.

Getting the fish: The importance of secure access

The opportunity to access fish resources has undergone considerable change. A growing realisation that fish stocks, although fundamentally renewable living resources, are ultimately finite in their capacity to withstand exploitation has compelled States to intervene to limit access and share fish resources between stakeholders. Some fisheries remain open access and unconstrained, especially in the high seas and in some developing country EEZs, but they now represent the minority of annual landings.

Finding solutions to the problem of ensuring resources are exploited sustainably while minimising the impact of resource constraints on opportunities to fish has been tackled in a variety of ways over the last five decades. Only in a very small number of cases, and in the main associated with rights-based access and allocation systems, is there strong evidence of vertical integration of the catching and processing sectors.

The products of the catching sector are frequently "globalised" by being consumed by people in other countries than those that fished them, but the fishers themselves are more often than not disconnected from the rest of the value chain. They operate inside domestic boundaries and in supply relationships that frequently discourage further value adding beyond the simplest stage capable of ensuring the product can be transported internationally for further value adding and final consumption.

Vertical integration of the catch sector with the rest of the value chain can be more readily identified in States where the resource access arrangements provide sufficient security of access to incentivise investment. This may, for example, be through harvesting rights-holding companies operating vessels to supply their processing operations and marketing operations. Another approach can be where fishers retain rights individually or collectively and form themselves into commercial entities to capture a greater share of value added through diversification into the processing and marketing value chain. Examples of the former are integrated companies like American Seafoods, Nippon Suisan Kaisha and Sanford Ltd. Examples of the latter are in fisheries cooperatives.

Annex 3.1 notes the ten largest publicly listed companies by market capitalisation in each of The Americas, Asia and Europe. In the Americas, six companies are directly involved in the catching sector with four of them headquartered in either Chile or Peru. All six have relatively secure rights to access the fisheries they are involved in, with most investment concentrated in pelagic fisheries. In Europe, only four out of the top ten are involved in catching – one headquartered in Spain and two headquartered in Norway and one in Iceland. The two Norwegian companies and the Icelandic company have clearly secure access rights. In Asia only four of the top ten companies have an involvement in the catch sector – one headquartered in China and the other three being Japanese headquartered multi-national corporations with significant investment in the catch sector outside Japan, particularly in fisheries in the Southern Hemisphere with relatively secure access conditions. Only two of the thirty top companies globally appear to have significant exposure to fisheries with relatively open and insecure access – they are based in Spain and China.

This investment pattern does not reflect the export and import trade flows between developing country producers and developed country consumers illustrated earlier. The lack of vertical integration between catching, processing and export marketing ensures that primary producers participate in the globalising market as suppliers of commodity fish, while the value adding that is entailed in getting commodity fish to food ingredient is captured separately and further up the value chain.

Globalising the catch sector

Work underway in the OECD investigating foreign investment issues in the OECD fisheries sector identifies widespread impediments to foreign investment in the catch sector. This is not uncommon outside the OECD too. Most of the largest publicly listed corporations in the seafood industry of the Americas and Europe identified in Annex 3.1 of this chapter that have a catch sector presence, are located in the fisheries of the countries in which they are headquartered. The Japanese Corporations and Pescanova from Spain that have significant catch sector investment in other fisheries have frequently done so through investing with business partners in the States where the fisheries occur, in order to overcome domestic investment restrictions, rather than attempt wholly owned subsidiaries.

The relative lack of exposure of major publicly listed corporates to catching fish suggests that investment in catching is not necessary to secure supply and may entail unacceptable risk. Investment risk is high if capital is committed to fisheries without secure access rights and even more risky where there is doubt as to the long term sustainability of fish stocks. Thus the interest that the OECD has to investigate impediments to investment, particularly in the catch sector, may be only of commercial consequence in a limited number of fisheries.

In recent years, international corporations have exited from investments in the catching sector. One example was Unilever's exit from its exposure to the catch sector in the mid 1990s, around the time it became a founding partner of the Marine Stewardship sustainable fisheries certification initiative. Another example is the marked reduction in the number of tuna purse seine vessels under the US flag fishing in the Pacific. A number of the vessels concerned have been sold to non-US companies and now fish under other flags. This dis-investment was coincident with the outsourcing of tuna canning to South-East Asia.

Access Agreements

The Law of the Sea[5] recognised the historic interests of distant water fishing nations in the newly "nationalised" fisheries of Coastal States by providing under Article 62 that

5 United Nations, 1997, *The Law of the Sea*, United Nations, New York.

coastal States are "...required determine (their) capacity to harvest the living resources of the exclusive economic zone. Where the coastal State does not have the capacity to harvest the entire allowable catch, it shall ....give other States access to the surplus ...". So, foreign fishing access agreements were born. Without going into detail, access agreements, particularly the models that date back to the 1980s, run counter to modern forces of globalisation The resource may be globalised – in that catches are consumed in the flag state of the distant water fisher rather than in the coastal state – but the context of consumption is as a "domestic" product of the flag State. The coastal State receives an access fee (in effect a resource rental) and remains disconnected from the value chain leading to final consumption. The value chain that follows from the act of catching by the distant water fisher is within the home state of the distant water fisher – as though the fish had been caught in the distant water fisher's home waters. Traditional access agreements carry over the conduct of fishing before UNCLOS. They are less globalised than the southern European cod fishers of the 500 years ago "making fish" in Newfoundland for transport back to, and consumption in, southern Europe at the end of the fishing season. In that historical case there was trade of the bacalao after it was landed in Europe.

It is acknowledged that Government to Government access agreements have evolved to so-called partnership agreements that provide development assistance that, inter alia, may also assist the development of coastal state fishing interests. Tying development assistance to fishing access can be a disincentive to coastal state fishing development, particularly development into fisheries also being accessed by the distant water fishing interests. It is likely that the distant water operators will be more efficient and on a larger scale than the coastal States' activity and crowd out coastal State fishers. If the coastal State enterprises then try to trade their products in the markets of the distant water fishers they will enter those markets as exporters rather than "domestic suppliers" and have to overcome the market access conditions that apply to imports. The coastal State has to assess whether the domestic fishing sector development will provide an equivalent or better net national benefit than the aggregate benefit of the fees for access and the related development assistance received, especially if coastal State development has potential to exclude distant water fishing activity and thus result in the elimination of the development assistance tied to fishing access.

Modern access/partnership agreements linked to development assistance perpetuate the "domestic supply" character of distant water fishing. The catch invariably ends up being shipped back to the flag States of the vessels and enters their markets free of any market access restrictions as domestic product. As has already been noted, the same products fished by enterprises based in coastal States' EEZs and exported to the markets of the distant water fishers will face normal MFN market access barriers, unless preferential tariff and access arrangements have been made.

Access agreements linked to development assistance can also crowd out private sector access contracts. Foreign, private sector ventures seeking to gain access to distant water fisheries under commercial terms are unable to offer the inducements of development assistance available to governments. However coastal States are likely to regard the total of transfers received under access agreements that also include a significant aid component as a basis for negotiating fees for private sector arrangements. This is a reality that is starting to impact New Zealand tuna fishers negotiating to maintain private access agreements in the Western Pacific fisheries[6].

6 Personal communication.

Access agreements and coastal State development into the globalising seafood business can coexist if the fees transacted are a fair market-related reflection of the resource's value and are sustainable both environmentally and within the cost structures of the fisheries and are de-coupled from development assistance. Coastal States can then make rational and unbiased decisions as to where their national interests lie in fisheries development.

Chartering

Private sector chartering of fishing services can be an alternative to government to government access agreements. They can also provide an alternative for enterprises in coastal States to minimise the need for investment, for example in seasonally redundant fishing capacity. This approach has been applied in New Zealand over the last 25 years. The strategy can be especially relevant when coastal state enterprises are at a start up stage and short of capital.

In the 1980s and 1990s, chartering of vessels and catching services from companies recently excluded from distant water fisheries with the establishment of EEZs, enabled coastal State based companies to get the use of vessels in relatively good repair, with skilled crews and technical capability to produce wholesome seafood products of merchantable quality. However, chartering fisheries services was only taken up as a commercial arrangement in only a few cases, New Zealand being one.

Chartering has largely failed as a rational strategy to make use of otherwise excess global fishing capacity and has gone into decline. Restrictions by coastal States of fishing access to domestic flagged vessels and incentives to expand domestic fishing capacity over the years have prevented and discouraged the private contracting of charter fishing services in all but a handful of fisheries. Today the remaining charter fleets are ageing and declining in number as vessels are progressively scrapped though old age. While those that remain may continue to be flagged to the historical distant water fishing nations, they are not being crewed there. Instead they are increasingly being crewed by people from some of the poorest and lowest wage nations in the world in an effort to keep down costs. Many vessels have been reflagged to other so-called flag of convenience states and this has further enabled crew sourcing and working conditions to migrate to the lowest cost sources. There is little evidence of new building of vessels to participate in the charter fishing market, with the notable exception of China. The remaining charter operations frequently only have capability to produce the simplest commodity form of frozen fish products. In WTO language, chartering of fishing services has become a now declining example of mode 4 trade-in-labour, as well as a declining example of trade in services. As an example, New Zealand companies contracting charter vessels have frequently found the vessels to be crewed by nationals from nations other than the vessels' flag states. There have been a number of instances of crew members jumping ship in New Zealand and attempting to immigrate illegally. This is despite New Zealand government requirements to ensure that the wages the crews receive in New Zealand are above New Zealand's ruling minimum wage and significantly higher than the wages that generally apply to the crews when they work outside New Zealand's EEZ. The vessels themselves often require significant maintenance and cleaning to meet food safety standards before being permitted to fish in New Zealand's EEZ.

Catch Sector summing up

Where fish stock levels are perceived to be under stress and States wrestle with overfishing, investment in the catch sector will be perceived as risky. Globalisation of the catch sector by major publicly listed corporates is an unlikely option both due to restrictions on inward foreign capital investment and perceptions that fish stock sustainability and uncertainty risks are too high.

However, where fish stock status can be reasonably assured by precautionary management settings and where there may be an option to obtain secure access rights, globalisation of catch sector ownership can be attractive to investors. Investors are likely to be companies already experienced in the catching sector in their coastal States and be familiar with or sympathetic to rights-based access regimes. They will favour investment in relatively under-utilised or developing fisheries operating in management regimes that either already have implemented, or are committed to implementing, rights-based management regimes. Examples of this are investments by New Zealand, European and Japanese companies investing in fishing operations in a number of southern hemisphere countries and the expansion of interest from European and Chinese fishers in fishing in the pelagic fisheries of the Pacific.

Aquaculture

Referring again to Annex 3.1, identifying the top thirty publicly listed companies in the seafood sector, only five are reported to have investment in aquaculture and none based in the Americas has direct exposure. This appears to run counter to the positive press that surrounds the very significant development of aquaculture round the world. One conclusion that could be drawn is that globalising investment in aquaculture is coming from private companies. If publicly listed companies are involved, they would have smaller market capitalisations than the ones in Annex 3.1.

However three of the five publicly listed companies that Annex 3.1 does identify as aquaculture investors are among the largest public companies in the seafood sector, and two are major Norwegian based investors in salmon aquaculture. While none of the major publicly listed companies in the Americas has direct investments in aquaculture, the Chilean and Peruvian companies and one company in the USA are strongly dependent on aquaculture as the consuming market for their products – fish meal, fish oils and aquaculture feeds.

Annex 3.1 indicates salmon aquaculture, particularly in Chile and in Norway, is a preferred area for globalising investment. The list also identifies one specialist investor in sea bream and sea bass aquaculture in the mediterannean.

The FAO's 2006 SOFIA report[7] confirms the continuing the rise of aquaculture, noting particularly the expansion of aquaculture in China. At the FAO's Aquaculture Sub-committee meeting in 2006, the FAO's Director of the Fisheries Resources Division, Mr Serge Garcia, noted that "…aquaculture currently accounts for 43% of global fish production used for human consumption and is expected to grow and compensate for the predicted global shortage of supply from capture fisheries and the demands of society"[8].

7 *Ibid.*

8. FAO Fisheries Report 186.

International trade in aquacultured finfish into the leading developed country fish import markets has been dominated in recent years by farmed Atlantic salmon. Shrimp has been the dominant aquacultured shellfish. But the trade in aquacultured finfish is undergoing significant change. Major expansion of farmed, tropical, omnivorous species – tilapia, catfish and lately barramundi – is bringing large quantities of low cost, farmed white fish to markets, particularly in North America and Europe.

Expansion of production and trade in fish species that can substitute for increasingly expensive capture fish species, produced in lower cost feeding systems and in low wage cost countries, for consumption in the major consumer markets is to be expected from a globalising sector. The diversification of the major importing markets of North America and Europe away from exposure to salmon should not be a surprise.

Examined on a global basis, and noting the huge production from freshwater aquaculture of carps and barbells in China to supply domestic demand, the top ten farmed seafood species groups globally by volume in 2003, according to the FAO[9] were:

Table 3.1. Top 10 Farmed Seafood Species Groups Globally by Volume, 2003

	Volume MT '000
Carps, barbels, cyprinids	17 215.1
Oysters	4 496.7
Other freshwater fish	4 250.1
Clams and Cockles	3 788.3
Salmons, trouts, smelts	1 828.8
Shrimps and Prawns	1 804.9
Tilapias and cichids	1 677.8
Mussels	1 589.5
Other marine molluscs	1 232.3
Scallops	1 178.5

Source: FAO.

9 FAO yearbook Vol, 96/2.

By unit value, the groupings are a better reflection of their relative status in developed country markets:

Table 3.2. Top 10 Farmed Seafood Species Groups in Developed country Markets by Unit Value

	USD/kg
Shrimps and Prawns	5.17
Salmons, trouts, smelts	3.06
Scallops	1.44
Other freshwater fish	1.33
Clams and Cockles	1.13
Tilapias and cichids	1.21
Carps, barbels, cyprinids	0.89
Oysters	0.84
Mussels	0.63
Other marine molluscs	0.51

As aquaculture steps up to the challenge of supplying growing consumer demand for seafood at volumes that sustainably managed capture fisheries can never meet, the farming models will be increasingly challenged to supply fish products that are attractive to consumers as an eating experience and profitable at mainstream prices for "centre of the plate" competitor products – chicken, pork and beef. It already appears more likely that tilapia, catfish, barramundi and their ilk will be more profitable in the mainstream of the market in the long-term than salmon, cod, halibut or tuna. Whereas the salmon story is one of gradually slipping down the prestige continuum as volumes increased, the tropical aquaculture fish species enter the market at a low cost advantage and the possibility that they might in time re-position up the prestige continuum.

Jim Cane, Group Commercial Director of Youngs was quoted in the trade magazine *Grocer* in 2001 as the inventor of a new word – "chickenability"[10]. He was referring to the gap in the food market for a fish species or product with similar flexibility for food processors and acceptability to consumers as chicken. Mainstream consumer markets not only in the West but also in the developing world are moving down a path towards greater spending power and less time to spend preparing meals. Peeled and cooked shrimp has already found a place in the convenience market, but there is not yet a finfish species that can take on the mantle of "chickenability", but perhaps that time may not be far off for tilapia or basa catfish.

Aquaculture in a globalised seafood market is at a cross-road, with strong indications that low cost, high volume production will be the way forward. Aquaculture has been identified by the FAO as already supplying close to half of the world's human consumption fish product needs. Several of the major globalised publicly listed corporates in Annex 3.1 are heavily exposed to salmon farming and its attendant risks but

10 www.wordspy.com/words/chickenability.asp .

not apparently significantly invested in new areas of aquaculture potential. Other, lower cost fish species farmed in tropical conditions – or in enclosed, secure farming systems in temperate countries to which they are alien – are coming into the markets and meeting consumer demands for fish at price levels that compete profitably with other mainstream animal protein competitors.

Impediments to investment in profitable aquaculture are discussed later in the chapter.

Processing

All but two of the publicly listed corporations in Annex 3.1 are invested in processing of fish products. For ten of the thirty, it is the basis of their involvement in the seafood sector. Processing, rather than primary production is the common factor for globalisation of the seafood industry. In addition to the publicly listed companies identified for illustrative purposes through this chapter, two of the largest processors and retail brand owners have been purchased by private investment funds – Unilever's Frozen Fish International is now owned by Permira and Youngs by Cap Vest. This is in line with a trend for capital fund investment in other staple food processing sector stocks.

Investment in fish processing has undergone significant change over the period since the GATT Uruguay Round of trade negotiations that concluded over a decade ago brought a new, risk and science based approach to regulating food safety. Adoption of risk-based food safety management systems has opened up opportunities for seafood products processed in developing countries to gain relatively ready access to markets in developed countries, once the processors can demonstrate compliance with food safety requirements. Risk management based food handling systems and the regulatory frameworks and institutional structures required to provide assurances of safety, wholesomeness and truth in labeling have been costly to develop and implement. But today consumers in Europe, the USA or Japan have confidence that seafood processed in Africa, Latin America or Asia will be safe to eat. Fish products are being packed under the most well known of European, American or Japanese household brand names in China, Thailand and Vietnam and consumers are buying them without obvious sensitivity to their countries of origin, but with confidence in the integrity of the brands.

The adoption of risk-based, systems approaches to food safety management and assurance is frequently identified as a growing market access barrier by producers in developing countries and in other exporting countries seeking to do commodity trading seafood business in Europe or the USA. A hindsight examination would suggest the opposite, given the expansion of trade in seafood products sourced from, or processed in, developing countries. Customers expect that the food they purchase will be wholesome and safe to eat, and it is an expectation that has no allowance for exceptions. Consumer purchasing patterns suggest that customers are neutral on origin, provided that they can be assured about safety and integrity – as for example through the product being packed under a trustworthy brand – or being offered for sale through a supermarket chain that consumers trust implicitly.

Moving people, moving businesses

Processing fish is a low status job – it is messy, smelly and frequently conducted in uncomfortably cold conditions in order to safeguard product integrity. As people have become wealthier in the developed world, processing sector jobs have become harder to fill. The food processing sector in developed countries generally, and seafood is no exception, is increasingly unable to attract staff and match remuneration paid in more attractive jobs in the service economy in developed countries.

Three solutions are available. Traditional processing sectors, *e.g.* fish canning in Europe, have been able to retain competitiveness behind high levels of tariff protection from competitor products from low wage economies and thus retain an ageing workforce. But this protection is likely to go in time, as is the workforce. Already, operations are coming under pressure from canned fish originating from developing countries that have been able to negotiate tariff preferences in the context of fisheries access and co-operation agreements. Examples are tuna canned in Mauritius or American Samoa that can respectively enter the markets of the EU or USA tariff free.

The second solution is to bring in temporary, seasonal migrant labour from low wage countries under work visas that compel the people to return to their original countries at the end of the contract. Minimum wage rates in developed countries can be attractive. Examples are seasonal processing workers in the USA from Central America and Eastern European origin workers in the processing operations in Iceland or the UK.

The third choice is to relocate processing to developing countries, confident now, in the post Uruguay Round climate, that the food produced will be safe and fit for purpose and will be able to gain access to markets. Thailand's tuna canning sector development reached a critical mass of supply dominance to the USA before the Uruguay Round negotiations were completed, so this is not a new phenomenon. Two of the leading ten publicly listed companies in Asia set out in Annex 3.1 are Thai based canners and processors. Today, China has become a leading location for processing imported fish raw materials in Customs free zones for re-export to developed country consumer markets. In the period 2002 to 2004 more than USD 580 million of seafood products were exported on average annually from North America to China and more than USD 1.4 billion imported[11]. Much of that trade was relatively unprocessed frozen fish products exported to China for further processing and re-importation back into the USA.

Business to business relations

Trading

The classic entry level to globalising of seafood is trading. Traditional trader-based trade remains a core part of the business mix. Margins are tight, returns vulnerable to supply surges and product attributes downplayed in favour of price. But much of the supply chain in many markets is initially supplied by trading. Trading enables small producers to enter export markets at low cost (as well as potentially uncertain returns).

For example within the United States, a significant proportion of seafood business is still conducted by brokers acting as middle-men between exporters to the US market and buyers of seafood products – particularly in the food service sector. Brokers may be denigrated as "order takers", but their strength in the US markets situation is often their capacity to provide a full range of seafood products to food service buyers who want to purchase a mix of seafood products – shellfish, shrimp and a range of popular fish species – and who want "just-in-time" delivery that will minimise their inventories. Selling through brokers is attractive to exporters who do not have a capacity to provide

11. FAO (2006), *Fishery Yearbook, Volume 99 Ibid.*

the full market mix that customers require. The brokers undertake this service by sourcing seafood products from a variety of countries of origin and suppliers in order to be able to meet customer expectations of product mix.

Markets in Japan and Europe are supplied through wholesale markets. They not only receive products from domestic fishers, they are also supplied by exporters selling relatively, unprocessed frozen fish, and also high quality, chilled and live fish and seafood. Logistics of getting imported product to the markets is normally handled by local agents. Large scale wholesale markets supply fish on a daily basis to wholesalers and traders supplying processors, food service and retail. In the case of Japan, products purchased at auction may pass through several traders before final sale and all within a matter of hours from auction.

Contracting

Brand owning processors and suppliers of processed fish products to brand owners need to ensure consistency of supply, quality and price on longer time frames than are provided by brokers or auction arrangements. While processors are extremely cost focused, they offer sales security provided that suppliers are able not only to meet price expectations but can also supply volume and meet specifications. Smaller producers can be challenged by the scale of processors' needs. If they are to supply, it may require that they do so through joint contract arrangements with other suppliers.

Investment

Contracting relationships between processors and suppliers can lead to equity investment as both sides seek to assure their needs. As discussed earlier, vertical integration of the supply chain is unlikely if there are doubts on sustainability of the resource or the rights or opportunities to access it or where investment is constrained by government intervention. This may discourage investment by processors in the harvest sector. However stakeholders in the harvest sector that wish to gain a share of the value chain are taking equity interests in processing, most recently in processing located in developing countries. Examples can be found in China and South East Asian countries like Vietnam.

Retail influences

Supermarkets are exerting major influence in retail markets for fish products. The independent retail fish shop is struggling to survive in many developed countries. Supermarkets offer the convenience of one-stop shopping. Retailing seafood is operationally expensive. It demands personal service, while supermarkets offer self service store environments. Seafood is both a cost centre and an opportunity for supermarkets to add value to the one-stop shopping experience. However, as generalists, supplying quality seafood is a challenge for supermarkets. Seafood requires a relatively large commitment of display space, quicker stock turn and higher labour input than is necessary for most other sections of a supermarket. This means that retail margins required are consequently higher and can reinforce consumer misconceptions that seafood is an expensive option. Supermarket buyers seek to correct that perception through seeking to drive down prices that they are prepared to pay for raw material.

As a high profile sector of the store, seafood has been required to contribute to the brand positioning of the supermarket operator. In some markets, notably northern Europe and the UK and increasingly the USA, supermarkets are promoting seafood in the context of supermarkets' social responsibility brand positioning. The brand positioning is manifested at two levels. One is in the general positioning of the supermarket brand as socially responsible. Supermarket chains in the USA such as Wild foods and Wal-Mart and in Europe Wal-Mart's subsidiary Aldi, plus UK chains including Sainsbury, Waitrose, Marks and Spencer and Tesco are publicly committing themselves to "responsible" sourcing policies for food generally and using seafood as a lead product sector in that positioning. This takes the form of demands for independent certification to confirm that fish was sourced legally and from independently verified well-managed fisheries. The second area is in "house-branded" products. Supermarkets such as Wal-Mart and Marks and Spencer are requiring that capture seafood products in their house brands are sourced from fisheries independently certified as well-managed. At present they are favouring seafood products from fisheries certified to the Marine Stewardship Council's standard. There is no equivalent reference standard for aquacultured fish products and, to some extent, organically labelled aquacultured seafood is taking that positioning.

Supermarket buying strategies differ round the world. In the USA supermarkets are major users of the services of brokers, as noted earlier. In Europe they contract directly with processors for supply and may also source wet fish from wholesale markets. In the UK, Waitrose as a smaller chain with a strong social responsibility positioning, is sourcing wet fish directly from suppliers that can meet its social responsibility positioning. In so doing, it also avoids at least one stage of margin addition in the value chain. In Japan, supermarkets source directly and indirectly from public auctions. In Australasia there is a mix of buying from auctions and contracting for supplies. The push for social responsibility branding, such as MSC, requires the supermarket to undertake chain of custody certification of its own purchasing processes. A result of this is likely to be an increase in sourcing fish from contracted suppliers rather than wholesale markets and to manage supplies through centralised warehousing and distribution. Wholesale and auction markets will continue to supply the food service sector and smaller independent retailers. Restaurants lack buying power and often favour buying fresh, whole fish. For them, auction and wholesale markets are an attractive option.

Branded products in supermarkets, including house brands, are more frequently found in the freezer case, rather than as chilled products. Packaged, frozen seafood products provide an opportunity for brand owners to place collateral statements related to their social responsibility positioning emphasising origin and management of the products. The fact that the product may have been processed in a third country is recorded at most in the fine print and is neither featured nor addressed as potential issue of concern to consumers.

External Influencers on Globalising Business

Governance issues – Capture fisheries

This chapter has asserted a point of view that vertical integration of the catch and processing sector is more likely where access and management regimes maximise opportunities to be assured of catch. Assurances are required for at least two elements – appropriate catch limits to assure abundance and access conditions that provide secure opportunities to fish. The evidence supplied from the thirty largest companies by market capitalisation suggests that vertically integrating investment in the catch sector is more likely in fisheries with secure rights of access. Significantly, investment in the catch sector by the identified companies is oriented towards pelagic stocks – with inherently less biological risk of over exploitation.

Furthermore, while the significant public companies in the sector are predominantly anchored in the processing sector, exposure to the catch sector is in the minority. Only one of the thirty companies has clearly invested in fishing activities with virtually no secure access rights – that is China Fisheries Group, which reports a strong reliance on high seas fishing in the Pacific Basin. The status of Pescanova is less clear.

Commercial fishing the world over continues to be dominated by smaller enterprises. Their products enter international trade, without doubt, and are globalised by the processing sector. But only in a very small number of cases, as in New Zealand, Chile and Iceland, has fishing aggregated into larger access rights-based enterprises and become vertically integrated into globalised value chains.

The default position for fisheries and governments responsible for them in most jurisdictions – regardless of relative wealth of individual nations – is to consider fisheries as an economic and social activity first and maximise the opportunity for employment. Not surprisingly, stakeholders exhibit strong resistance to pro-active efforts to reduce fishing to match stock sustainability – and change comes about as a result of crisis. Despite efforts to put an end to the "tragedy of the commons", the tragedy continues. This is not a basis for encouraging new investors and thus the trade relationships that stakeholders have with the markets for their fish are likely to remain commodity trade based.

National attitudes towards fisheries governance become even more clearly exposed in the operation of regional management arrangements. Based on well established practices of international law favouring consensus decision making wherever possible, RFMOs characteristically move at the pace of the slowest member. Thus States that domestically may favour precautionary stock management settings, or favour clear access and allocative mechanisms that attempt to address the need to balance catch efforts with sustainable abundance, acquiesce in RFMOs that avoid addressing such fundamental issues for as long as possible due to inability to achieve consensus.

Where RFMOs attempt to address allocation of the opportunity to fish between parties, it seems unavoidable that States become advocates on behalf of the fishers under their jurisdictions and compromised between demands to maximise opportunities to fish for their nationals and the need to take precautionary measures for stock sustainability reasons. This frequently translates into allocating access by determining the number of vessels that can participate and separately determining a catch limit, but avoiding linking the two together through individualising the catch limits and confronting vessel operators with the prospect that the outcome may be uneconomic. As a result vessel operators are incentivised to race for fish and cheat in order to be profitable.

It should be no surprise then that brand owners who wish to be in the business of supplying fish to their customers, recognise the threat to their brand integrity positions from being associated with supplying fish from failing resources. They show little interest in investing directly in the catch sector. But they are expressing their interests in ensuring that sources of supply are secure into the future. The wave of retail brand owners – both independent brands such as Youngs and Iglo and supermarkets such as Tesco, Aldi and

Wal-Mart – are demanding independent verification and certification that the capture seafood that is offered under their brands is sourced from sustainable fisheries. This may be dismissed as "grandstanding", as they have no capital at risk in the catch sector, and can afford to take a high moral stance. But they also have coherent brand positionings to safeguard against charges of consumer deception. Brand owners cannot tolerate undermining fundamental values of their brands – they are real capital assets.

The question remains open as to whether the scale of their demands in terms of the volume of sales they represent can be met or whether the brand owners will switch to sourcing alternative aquacultured supplies with equivalent assurances of sustainability, or whether they must take the decision and abandon the capture seafood parts of their businesses.

Aquaculture

The illustrative sample of public companies used throughout this chapter has a minority of companies invested in aquaculture or business activities strongly linked to aquaculture. The investment is overwhelmingly concentrated in salmon aquaculture or feed associated with salmon aquaculture. If nothing else, this is confirmation of the current success and scale of this element of the aquaculture sector.

It also reflects the history of salmon aquaculture development. The major developments in Norway and Chile, took off in benign regulatory environments in the 1970s and 1980s. Water space was relatively easy to secure and expand and both locations were physically favourable to the species being farmed. By comparison, salmon aquaculture in New Zealand and Australia started at about the same time, including with some of the same establishing investors, but both locations have failed to attain similar scale.

Aquaculture requires secure access to coastal space on a long term basis. Security of tenure requires not only a legally defensible position for the aquaculture operation but also that any conflicts related to the alienation of space have been addressed.

Marine aquaculture demands that societies agree to the alienation of space to the activity. This does not appear to be a cultural problem in some societies, but in others it poses challenges to perceptions of what is acceptable activity in coastal areas, especially in sight of coastal communities. The potential for aquaculture will be constrained by its social acceptability and thus, the future location of aquaculture will not necessarily be where it makes most "sense" in a physical sense. It is unlikely that aquaculture will be allowed to thrive in spaces that have high amenity value for coastal communities that have the wealth available to choose to exclude it.

Earlier, this chapter identified "chickenability" and the place for low cost aquacultured fish products that can compete equally with the volume animal proteins on functionality and price. Currently a small number of tropical omnivorous fish species appear attractive candidates. But where will production be located? Will it be concentrated only in those places where they occur naturally – or have been successfully acclimatised already? Or will societies be prepared to confront the biosecurity risks of permitting aquaculture of new, alien species?

The late 20th century marked a shift in social attitude to alien species introductions, especially in developed countries colonised and settled in the last 200 years with accompanying major impacts on the native ecosystems from the introduction of farming and urbanisation. For hundreds of years there was little or no social resistance to, and in fact active encouragement of, the transfer and establishment of species from one place to another, despite growing evidence of negative impacts on some native species and habitats from alien introductions. By the end of the 20th century, biosecurity impact concerns have encouraged some governments to enact more and more stringent controls to limit risk. This new risk aversion has popular support where societies have a growing sense of guilt over what has been lost in the course of the last century or two.

A consequence for aquaculture development in societies that have the economic power to afford to make decisions to protect environmental amenity values rather than encourage further wealth generation, is to limit location to places that are out of sight and to species that are already available in the local environment. For globalising aquaculture business development, location will be driven by security of access and potential for profit. Societies with heightened sensitivities and resistance to perceived amenity impact will be avoided, as the cost consequences of accommodating such concerns will redirect investment to more accepting locations.

Governments as technical regulators

Globalisation of the food business, including the processing and marketing of seafood, has found a fundamentally synergistic relationship with the new post-Uruguay Round risk-based approach to food safety and technical regulation. The WTO's Agreements on technical issues, including food safety and plant and animal health are built on business models of minimal but demonstrably necessary intervention and fitness for purpose.

The globalised, multi-national seafood processing and marketing sector has had little significant difficulty in working in the new environment. Frequently demands to assure brand integrity will operate more risk-averse settings for food safety than are required by regulatory intervention. And the same can hold for technical specifications and labelling requirements.

In the new risk based environment, governments are challenged to find ways to negotiate and implement mutual recognition and equivalence agreements that will enable verification and certification systems between States to communicate and thus facilitate trade. Trust and understanding between States has to be matched by integrity within them. Corruption of that trust leads to trade being stopped.

It is clear also that producers that are not integrated into a global supply chain, but simply trade their products into it, whether in developed or developing countries, have been challenged to meet the new environment. For many, the move to self-disciplined, quality management systems has been culturally as well as economically challenging and demanded significant internal change to management and staff relationships. The new risk based systems fail if staff are not empowered to take responsibility. When the food safety regulation was based on command and control, the incentive was to cheat the inspector. When the rules changed to require continuous performance and the threat was exclusion from markets or customers, attitudes and performance have improved.

There is frequently a marked difference in the attitude of officials towards stakeholders between regulatory regimes for food safety and fisheries management systems. Food safety regulatory regimes in the last ten years have been restructured away from command and control towards performance auditing of self managed food safety systems that the food producers own. Fisheries management regimes remain firmly wedded to command and control, even in rights-based management systems. Decisions on harvest strategies, stock sustainability reference points are more often that not the realm of government employed scientists, officials and politicians. In food safety management, it is up to the producer to identify food safety hazards and appropriate controls to manage them. The role of the regulator is to set the standards for performance outcomes expected and audit the producer's performance against the standards.

Globalisation of Civil Society – Relationships with the NGOs

The rise of the environmental movement since the 1950s and 1960s has had a patchy impact on the seafood sector. As discussed earlier, the impact has been significant in developed countries with the wealth to make choices to favour amenity values over economic values.

NGOs have had only limited impact directly on government policy settings in relation to fisheries and aquaculture. Success appears to have been proportional to their strength or influence at the ballot box. Occasionally there have been major victories, as in the case of the "dolphin safe" campaign that impacted heavily on consumer purchasing decisions in the USA and consequently on national and regional management measures in tuna fisheries. The commercial sector responded, driven by the need to restore and safeguard sales, and provided the political support for regulatory change to be implemented at national and multi-lateral levels.

NGO strategies are increasingly following the model of trying to shift public opinion through campaigning while playing on the vulnerabilities of business to loss of markets if their campaigns are successful in shifting public attitudes against the businesses concerned. Their objective is to bring together a social consensus for change that enables governments to take actions NGOs advocate with minimal ballot box impact.

The Marine Stewardship Council standard and eco-labelling scheme was initially a joint initiative of NGOs and globalised seafood processors and retail brand owners concerned to safeguard brand integrity. It sought to by-pass governments and fisheries managers and appeal directly to catch sectors to promote changes to management settings that would be rewarded by market demand for those fisheries' products. Up to now the fisheries that have been MSC certified have been on the basis of current performance and sustainability settings with only relatively marginal or incremental changes required over the period of certification.

There is no clear example yet of stakeholders in a fishery voluntarily promoting changes to management settings in order to obtain certification at some future date. However the uptake by key retail brand owners of policies to stock only independently certified seafood may be of sufficient market impact that it will induce the changes that the MSC's founders foresaw.

Conclusion

This chapter has sought to discuss globalisation of the fisheries sector through catching, aquaculture, processing leading to vertical integration through the value chain from production to plate. It shows the concentration of demand for imported fish products in the developed country markets of Europe, North America and North Asia, while identifying that developing countries are now supplying the majority of fish that is exported.

The chapter recognises that a large proportion of fish production is traded internationally. Setting aside trade between European Union member states, more than 60% of fish exports by value originate from developing countries. However much of that fish is traded in commodity trade arrangements and little is traded in more secure supply contracts or conducted as a result of transfer trading between companies that relate to each other through shared equity.

The 30 largest publicly listed companies by market capitalisation have been examined as examples of globalised businesses in the sector. The chapter has identified strong evidence of globalisation forward in the market place from processing through to markets, but weaker evidence of integration between the catch and processing sector and a concentration of investment in aquaculture only where it is proven to be commercially successful. The chapter suggests that evidence of vertical integration is stronger where investors can have assurance of resource access and resource abundance. Fisheries that are managed with precautionary settings, or fish stocks that are naturally resilient as in the case of small pelagics and fisheries that have secure access rights are attractive for globalising, vertical integration.

In regard to aquaculture, the chapter identifies a concentration of public company investment in salmon aquaculture and businesses servicing the needs of that sector. It notes that there is opportunity for diversification into aquaculture systems that can operate at lower cost structures and offer products that may compete better in mass markets supplied by other farmed animal protein products.

The chapter discusses the role of governments and the influence of society values on fisheries management, aquaculture development and regulation of food safety.

In regard to fisheries management, it concludes that continued failure to address fisheries sustainability and secure rights to access fish discourages investment and perpetuates fisheries catch sector stakeholders in entry-level commodity trading relationships with international markets. Vertical integration between the catch sector and the rest of the value chain is taking place in fisheries that can assure stock abundance and secure access.

In regard to aquaculture, the chapter notes the need for secure access to water space as a key factor for aquaculture developments. It discusses the clash of values between societies that place amenity value of coastal space ahead of economic development and concern to protect native biodiversity with highly risk-averse policy settings is likely to lead to aquaculture development locating to places that are less sensitive to such values.

In regard to regulation of the processing sector, the chapter suggests that rather than discouraging trade, investment and globalisation, the post Uruguay Round science and risk-based approaches to food safety assurance have facilitated the development of the processing sector and its relocation to developing countries to take advantage of lower costs while delivering safe and wholesome food. The chapter contrasts the enabling relationship between regulators and processors that lie at the heart of performance based food safety regulation, with the continued adoption of top-down command and control fisheries management regimes that dis-empower stakeholders.

The chapter concludes with some observations on the influence of globalising environmental NGOs and discusses the Marine Stewardship Council as an example of civil society and brand owning corporates joining to provide market based incentives to catch-sector stakeholders to promote less risky fisheries management settings than those implemented by governments.

The New Zealand Seafood Sector as a Case Study of Business to Business Relationships

The New Zealand seafood sector is the fifth largest goods exporting sector in the New Zealand economy[12]. Total seafood exports in 2006 amounted to NZD 1.35 billion f.o.b. International markets provide about 90% of total revenues for the sector. New Zealand is a small producer – supplying less than 1% of global seafood production and less than 2% of global seafood trade. However it is a microcosm of the globalising seafood business.

The fishery is managed under a quota management system with individual, perpetual, transferable property rights. Rights are overwhelmingly domestically owned and there are restrictions on foreign investment in fish quota rights. There are no foreign investment restrictions on aquaculture developments.

There are over 2 000 rights owners, but 66% of primary landed value is concentrated in ten companies. The top ten companies, measured by the value of their rights, are

- Sanford Sustainable Seafood Ltd
- Sealord Group Ltd
- Talley's Fisheries Ltd
- Te Ohu Kai Moana Trust Ltd
- Aotearoa Fisheries Ltd
- Vela Fisheries Ltd
- United Fisheries Ltd
- NZ King Salmon Ltd
- Independent Fisheries Ltd
- Ngai Tahu Seafood Resources Ltd

Nine of the companies are active fishing and/or aquaculture companies. Te Ohu Kaimoana Trustee Ltd owns quota on behalf of Maori tribes and is going through a process of divesting that quota to those tribes. While it continues to own quota, it allocates it to tribes who in turn can fish the quota or annually trade the catch entitlement that it represents to other fishers.

The table in Annex 3.2 sets out the key attributes of each company and its involvement in the value chain.

Only Sanford Ltd is a publicly listed company. Sealord Group Ltd is 50% foreign owned by Nippon Suisan Kaisha Ltd and 50% owned by Aotearoa Fisheries Ltd, but operated independently of its two shareholders. It is the most internationally integrated of all the companies in the seafood sector. NZ King Salmon Ltd is the only solely aquaculture based company in the group and the only company that is 100% foreign owned. Of the remaining privately owned companies, four are family owned and the others are owned by Maori. Aotearoa Fisheries Ltd is owned by all Maori tribes and Ngai Tahu Seafood Resources Ltd is owned by one tribe.

A total of 138 exporters of fish and fish products are registered on the New Zealand Seafood Industry Council's web site www.seafood.co.nz . They breakdown as follows:

---

12 The primary source for the information in this Case Study is the NZ Seafood Industry Council's data bases and records.

Table 3.3. New Zealand Registered Exporters of Fish and Fish Products

Brokers	57
Fishing Companies	22
Aquaculture Companies	12
Vertically integrated processors	31
Specialist processors	17

New Zealand seafood products are exported to over sixty countries, but the proportion of trade to the European Union, Japan and USA are similar to global patterns. However there are two significant exceptions. Australia is the largest market on an individual country basis, and ranks second if one aggregates sales to all European Union member states. Australia took 16% by value of all seafood exports in 2006. The second exception is the proportion of trade to China and through Hong Kong to China. Collectively these two markets accounted for 21% of New Zealand's seafood exports and were the leading export markets. The top ten country markets in 2006 were:

Table 3.4. Top Ten Country Markets in 2006

Australia	NZD220 million	16.3%
USA	NZD209 million	15.5%
Hong Kong	NZD165 million	12.2%
Japan	NZD143 million	10.6%
China	NZD126 million	9.3%
Spain	NZD 92 million	6.8%
Korea	NZD 68 million	5.0%
Germany	NZD 37 million	2.7%
France	NZD 31 million	2.3%
Singapore	NZD 25 million	1.9%
EU collectively	NZD253 million	18.7%

The market concentrations can be explained as follows:

- *China and Hong Kong*: Close to 70% of sales to China are headed and gutted frozen fish and relatively unprocessed frozen squid for further processing and re-export to markets in North America and Europe. Much of that processing and re-export is through operations partly owned by the two largest seafood companies in New Zealand. About 80% by value of the sales to Hong Kong are of two products – live rock lobster and processed abalone. Most of these products are further exported from Hong Kong into China where they are consumed by wealthy diners.

- *Australia*: Australia has become an extension of the New Zealand domestic market and takes a full cross-section of all seafood products. It is the primary market for added value, branded seafood products manufactured in New Zealand. While New Zealand's 4 million people can make a limited consumption impact on the total production of seafood Australia, with per capita consumption levels similar to levels in Northern Europe or the USA, offers a market five times larger than New Zealand. Australia has a reliance on imported seafood proportionally similar to the USA's. New Zealand's supply relationship to Australia is similar to Canada's relationship with the USA.

- *European Union*: New Zealand's fisheries are predominantly temperate to sub-Antarctic. These environments favour fish products with functional properties that can substitute for fish resources that the EU's fisheries are now unable to supply. The leading markets in the EU are Spain, which has become a leading market for squid and for hake and ling. Germany and France's white fish processing sectors are the leading users of frozen fillets and fillet blocks of hoki.

- *Japan*: Twenty years ago Japan purchased and consumed more than 40% of New Zealand's fish exports and was an active distant water fisher in the New Zealand EEZ. The decline in sales to Japan has come about as New Zealand exporters have diversified their market risk. It also reflects a cycle of development in New Zealand towards greater value adding in the 1990s that resulted in sales in the USA, Australia and Europe at the expense of Japan. A significant proportion of that value-adding processing is being moved to China.

- *USA*: The USA has been New Zealand's leading market for processed white fish products for most of the last twenty years. It is the key market for orange roughy and a leading market for frozen filleted hoki. The USA has also been developed as the lead market for aquacultured Greenshell™ mussels. Figures over the last three years would suggest that sales to the USA of hoki have declined as sales of headed and gutted hoki to the China have increased. However, this reflects the outsourcing of processing to China, as the final consumer market for that fish remains the USA.

Conclusion

New Zealand represents a microcosm of globalisation in the seafood business today. The largest companies are diversifying their businesses and moving their marketing control up the value chain, principally by developing deeper in-market relationships and by outsourcing processing to China through processing operations in which they have some ownership.

A significant proportion of companies use foreign chartered fishing vessels to fish in seasonal and lower value fisheries – thus focusing their own vessel investment in vessels that are actively fishing throughout the year. A small number have developed distant water fishing operations – in tuna fishing in the Pacific and deepwater trawl and longline fisheries in the high seas.

Most companies have invested in aquaculture as well as capture fishing, and most are vertically integrated, processing companies. Only one company has actively diversified into other food related businesses.

Sales are achieved by a mix of trading and through agency arrangements. A small number of companies are investing in their own sales networks, particularly in Australia.

References

FAO (2007), *The State of World Fisheries and Aquaculture* (SOFIA), FAO, Rome

FAO (2005), *Fishery Yearbook, Fishery Statistics, Aquaculture Production*, Vol.96/, 2003, FAO, Rome

FAO (2006), *Fishery Yearbook, Fishery Statistics, Commodities*, Vol.99 2004, FAO, Rome

FAO (2006), *Fisheries Report 186*, FAO, Rome

Glitner Bank(February 2007), *Stockwatch Seafood Index,* Intrafish, Vol.5, Issue 2

United Nations, 1997, *The Law of the Sea*, United Nations, New York

Washington Post, 13 November 2006

Wright Reports, http://wrightreports.ecnext.com/
www.wordspy.com/words/chickenability.asp .

Annex 3.1. Major Globalised Publicly Listed Corporates

Ten Largest Companies: Americas	Country	M. Cap USD	Fishing	A'culture	Fish Feed	Processing	Trading	Retail	Multi-national	Diversification
Connor Bros Income Fund	Canada	504				√	Global	Retail brands		Shelf stable meat products
Sociedad Pesquera Coloso SA	Chile	329	√			√	Global			
Pesquera Itata SA	Chile	254	√		√	√	Export			
Pesquera Iquique-Guanaye SA	Chile	180	√			√	Export			
Clearwater Seafoods Income Fund	Canada	124	√			√	Global			
Omega Protein Corp	USA	107			√	√	Global			Shipyard
Fishery Products International (FPI) Ltd	Canada	96	√			√	Export		USA, Europe, Japan, China SE Asia	
Copeinca SA	Peru	90	√		√	√	Export		Norway	
High Liner Foods Ltd	Canada	87				√	North America			
Vita Food Products Inc.	USA	9				√	North America			Honey, processed foods

Annex 3.1. Major Globalised Publicly Listed Corporates (cont.)

Ten Largest Companies: Europe	Country	M. Cap USD	Fishing	Aquaculture	Fish Feed	Processing	Trading	Retail	Multi-national	Diversification
Marine Harvest ASA (former Pan Fish)	Norway	3 712		√		√	Global		Europe, North America, Japan	
Cermaq ASA	Norway	1 470		√	√	√	Global		Europe, North America, Chile	Byproducts processing
Austevoll Seafood ASA	Norway	1 299	√		√	√	Global		Chile, Peru	
Leroy Seafood Group ASA	Norway	778				√	Global		Europe, North America	
Pescanova S.A.	Spain	526	√	√		√	Global		Chile, North America, Australia, Europe	
Biomar Holding A/S	Denmark	477			√		Global		Europe, Chile	Pharmaceuticals
Alfesca HF	Iceland	420				√	Global	Retail brands	Europe	Processed foods
Icelandic Group HF	Iceland	305	√			√	Global		Europe, North America. Korea, Thailand, Japan	Food services
Aker Seafoods ASA	Norway	241	√			√	Export		Europe	

Annex 3.1. Major Globalised Publicly Listed Corporates (cont.)

Ten Largest Companies: Asia	Country	M.Cap USD	Fishing	Aquaculture	Fish Feed	Processing	Trading	Retail	Multi-national	Diversification
Nireus Aquaculture S.A.	Greece	232		√		√	Europe	Retail brands		Dairy and confectionary products, fish farming equipment
Nippon Suisan Kaisha Ltd	Japan	1 591	√	√		√	Global	Retail brands	Global	Pharmaceuticals, Marine engineering cold storage and transportation
China Fishery Group Ltd	China	989	√			√	Global			
Maruha Corp	Japan	740	√			√	Global		Global	Meat, byproducts, pharmaceuticals, storage and logistics
Thai Union Frozen Foods Group	Thailand	553				√	Global			
Pacific Andes	China (Hong Kong)	359				√	Global		North America, China, Japan, Europe	Shipping services, cultivation & processing of vegetables, property
Nichiro Corp.	Japan	303				√	Global		Global	Hotels, packaging machinery
Sea Horse PLC	Thailand	276				√	Global			
Kyokuyo Co Ltd	Japan	239	√			√	Global		USA, Panama, Thailand, China	Other processed foods, storage and transport, insurance
Uoriki Co Ltd	Japan	171				√	Domestic Japan	√		Restaurants
Chuo Gyorui Co Ltd	Japan	129					Domestic Japan			Storage, transport, property

Annex 3.2. Leading New Zealand Seafood Companies

Company Name	% of sector	Types of rights owned	Rights Trader	Own fleet	Distant water fisher	Foreign vessel Charterer	Aquaculture	Domestic processor	Offshore processor
Sanford Sustainable Seafood Ltd	19%	quota, a'culture	√	√	√	√	√	√	√
Sealord Group Ltd	14%	quota, a'culture	√	√	√	√	√	√	√
Talley's Fisheries Ltd	10%	quota, a'culture	√	√	√		√	√	
Te Ohu Kai Moana Trustee Ltd	8%	quota	√				√		
Aotearoa Fisheries Ltd	5%	quota, a'culture	√				√	√	
Vela Fisheries Ltd	3%	quota	√			√			
United Fisheries Ltd	2%	quota, a'culture	√	√		√	√	√	
NZ King Salmon Ltd	2%	a'culture					√	√	
Independent Fisheries Ltd	2%	quota	√			√		√	
Ngai Tahu Seafood Resources Ltd	1%	quota	√	√		√	√	√	

Annex 3.2. Leading New Zealand Seafood Companies (cont.)

Company Name	Seafood Trader	Supplies brokers	Established Agents	Own offshore Marketing structure	Ownership	Off shore investment	Diversi-fication
Sanford Sustainable Seafood Ltd	√		√		Public	√	
Sealord Group Ltd	√		√	√	Private, 50% foreign	√	
Talley's Fisheries Ltd	√	√	√		Private, family		Meat, Dairy, Vegetable processing
Te Ohu Kai Moana Trustee Ltd					Tribal		
Aotearoa Fisheries Ltd	√	√	√		Tribal		
Vela Fisheries Ltd	√	√	√		Private, family		
United Fisheries Ltd	√	√	√		Private, family		
NZ King Salmon Ltd			√	√	Private, 100% foreign		
Independent Fisheries Ltd	√	√	√		Private, family	√	
Ngai Tahu Seafood Resources Ltd		√	√		Private, Tribal		

PART 2

Retailing and Distribution of Fish: Trends and Issues

Chapter 4

Global Seafood Sourcing: A Restaurant Perspective[1]

Roger Bing, Vice-President, Protein Procurements, Darden Restaurants, United States

Darden restaurants source seafood from over 30 countries and seafood comprises 70% of their protein purchase. Darden's strategic focus is to shorten supply chains by contracting directly with the producer. The company's Quality standards are more stringent than national regulations and overseen by the employment of 50 employees in Darden's Quality Assurance Department alone, 10 of whom are based at the point of origin of seafood purchases. Darden is proactive in addressing social and environmental laws in addition to implementing food safety measures: By contract, all suppliers must adhere to all laws in the country where they operate. If the sustainability of species is questioned, Darden will remove it from its List of Approved Species.

Darden Environmental Trust provides funds to projects that address sustainability. In addition, Darden Restaurants Inc. was the first American supply company to adopt Best Aquaculture Practices guidelines established by the Global Aquaculture Alliance and certified by the Aquaculture Certification Council (ACC).

Darden's research shows that to maintain per capita consumption of 7.35 kg per annum by 2025 in the United States, approximately 400 000 additional tons of fish and fish products (edible weight) will be required. This represents a 15-20% increase. As wild fish production has been stable for the past decade, capture fisheries are unlikely to satisfy this demand, so supplies will have to come from aquaculture. As well as offering a competitive advantage, aquaculture also provides easier traceability and a more consistent supply compared to wild fisheries.

Darden Restaurants is the largest casual dining restaurant company in the United States. The Company owns and operates four distinctive brands – Seasons52, Olive Garden, Bahama Breeze and Red Lobster - where seafood has a strong presence.

1 Thanks to the Food and Agriculture Organization (FAO) of the United Nations and the Organisation for Economic Co-operation and Development (OECD) for their leadership role in co-sponsoring this Workshop on the Opportunities and Challenges of Fisheries Globalisation.

Figure 4.1. Darden -- A Global Seafood Purchaser

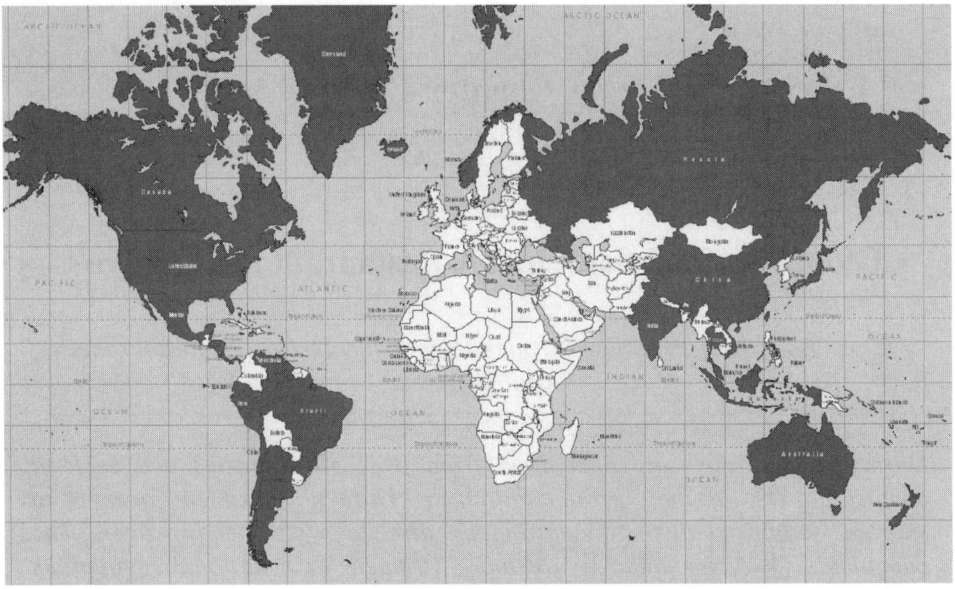

Darden sources seafood from more than 30 countries by contracting directly with the producer in the country of origin (see Figure 4.1.), a procurement practice that is unique in the U.S. restaurant industry. The Total Quality Assurance Department, which includes staff in Orlando, Fla., and other key locations throughout the world, supports the process by ensuring – first and foremost – that food safety and our quality product specification requirements are met. Darden consistently meets and exceeds stringent U.S. regulations.

Figure 4.2. U.S. Restaurant Growth (1)

(current USD billion)

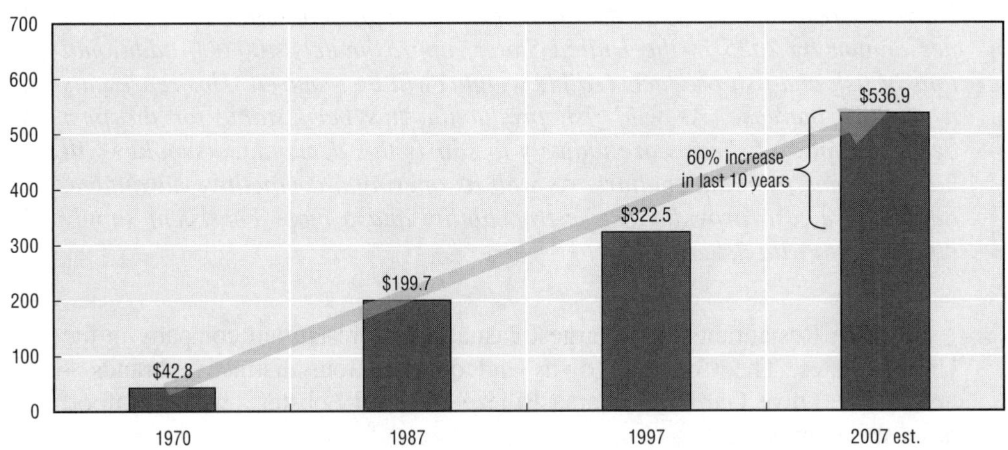

(1) 935 000 restaurants will meet USD 539.9 billion in sales in 2007.
Source: National Restaurant Association 2007 Restaurant Industry Forecast

All of us are familiar with the information published by the Food and Agricultural Organization (FAO) of the United Nations that tells us wild capture fisheries are static, so when the National Restaurants Association of the United States published details about the significant growth of the industry for the past 25 years – a growth that mirrors Darden's trajectory – we conducted a research exercise that demonstrated a deficit of 400 000 metric tons of seafood in the United States alone by the year 2025.

Figure 4.3. Future U.S. Food Supply (1)

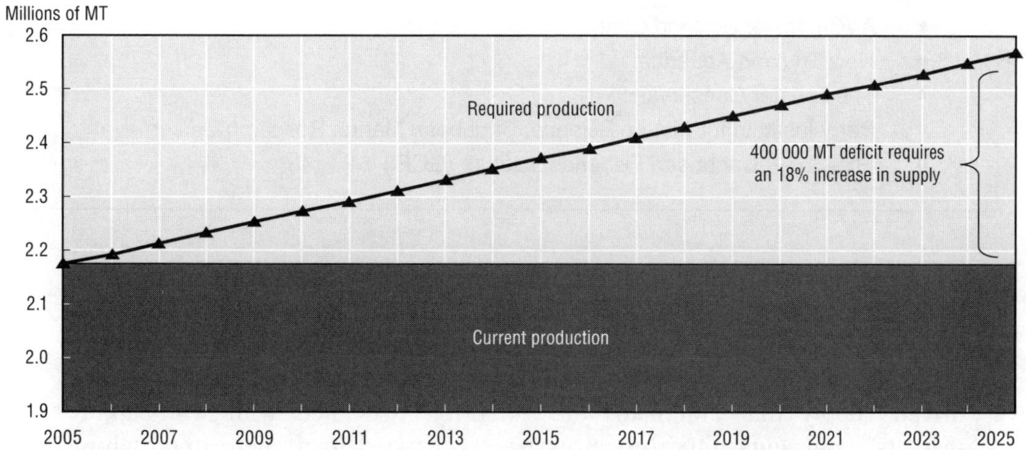

(1) To maintain per capita status quo (7.35 kg) will require approximately400 000 additional metric tons (edible weight) by 2025.

Source: Data from U.S. Census and FAO. Assumes constant seafood consumption of 7.35 kg/capita and population growth of 0.9% per annum (US Census).

We anticipate this deficit will be global, to varying degrees, over the next 10-15 years. Clearly, this increase in demand can only be satisfied by an increase in aquaculture production.

Seafood sustainability is very important to the long-term success of our business, and Darden has taken steps to address this concern. For wild-capture fisheries, the company has removed species such as orange roughy and Chilean sea bass and is using the New England Aquarium's ocean conservation expertise to assist the company in assessing species under consideration for its approved purchase list. In addition, the Darden Environmental Trust of the Darden Restaurants Foundation funds and supports many seafood sustainability projects that aim to conserve and protect the ocean's resources, including spiny lobster stock assessment in Belize; the Gulf and Caribbean Institute's workshops on queen conch, spiny lobster and large pelagic species; and the annual U.S./Canadian Lobstermen's Town Hall Meeting, just to name a few (see Box 4.1).

Box 4.1. Darden Environmental Trust

- Belize Fisheries Department
- Hubbs-Seaworld Research Institute
- Rhode Island Sea Grant
- The Department of Marine Science, University of Southern Mississippi
- Lobster Institute
- AVC Lobster Science Centre
- New England Aquarium
- The Lobster Conservancy
- Perry Institute for Marine Science, Caribbean Marine Research Center
- The Gulf & Caribbean Fisheries Institute (GCFI)

Aquaculture production is very important to Darden; in fact, significant amounts of the seafood products we purchase are aquacultured. Not only would aquaculture enhance wild fish stocks, aquacultured products are highly desirable because they provide product traceability and a consistent, quality-driven supply that eliminates the volatility associated with wild-capture markets. But aquaculture practices have generated many detractors, whose negative seafood messaging with regards to food safety and environmental and social-justice concerns have the potential to affect consumer attitudes toward aquacultured products. We ask the aquaculture industry to be proactive in addressing these issues.

Darden is doing its part. By contract, our suppliers must adhere to the laws of the countries where they operate. Darden also requires that all its suppliers of aquacultured shrimp adopt the Best Aquaculture Practices (BAP) developed by the Global Aquaculture Alliance and be certified by the independent Aquaculture Certification Council to ensure their compliance with BAP standards. This is a first for the restaurant industry of the United States.

At Darden, we believe standards and labels have many potential societal benefits. They can protect consumers from unsafe food, the environment from pollution and the exploitation of resources, and workers from social injustices. But their proliferation and inconsistencies are creating a host of other problems for Darden, as we seek to identify standards with the clearest and best guidelines for our business; for consumers, potentially our guests, who want to understand the link between purchasing decisions and effective management of seafood resources; and for emerging nations that would like to enter the aquaculture industry but don't know which standards to adopt. Darden urgently asks for a harmonisation of standards and labels as soon as possible (see Box 4.2).

Box 4.2. Seafood Industry Standards and Labeling

- Alaska Quality Seafood (AQS), US
- Alter-Trade Japan, Japan
- Australian Prawn Farmers Association, Australia
- BioGro, New Zealand
- BioSuisse, Switzerland
- Blue Angel, Germany
- Carrefour Quality Line (CQL) Shrimp
- Dolphin Safe, Department of Commerce, US
- Dolphin Safe, Earth Island Institute, US
- Environmental Justice Foundation (EJF)
- EurepGAP
- Fundacion Chile, Chile
- Food Certification Scotland, UK
- Freedom Food/Royal Society for the Prevention of Cruelty to Animals, UK
- Friend of the Sea, Earth Island Institute Project
- Global Aquaculture Alliance (GAA)/ Aquaculture Certification Council (ACC)
- International Federation of Organic Agricultural Movements (IFOAM)
- International Principles for Responsible Shrimp Farming
- International Standards Organisation (ISO)
- KRAV KRAV, Sweden
- Label Rouge, France
- Marine Stewardship Council (MSC)
- Naturland, Germany
- Organica Food Federation, UK
- Safe Quality Food (SQF) / Salmon of the Americas (SOTA) Farmed Salmon Cert.
- Shrimp Seal of Quality (SSOQ), Bangladesh
- Soil Association, UK
- Tartan Quality Mark (TQM), Scotland

In summary, Darden knows there are tremendous opportunities for the global seafood market. But along with the opportunities, the challenges of the not-so-new globalisation on fisheries must also be addressed. They include:

- The sustainable management of wild-capture resources;
- Sustainable and responsible aquaculture;
- Processed products that meet buyers' health and food safety requirements; and
- Harmonisation of meaningful standards that are based on current best science of the day and practical application

It is incumbent upon all of us – governments, international institutions, non-governmental organisations, fishers, processors, distributors and retailers – to work together to overcome these challenges so we can take advantage of the many opportunities offered by fisheries globalisation.

Chapter 5

Buying Power as an Instrument of Change

Peter Redmond, Vice-President, Wal-Mart, United States

Wal-Mart is a large retailer and owns more than 6 700 stores worldwide and sources its seafood from 70 countries. Sustainability is crucial to Wal-Mart because "it is the right thing to do"*, and because it is critical to ensure continued supplies for Wal-Mart stores. Many consumers hold businesses, in particular retailers, accountable for addressing local and global societal needs. In addition, food safety plays a large role in the US. Therefore, Wal-Mart has adopted a risk-minimising policy of only selling seafood that has been certified and labelled as sustainable and expects this to be widely in place in its North American stores within 2-4 years. The Wal-Mart seafood certification initiative is, volume wise, the largest certification initiative ever undertaken.*

Shrimp is the primary seafood product sold in the United States. By January 2008, all Wal-Mart shrimp will be farm-raised shrimp. ACC certification for shrimp was granted in October 2005 and 65 facilities were certified by the summer of 2006.

As a large, multi-national company, some might assume that Wal-Mart is in favour of globalisation; they would be correct. Wal-Mart is a very large company and getting larger every day. Wal-Mart has 1.8 million associates around the world and a total of 6 800 stores. In addition, 28 source offices cover 70 countries. In 2008, between 625 and 660 stores will open; 270 within the US alone.

So the question is: why is environmental sustainability important for Wal-Mart? One of the biggest issues Wal-Mart faces is a continued license to grow. Consumers want to know that Wal-Mart is 'doing the right thing' to protect the environment. This is also essential in building trust with influential stakeholders. Continuity of supply is critical for Wal-Mart and sourcing products for 600 stores is incredibly difficult. Wal-Mart will sell 50 million pounds of shrimp in 2007 in the Wal-Mart division alone. By the end of this decade it will be over 100 million pounds of shrimp. Sourcing that product requires a tremendous amount of planning.

A company is built around a strong value system and this is completely in line with Wal-Mart's environmental sustainability programme. Wal-Mart's associates are proud to work with a company that is committed to sustainability. Sustainability is also a competitive advantage and provides a license to grow from the consumer. As consumers become more aware of the choices they can make, their purchasing power will drive that further and faster.

Sustainability in Wal-Mart is broken into three categories. The goals are to be 100% reliant on renewed energy, to create zero waste and to sell products that sustain resources and the environment. How is Wal-Mart managing that goal? Within Wal-Mart, sustainable value networks have been created that will bring added value for customers and shareholders. A sustainable value network is a group of people that Wal-Mart has assembled, consisting of internal associates, domestic or international suppliers, academics and NGOs. The seafood value network has over 70 members, of which a dozen are NGOs.

Macro-trends cut across the networks and set contexts for strategic direction. Across the 14 networks, natural resourced demands and overexploitation are issues of concern, whether the product sold is paper or shrimp. Moving from centralized to de-centralized production of energy and products is increasing efficiencies and dependability. Technical advances create the ability to measure and reduce hidden costs. These are critical for Wal-Mart and the business will become accountable for addressing local and global social needs. People look to large, multi-national companies such as Wal-Mart for change.

The seafood sustainability network is divided into five categories. One is wild fisheries and one is farm raised aquaculture. Certification is the network's biggest initiative. Wal-Mart is aligned to the MSC (Marine Stewardship Council) programme and the GAA and ACC. One of the main roles of the network is to work closely with NGO partners. The vision of a sustainable future is stable/restored ocean fisheries that meet present and future needs. This area is of particular interest as a result of Wal-Mart's sourcing requirements. Wise global management of oceans, certified-sustainable fisheries and aquaculture worldwide, and seafood practices that do not threaten species to extinction are crucial.

In 2004, Wal-Mart collaborated with the NGO, Conservation International, to develop Wal-Mart's sustainability programme. In January 2006, Wal-Mart announced that it was adopting the MSC programme. All suppliers met with Wal-Mart and universally embraced this challenge. 2007 sales of MSC-certified products are projected to reach USD 56 million. The initiative includes Wal-Mart's US division, the Sam's Club division (a warehouse cash and carry division), and in the UK, ASDA, Wal-Mart's sister company. The goal is for Wal-Mart to only buy wild caught product from MSC-certified sources within two to four years. Fisheries products have been divided into three main groups: short-term, medium-term and long-term. Short-term fisheries are those that can receive the certification in a straightforward manner as they are very well managed. Medium-term fisheries (around 11%) require some work. 67% of fisheries are long-term; they have serious problems. Part of the two to four year strategy is to have agencies such as WWF and Conservation International involved in the fishery and helping the fishery to achieve the standards of certification.

BAP standards for Wal-Mart's shrimp programme were recently adopted through the GAA and the ACC. In 2004, Wal-Mart worked again with Conservation International, who assisted the company on seafood sustainability issues and across the value chain. In October 2005, Wal-Mart announced that it was adopting the BAP standard to be certified by the Aquaculture Certification Council. The next step was to ensure processing facilities were ACC certified by mid-2006. There are now 65 facilities that have the processing certification. The next phase, which is the most difficult, is the certification of farms. At the farm level, a fragmented industry exists. The intention of certification at the farm level is to coordinate supply, ensure strategic partnerships with suppliers, processors and companies and countries they source from. The other tangible benefit of farm certification is food traceability and food safety. Food traceability and food safety are clearly issues that require risk mitigation. By the end of January 2008, only farm raised shrimp from ACC certified sources will be purchased. This is a clear barrier to entry for both wild caught species and aquaculture within Wal-Mart.

However, Wal-Mart also buys significant amounts of domestic products. The Wild American Shrimp Programme is a non-profit organization similar to organisations Wal-Mart already works with, such as the ACC and MSC although WASP is strictly US-only. Shrimp is the number one seafood in the US. Country of origin legislation recently passed has made a strong point of educating the consumer. Consumers know what they are looking for and where the shrimp they purchase comes from. Given a choice, 94% of consumers prefer to purchase wild caught USA shrimp rather than an import. Wal-Mart puts a lot of effort into educating consumers, including TV advertising. POP (point of purchase) advertising identifies wild caught shrimp on wet counters in frozen food areas. It highlights that the certification programme sets a quality standard that is similar in vein to the ACC and MSC. It is increasingly important to know not just where shrimp is from, but further information on sustainability.

Shrimp from the Gulf of Mexico increased by 108% in 2006. This speaks to the fact that customers not only identify with wild caught compared to farm raised, but also country of origin. In 2007, the figures are expected to reach five million pounds of shrimp from the Gulf; almost 10% of total shrimp volume. At some point, a natural limit will be reached of what the Gulf can produce.

Domestic products command a premium with the customer, not from a price but from an acceptance and desire point of view. It is therefore in Wal-Mart's best interest to point out when shrimp in coastal areas are being sold in coastal areas, using packaging.

Sustainability within Wal-Mart is a way of life and a road map for the future. However, difficulties are raised by the number of certification programmes, although the programmes themselves are good things. However the number confuses retailers and confuses suppliers. There will always be differences of opinion on what makes a good or bad sustainability or certification programme. However, at some point, rationalisation is inevitable. Current generations are the stewards of the oceans. Therefore, change is essential.

Chapter 6

Meeting Customer Needs for Seafood

Peter Hajipieris, Group Policy Manager - Seafood, Tesco Stores Ltd, United Kingdom

Shareholders demand a clear, long term strategy, good governance, transparency, regular reporting, strong performance and a good return on their investment. Market research has shown that customers consider the environmental impact of harvesting seafood when making a purchase decision. These two views have informed Tesco's seafood policy.

Sustainability forms a part of the Tesco Community Plan which sits within the corporate balanced scoring card – the Tesco Steering Wheel.

Taking UK Fish counters as an example, Tesco sources its seafood from 38 countries worldwide. In order to meet demand for sustainable seafood, Tesco has looked to source seafood from approved sources. In identifying these approved sources, Tesco has focused primarily on the governance regimes in place as well as regulatory frameworks, in particular with regard to seafood quality. These include in particular highly visible and active inspection control regimes. Tesco's supply criteria aim to mitigate the potential damage to existing and future markets from unsustainable products.

In the future, aquaculture will continue to be a dominant feature of the value chain, as it continues to provide a secure and sustainable supply of seafood, as well as guarantee quality and traceability.

The key to future sourcing will be to form strategic cross-border alliances which can guarantee supplies, quality and traceability. By the same token more emphasis will need to be placed on collaboration between processors and retailers with regard to product development and innovation.

This chapter will provide some insights into the challenges and facets that go into running a supermarket operation. In particular, this will include some aspects of shareholder demands, what shapes in-store consumer demand (for fish counters and pre-packed seafood), emerging consumer attitudes and finally, the impact of own-label and sustainability on customer behaviour.

Shareholders make a number of demands on supermarkets. The first thing shareholders require is a clear long term strategy. TESCO has a governance model called the 'steering wheel', which governs the way the company operates and communicates to customers and shareholders the way in which the business is run. The steering wheel brings good governance through aspects of the steering wheel, including transparency, regular reporting of the financial performance strategy and clear commitment to a growth strategy. A strong performance on the return of an investment is also vital. Central to this plan is that TESCO listens a great deal to customers. For example, the issue of sustainability may not register to the same extent in every socio-economic group, but sympathy exists for social, environmental integrity and sustainability issues. This is likely to continue to grow in the public mind, aside from sustainability also being the 'right thing' to do.

How does TESCO listen to its customers? The TESCO slogan is "every little helps", which includes one value of vital importance: "I can get what I want". This means "is the product the correct quality?", "Is it fresh?", "Can I find the range I want?". "I can get what I want" is a very simple statement that drives a great deal of business.

In terms of what shapes consumer demands for seafood, in the UK, roughly 450 fish counters exist across the country, with pre-packed seafood in 1 300 stores. For counters, some of the key areas that affect the way consumers behave are the type of store and concept design. For instance, a convenience format is suitable for pre-pack, a superstore for counters and hyper-markets justify a larger counter. In all stores, product quality is crucial. Each range is targeted for the consumer groups that exist within that locality. In addition, training staff is absolutely key. In the UK, the labour market is particularly acute and acquiring skills to serve seafood products across a counter is extremely demanding. Customer service is crucial and obviously, so is marketing. Underpinning all this is that suppliers and sourcing is fundamental.

Customer behaviour can be influenced and fish consumption increased in many ways. Some examples are through advertising, branding, promotion, in-store sampling demonstrations etc. From customer insight, the primary reason why customers shop at a counter rather than buying pre-packed or frozen fish is for freshness. Secondly, it is to service the customer's exact shopping needs, such as size, cut, etc. This third reason is the range and variety. In the early 1980s, supermarkets in the UK sourced most of their fish from UK waters. The situation now has completely altered. TESCO sources from over 38 countries to satisfy the demands of counters and pre-pack alone.

Consumer drivers for pre-pack are dominated by the desire for convenience foods and a different customer profile. For example, the UK consumes half as much seafood on a per capita basis as countries like Portugal, Spain and France. Therefore, consumers do not have the same attitude to seafood and extra encouragement is required to advertise fish counters. Those customers with less confidence are likely to stick to pre-pack.

On-pack information such as the origin of the product is now standard. TESCO is also working on including sustainability certification on the pack. Pre-pack also allows purchases of a particular product on a promotional basis. This has two effects: it puts demands on supply but also brings new consumers into the fish arena. In the last year alone, salmon sales have risen by 35% due to successful promotions.

Products are categorised according to a tier system. For example, the bottom of the smoked salmon packs is the TESCO Value line. After the next level, the standard, comes organic, and then finally, a premium brand. Each product represents a different level of quality and a different level of smoking and flavour profile. They appeal to different socio-economic groups and different consumer occasions.

How will emerging consumer attitudes affect the seafood chain? Four dominant store formats exist at TESCO: Express, Metro, Superstore and Extra. Express is for convenience shoppers, Metro is for metropolitan areas, the Superstore is a standard supermarket, and Extra is a hyper-market. Regardless of the store, all consumers expect their needs to be met and expect value for money in terms of price and quality. In addition, consumers want expertise, everything under one roof, and provenance i.e. social responsibility, especially in how seafood is procured.

Governance regimes are crucial in promoting long-term planning for sustainability. In particular, a highly visible and active inspection control regime is essential, both from an industry perspective and a consumer viewpoint. Infrastructure development is also a requirement to meet the demands of a fast moving retail sector. Collaborative arrangements for sustainability are required as well as knowledge and capability improvement for all sectors and stakeholders. It is not a given that suppliers understand the need for sustainability anymore than anyone else. More innovation is required and more raw material utilisation. Also, understanding demographics and the different retail business models are important on the supply chain side.

At a supply chain level, there is no question that there is increasing pressure to source seafood from approved sources. The word 'approved' is crucial, with so many available schemes in existence. In the future there are likely to be more strategic cross-border alliances, particularly to gain security of supply, quality and traceability. Innovation is likely to play a role in the future in obtaining raw material. However, occasionally, an inability exists on the supplier or retailer side to grasp partnerships.

TESCO own-labelling markets include the UK, Europe and America. Own-labels are not new but do contain consumer expectations. Own-labels are trusted and recognised, the brand is very important and it signals a difference in quality. Product consistency exists due to stringent code of practice specifications that are aligned to own-label products. Customers need clear, consistent and transparent messages as shopping time is limited and decreasing.

In the future, all brands will need to demonstrate a sustainability policy and clear labelling on sustainability will be essential. Retailers owe it to the fishing industry and the communities the industry supports, to actively support a reduction in negative environmental impacts. At TESCO, a structured and transparent decision making process for wild fish procurement called the 'wild fish stocking tree' exists to ensure this. Sustainability is also about risk management: existing markets, future industries and their communities. The aim of course is to ensure fish stocks for future generations.

PART 3

Processing Fish: Where, Why and How?

Chapter 7

Seafood Processing – Local Sources, Global Markets

Alda Möller, Seafood Consultant, Iceland

Alda Möller introduced a number of case studies showing changes that have taken place in the fish processing industry. Demographic and lifestyle trends, consumer sophistication, knowledge and changing consumer preferences have all driven seafood consumption, hence trade and processing. Meanwhile, key considerations include seafood safety and hygiene, which are non-negotiable in the seafood business. Demand for safe food is bound to augment partly due to better detection techniques that are becoming more and more sophisticated.

The case studies considered include the Pangasius fish species in Vietnam, Cape hake from Namibia and the Chinese re-processing industry. Each of the cases suggests different challenges in terms of how they have benefited and dealt with the globalisation of the international fisheries markets. One of the key issues to be dealt with in the near future is the emergence of labour shortages and wage increases in China. Further consolidation in the fishing and seafood processing industries is likely to take place to balance the influence of the large retail chains, which presently are powerful players.

Disposition of the World Fishery Harvest

The world fisheries production in 2005 was close to 142 million metric tonnes (mt), comprising a stable supply from the capture fisheries of about 94 million mt and a growing supply from aquaculture, amounting to 48 million mt. Of the total production about 105 million mt were used directly for human consumption.

Utilisation in the fisheries seems to have changed significantly in recent years, although it should be acknowledged that the statistics are fraught with many uncertainties. FAO statistics show utilisation in several categories that include fresh disposition and also the traditional conservation categories (FAO 2006c).

The following figure shows that fresh utilisation has increased rapidly in recent years. In 1994, about 35 million mt were used for fresh production, but ten years later this was close to 55 million mt. About half of the world's total production for direct food is therefore used for fresh produce. Disposition by freezing, on the other hand, has remained stable in the past years with the industry processing about 27 million mt, or a quarter of the world production for human consumption. Curing methods, such as salting and drying, have also remained stable at close to 10 -11 million mt and processing by canning at similar levels.

Fishmeal processing has traditionally been quite variable, depending as it does on large fish stocks that are subject to sudden changes in production.

Figure 7.1. Utilisation of world fisheries harvests 1991-2004

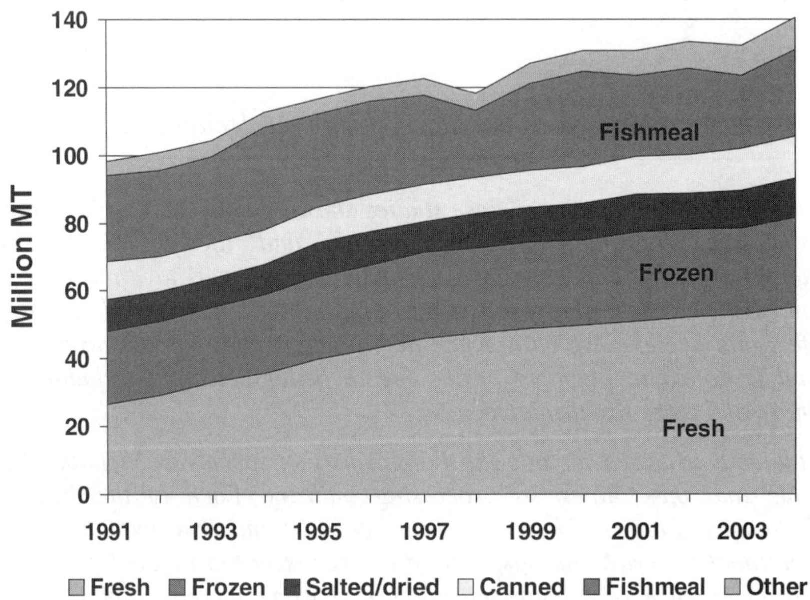

Looking more closely at the statistics it seems that all the significant changes in disposition have taken place in developing countries, where fresh production has increased by almost 20 million mt in the past decade[1]. It seems clear, however, that in the developed countries fresh production is increasing relatively slowly, and that freezing is still prominent as the primary production. This is illustrated in the following table. Fresh and frozen disposition of world fisheries production.

1 It is logical to conclude that this is mostly due to fresh utilisation of farmed products but disposition statistics do not specify production by origin.

Table 7.1. Fresh and Frozen Disposition of World Fisheries Production 1994 and 2004 - live weight basis

	1994 Million mt	2004 Million mt
Fresh - developing countries	34.5	52.9
Frozen - developing countries	10.0	14.7
Fresh - developed countries	1.0	2.0
Frozen - developed countries	13.0	12.0
Fresh - world	35.5	54.9
Frozen - world	23.0	26.7

International Trade

Fishery products have for a long time been among the most international of food categories, in the sense that they have been widely traded across borders. Forty years ago just under a third of the world production entered international trade (Kurien 2005), and in 2004 the proportion was 38% (FAO 2006c). The quantity traded internationally has, however, increased by almost 5 million mt (calculated back to live weight) in the past ten years and the industry is today a truly global business. The following table also shows that the increase in trade is across the board - export from developed countries has increased by almost 20% and from developing countries by 7% in the past decade. In 2004, the export value of fisheries commodities was close to USD 72 billion. Trade in world fisheries production, 1994 and 2004 - live weight basis.

Table 7.2. Trade in world fisheries production, 1994 and 2004 - live weight basis

	1994 Million mt	2004 Million mt
Production - developing countries	79.7	110.3
Exports - developing countries	28.1	30.1
Production - developed countries	33.2	30.2
Exports - developed countries	19.2	22.7
Production - World	112.9	140.5
Exports - World	47.4	52.8
Proportion for export	42%	38%

A recent report found that international trade in fishery products has not had a detrimental effect on domestic fish supply, in eight of the eleven developing countries studied. On the contrary, increases in production, coupled with the import and export of fishery products, have meant continuous availability of fish for domestic supply and increased earnings. On the other hand, the report warned that international trade had been found to have a negative impact on fish resources for all the countries, highlighting a need for more effective fisheries management (Kurien 2005).

Twenty countries export annually more than USD 1 billion worth of seafood products. The top-three exporters in 2004 were China, Norway and Thailand, which are all prominent capture and aquaculture countries. For many years prior to 2002, Thailand had been the largest exporter.

Seventeen countries import annually more than USD 1 billion worth of seafood. The top three importing countries in 2004 were Japan, USA and Spain, but as a whole the European Union is by far the largest importer with 38% of all imports, compared to 20% for Japan and 15% for the USA (FAO 2006c).

Largest International Product Categories

International trade is categorized extensively in FAO publications. From the standpoint of someone with interests in the international trade of chief seafood categories, it comes to light that four large categories account for over 40% of the international trade. These are the categories of shrimp (prawn), salmon and trout, tuna, and fish fillets. Three of them are based on groups of related species while the fish fillets are a product category, defined by processing. The following table gives a further description (approx figures (FAO 2006c).

Table 7.3. International trade - Four Important Seafood Categories

Category	Value USD billion	Category	Value USD billion
Shrimp	12.2	Tuna	5.5
Frozen	8.6	Frozen	2.0
Canned	2.8	Fresh	0.8
Other	0.8	Canned	2.7
Category	**Value USD billion**	**Category**	**Value USD billion**
Fish fillets	7.3	Salmon/trout*	5.7
Fresh	1.7	Fresh	3.6
Frozen	5.6	Frozen	2.1

Note: Salmon and trout fillets included.

Each of these categories is worthy of extensive studies on recent developments in processing and markets, and the same goes for a number of other less prominent categories. *For this paper, however, it was decided to focus on major developments within the fish fillets category, with only a brief and general mention of developments in the other "champion" categories.*

Shrimp

The shrimp category is large and heterogeneous as regards species, and the capture production, in particular, is widespread. About 6 million mt are harvested annually, with capture production close to 3.5 million mt and farmed production 2.5 million mt

(FAO 2006a, 2006b). Shrimp has a short shelf life in the fresh state and it is therefore to be expected that frozen products account for most of the traded value, but the canning sector is strong in Thailand and increasingly so in China.

Three producers, Vietnam, Thailand and India, supply about a third of the total international trade and two markets, the US and Japan take over 50% of the traded volume. Industry sources (Bush 2007) indicate the dominance of farmed products in international trade, and that two species are by far the most important on the market, i.e. whiteleg shrimp (*Penaeus vannamei*), and black tiger shrimp (*Penaeus monodon*).

Interestingly, production has shifted significantly from black tiger to white shrimp in the past few years, especially in Thailand, where black tiger used to dominate production but white shrimp is now thought to constitute 95% of production. Vietnam is also increasing production of white shrimp quite rapidly. These changes have been driven by production issues rather than the market, and are due to fewer problems and improved productivity in white shrimp farming compared to black tiger production. Other marked changes in the Thai industry involve the use of new technology in processing, especially for grading shrimp automatically by size before packing, and also increased emphasis on value addition, by peeling, cooking and coating the products (Ingolfsson 2007)

The market (US) is a mixture of relatively low value headless, shell-on products that still dominate in the market, while higher value products, such as peeled, cooked and breaded products are gaining increased market share.

The chief issue of concern to both producers and the market are anti-dumping duties established unilaterally in the US in 2005. Brazil, China, Ecuador, Thailand, India and Vietnam were all found to have been selling shrimp at less than fair value and are still subject to duty and custom bond requirements, which restrict trade. All the affected parties challenged the decision. The duties are the largest for China and Vietnam (over 100% in some instances). Another issue, which is presently felt to limit trade with Europe, is a zero-tolerance for certain contaminants, notably chloramphenicol and nitrofurans (FnBnews 2007).

Salmon and trout

Salmon and trout are important and valuable resources that are harvested from temperate and cold waters, both in capture fisheries and aquaculture. The 2004 capture landings were 900 000 mt and aquaculture produced 2.0 million mt in this category (FAO 2006a, 2006b).

Farmed Atlantic salmon (*Salmo salar*) is the most valuable resource within the category. Latest estimates are that production in 2007 will be nearly 1.4 million mt, of which one million mt will be produced in Norway and Chile.

There are significant differences in salmon disposition among the major producers. In Norway, by far the most important product is fresh whole salmon, while in Chile most of the salmon (close to 80%) is sold as fresh or frozen fillets (Infante 2006).

Overall, the 2005 disposition for farmed Atlantic salmon was 52% for fresh whole salmon, 20% for fresh fillets, 16% frozen fillets, 8% frozen whole, and 4% other products, such as smoked salmon fillets (Groundfish Forum 2006).

The EU and the US are the traditional markets for farmed Atlantic salmon, and both are to a large extent, based on fresh salmon products. The US is the largest market for

Atlantic salmon from Chile, which is primarily exported as fresh fillets, frozen fillets and fillet portions. Chile's exports to the EU are based on frozen fillets and portions. The EU is by far the largest market for salmon from Norway but trade with Russia is increasing rapidly. Japan has also become an important market for Norwegian salmon for use in *sashimi* products (Leroy 2006).

Value addition of farmed salmon in the US and Europe is following the trend of traditional whitefish products in terms of coating and portion control, but there is more emphasis on marinated products. The distribution of fresh salmon fillets is based on steady supply and increasingly on a system of total supply chain management from the farm to the end user. The marketing of fresh (chilled) recipe dishes is especially evident in the UK and France. Value-added salmon products are important in foodservice distribution and are widely available in the retail trade.

Issues for the farmed salmon industry are connected to its image as a very large and global business. Newspaper reports on traces of persistent pollutants have caused temporary setbacks in trade, but the industry appears very solid and able to handle adverse publicity. Another issue for future concern is the limited supply of fishmeal and oil for salmon feed, but indications are that the industry will be able to replace fishmeal to a large extent with vegetable protein.

Tuna

The tuna category is large as regards number of species, and the resources are extremely widespread across the oceans - ranging from cold-temperate waters to tropical areas. The tunas are capture fisheries resources, with developments just emerging in aquaculture. Processing of tuna at source usually only involves chilling or deep-freezing in nitrogen, and is followed by worldwide transport - with Japan as the most frequent and lucrative point of sale.

The largest and most widespread resource is skipjack tuna (*Katsuwonus pelamis*), with 2.1 million mt landed every year. Its value in international trade is USD 400 million.

The largest resource, in terms of international trade, is yellowfin tuna (*Thunnus albacares*). Landings are close to 1.4 million mt and international trade is worth close to USD one billion. This is followed by bigeye tuna (*Thunnus obesus*), with harvests of 400 000 mt and international trade of almost USD 800 million.

The bluefin tunas are, however, the most valuable tuna resources in terms of value per kilo harvested, with prices for whole fish in fresh export trade over USD 12 per kilo on average and frozen products close to that price. Atlantic bluefin tuna (*Thunnus thynnus*), Pacific bluefin tuna (*Thunnus orientalis*) and southern bluefin tuna (*Thunnus maccoyii*) are harvested in rather small volumes - combined total of 60 000 mt in 2004, but overall trade in fresh and frozen products was USD 550 million in 2004. Most of the trade is in fresh products, or about 70% of sold value.

Bluefin tuna is under great fishing pressure and landings have decreased in all areas in recent years. Landings of Atlantic bluefin tuna in the Mediterranean have decreased by 40% in the past decade, which must be considered as clear evidence of overfishing.

On-growing of bluefin tuna in Spain (and of southern bluefin tuna in Australia) is becoming well established. Production in Spain was 6 500 mt in 2004. Small-size tuna from the French, Spanish, Italian, Turkish and Croatian quotas are landed live, transferred to enclosures and grown to market size. Most of the on-growing takes place in Spain, but increasingly also in Croatia and Turkey. The full-grown tuna is then slaughtered on-board Japanese motherships, and deep-frozen in nitrogen for the Japanese market. It is reported that 80% of Atlantic bluefin tuna imports to Japan are presently from farming sources (FAO fisheries report No. 779 2006).

Fish fillets

This large category was selected for a study of major processing and market trends. But even within this one category, it proved necessary to focus on selected industries and markets.

The case studies are:

- Pangasius fillets from Vietnam - a story of rapid development and market acceptance of a new whitefish species from aquaculture.
- Hake from Namibia - a story of technological transformation, and a production shift from commodities to customer-based products.
- Alaska pollock from the US - showing how fishery management and a market shift have transformed the industry.
- Alaska pollock in Russia - discussing briefly the operations of this large industry and its need for transformation.
- The re-processing industry in China - a considerable force in the global whitefish business.
- The Icelandic fresh cod and haddock industry - following market trends and responding to an influx of low-price products in the market.

Pangasius in Vietnam

Vietnam is one of the principal fisheries countries of the world, ranking 12th in capture fisheries with close to 1.9 million mt, and harvesting 1.5 million mt by aquaculture in 2005 (FAO 2007).

Aquaculture production in Vietnam has in the past ten years increased fivefold. A great variety of species is farmed but the most significant development in recent years has been in the farming of whiteleg shrimp and catfish. Production of whiteleg shrimp *(Penaeus vannamei)* started in 2003 and in 2005 it reached 100 000 mt. Production of blacktiger shrimp (*Penaeus monodon*) is, however, stable at about 180 000 mt. Production of the catfish species *Pangasius hypophthalmus* has gradually taken over in catfish farming. Production was reported at 500 000 mt in 2005, it increased to 800 000 mt in 2006 and in 2007 it is expected to reach the one million mt mark (Neubacher 2007). Most of the production is exported as frozen fillets.

The following shows the rapid development for pangasius exports in the past five years. Exports in 2006 were 286 000 mt, valued at USD 740 million (Vietfish International 2007), which may be a conservative estimate (Neubacher 2007).

Figure 7.6. Pangasius Exports from Vietnam (2001-2006)

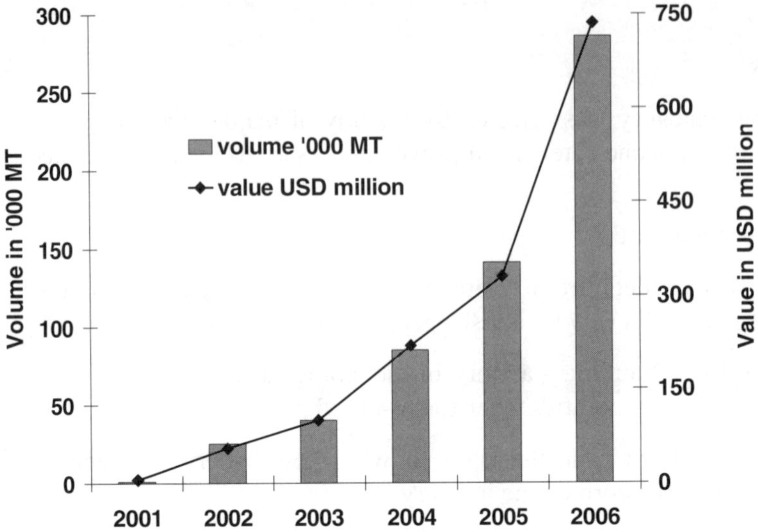

The Vietnam shrimp and catfish processors have shown a remarkable drive for export, their export association VASEP[2] seems to be an active promoter, and in the past few years Vietnamese seafood has become a force to be reckoned with in international trade. This is especially true for the introduction of catfish fillets to the European market in the past 3-4 years, which is the subject of this case study. In 2006, business partnerships with European customers were strengthened and the industry diversified its exports in terms of market spread, indicating that the pangasius trade is about to become a global business.

Pangasius production and markets

Vietnamese catfish is known by various names in the market, but pangasius is the common product name on the European market. In Vietnam it is known as *tra* and *basa* and in the US it is referred to as *basa*.

Pangasius is a whitefish species farmed in the delta of the Mekong River, which is its natural habitat. The fish grows from a 3 cm juvenile to market size of 1.5-2 kg within six months. It can be herded in great density without cannibalism and there is very low mortality. The feed is mostly of plant origin and only has a small proportion (2-5%) of fishmeal. By far the most important pangasius species is *Pangasius hypophthalmus*, but *Pangasius bocourti*, the originally farmed species and the correctly named *basa*, is scarce and quite expensive.

The pangasius business is still quite fragmented. The fish is farmed in thousands of *quangs* and floating cages by small-scale farmers who bring the fish live by boats to processing companies for filleting and freezing. The top-five companies process about

2 Vietnam Association of Seafood Exporters and Producers, www.vasep.com.vn.

40% of the total volume for export, and the top-ten about 55% (Vietfish International 2007).

At the plants, the live fish is killed and bled, hand filleted and skinned. The fillets are trimmed, IQF frozen and graded for colour, and finally packed for individual customers. There is very little block production and while the production is essentially that of a commodity line, there is already export market segmentation, based on quality and packaging. The large plants employ thousands of workers, mostly for the trimming jobs, but the newest plants are also state-of -the art as regards processing technology and there is great attention to hygiene, which is inspected by their European customers. Prices for IQF fillets have been USD 2.50-3.00 per kilo fob, based on 100% net weight. Current raw material prices suggest that the margins for the plants are quite small, indicating that high throughput and attention to productivity and yield are keys to profits for the plants.

Table 7.4. Market Spread for Pangasius from Vietnam

(2006)

Market	Percent by value rounded figures
European Union	47
– including to Spain	10
– including to Poland	9
– including to the Netherlands	9
Russia	11
Southeast Asia	8
China	5
United States	10
Australia	4
Others	15

There are historical reasons for the dominance of the EU market and they serve to illustrate how trade limitations in one country may affect seafood trade globally. *Pangasius* species were first imported to the US in 2000 and US catfish farmers soon began to fight these imports that were labelled as catfish. In 2003, an Anti-Dumping Law, decided unilaterally by the US, came into effect and, as a result, Vietnamese exporters had to pay duty of up to 60% in the US market. Vietnam was at the time not a member of the World Trade Organization and was therefore unable to challenge the decision. Furthermore, Vietnam was not allowed to call the product catfish on the US market. Imports to the US subsequently decreased and Vietnamese exporters transferred their marketing effort to Europe with immediate success. (The sequel to this story is that Chinese producers are now exporting catfish to the US in large volumes, and in this case unarguably channel catfish, the only species allowed the catfish name in the US).

In continental Europe, pangasius is rapidly becoming a major source of low-priced whitefish, quite frequently further processed for foodservice operations and recently also for retail. Today, about 20% of the market is estimated to be retail (Kaptijn 2006). Historically, US producers of Alaska pollock and major Atlantic pollock producers have supplied this market with low-cost fillet products. Last year, however, pangasius seems to have exceeded pollock from these sources as shown in the following table. Interestingly also, the prices per kilo of product for pangasius and Alaska pollock appear quite similar.

Table 7.5. EU Imports 2006 - Frozen Whitefish Fillets

	Volume mt, Rounded figures	Value EUR million, Rounded figures
Pangasius fillets from Vietnam [1]	123 000	274
Pollock fillets sf. total [3]	119 000	293
Atlantic pollock fillets from Norway, Iceland and Faroe Islands [2]	42 000	117
Alaska pollock fillets from USA [2]	77 000	176

Notes:
1. Vietfish, 2007 (export figures).
2. Eurostat, 2007 (import figures).
3. Sf. denotes single frozen products, and therefore excludes imports from China.

Pangasius appears to be gaining the same market position in Europe, which tilapia has secured in the United States, where it is presently number six on the top-ten list for consumption of seafood (NFI 2007).

Most recently the Russian market for pangasius fillets has developed very fast - from near-zero imports in 2005 to a volume of 43 000 mt in 2006, and a market share of 16%. Vietnam and Russia have historical ties, including trade, and all indications are that this market will continue to develop rapidly as Vietnamese plants acquire Russian veterinary import permits. Just recently, it has been decreed that all shipments bound for Russia need certificates of hygiene and food safety in accordance with the Russian VPSS[3] requirements (VASEP 2007).

Challenges and drivers for growth

Drivers for the rapid market acceptance of pangasius are without doubt its abundance and moderate price of products, sensory quality and confidence in safety standards used by producers. There was obviously a need for supply of moderately priced whitefish fillets to complement pollock in Europe and pangasius has helped to fill that gap. It follows that pangasius is also likely to replace considerable volumes of Alaska pollock imported from China to the EU.

At the same time, there are also many challenges facing the Vietnamese producers, including:

- Anti-dumping taxes in the US,
- Cost inflation,
- Integration and consolidation,
- Possible shortage of labour,
- Food safety issues.

3 Federal Service for Veterinary and Phytosanitary Surveillance (VPSS) of Russia.

Anti-dumping taxes of up to 60% for some producers are clearly prohibitive to business, and it is difficult to see *basa* "taking off" on the US market unless this barrier is removed. No such tariffs exist for exports to the EU.

Cost inflation is already a major problem in the industry. Raw material costs have increased and margins in processing are insignificant unless fillet prices exceed USD 3 per kilo (fob). This will represent a major increase in fillet prices and may slow down market developments. Integration of farming and fillet production is already taking place, with the largest processors having invested in pond farming on a large scale. It is also likely that there will be consolidation in processing, with technologically advanced large plants gradually dominating production. Shortage of labour is likely to become significant unless there is increased automation ((Neubacher 2007; Hoe 2007).

Food safety issues become apparent when considering the pressure of production of almost one million mt in the Mekong River, although it is a vast ecosystem. These issues arise from possible use of illegal additives, such as antibiotics, or contamination of the environment, and they can lead to closure of the market without much warning. These issues therefore need constant vigilance by health authorities and in production.

Consolidation in the industry will undoubtedly bring mechanization in processing, (including the most labour-intensive task of fat-line removal). Mechanization will also serve to reduce costs and is likely to provide added safety in processing.

Cape Hake in Namibia

Namibia is among important fisheries nations, ranking 30th among principal fisheries in 2005 with close to 590 000 mt harvested. In 2005, the most important species ranked by landed weight were horse mackerel, 325 000 mt; cape hake, 158 000 mt; pilchard, 27 000 mt and cape monkfish, 11 000 mt. Smaller volumes of various other species, such as kingclip, red crab and rock lobster were also fished and three species of tuna (albacore, yellowfin and bigeye tuna) are of commercial importance. Aquaculture is still small-scale, with tilapia farming, oysters and abalone culture developing (FAO 2007)

The Namibian fisheries are export-driven and important to the national economy, both in terms of labour and foreign trade. In 2006 the export value of seafood products was NZD 3.9 billion (equivalent to USD 540 million). Hake, monkfish and tuna are the most important product categories in terms of export value. Tuna is primarily exported fresh to markets of Europe and Japan while frozen monkfish tails are mostly sold to Europe (Kuntzsch 2007)

The following study will focus on the hake industry, showing how it has adapted to the international groundfish markets and will attempt to define the drivers for these developments.

Hake production and exports

Cape hake (*Merluccius capensis*) is processed on-board factory vessels and in land-based factories. The frozen-at-sea sector receives today 30% of the TAC while 70% is processed on land. Utilisation of hake has changed markedly in recent years in a drive to increase product value and both sectors now emphasize fillet processing and value addition by making products that are tailored to individual customer needs, but this development is more significant for the land-based factories and varies between processors.

Figure 7.7. Cape Hake in Namibia - Utilisation 1996-2005

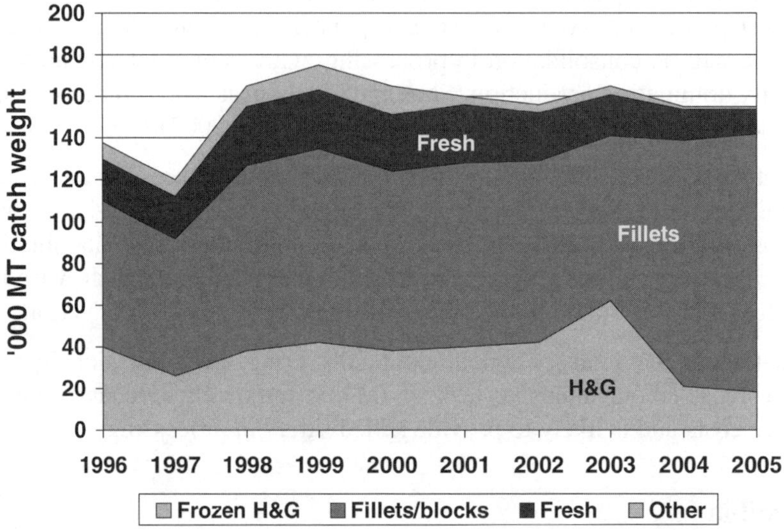

The graph shows the positive development for the frozen fillet industry, which in 2005 processed 80% of the total catch of 157 000 mt as opposed to 50% of the total catch of 138 000 mt ten years earlier. Furthermore, there has been a remarkable development within the industry towards production of fully trimmed fillets and fillet portions with the result that less than 5% of fillet products were industrial blocks in 2005, as opposed to 25% of the output ten years earlier.

The graph also shows that there is a small but significant proportion of the catch that is processed fresh. This is primarily long-line hake, which has a high price on international markets, especially in southern Europe. Similarly, the small proportion of the catch, which is frozen whole (H&G), has a good traditional market in Spain.

About 85% of hake fillet products are exported to the EU. Chief customers are in Spain, which is by far the largest market, with Italy, UK and Germany also developing as important markets (Industry sources, Namibia 2006).

The Namibian industry is still a primary producer with no production of coated products or recipe dishes. The industry has, however, made great changes towards value addition by modernization of the fillet industry. The land-based sector processes frozen retail and foodservice products. A significant proportion is made to individual customer requirements, and increasingly they are size-graded fillet portions or portions that are moulded from small hake fillets.

The industry states that Namibia needs to diversify its markets in order to achieve more flexibility and that it still needs to shift production more in the direction of value addition. This will also create a more predictable market environment with longer-term contracts and less price volatility. As in many other supplying countries the industry may have been too focused on production in the past and had too little focus on the market. This is partly because foreign partners, especially in Spain, have traditionally taken care of marketing. It is therefore acknowledged that the Namibian fish industry must become more market oriented and get engaged more deeply in the value chain by producing more value added products (Kuntzsch 2007).

The focus of the Government on employment and value-addition in processing has led some industry participants to build onshore facilities in order to pack seafrozen fillets in retail bags. This has to some degree also led to utilizing the H&G for other product types and increasing the value for Namibia. Markets in Asia are also developing and this may be driven by the Asian demand for aquacultural products such as oysters, with fish forming part of the portfolio that is being developed.

Keys to progress for the hake industry

Industry sources have identified the following keys to present and future success of the Namibian hake industry:

- Improved processing technology.
- Availability of skilled labour.
- Resource management.
- Vertical integration.
- Consolidation.

Industrial sources emphasize the need for increased productivity, improved yield and enhanced value addition for the Namibian industry. These challenges can only be met by employing new processing technology, which already is significant among some of the larger companies. Processing lines that include mechanical filleting, computerised portion cutting and size grading for frozen portions, together with incentive-based remuneration systems have been installed in some of the largest companies with the result that productivity (output per man-hour) in one company is reported to have increased by approx. 15% in the first year of running the new processing lines . The company still employs 1 100 people, the same as it did prior to installment of the new lines, but the value of the product mix calculated on the basis of sales price per kilo of raw material has increased by 40% (excluding exchange rate improvements) by a combination of general price increases, more value-added emphasis, and improved yield and accuracy. The increase in contribution amounts to 15 – 20% (Kuntzsch 2007).

Skilled labour becomes more important in a technical processing environment. This applies to engineering skills, on-line quality control, production planning and performance assessment, to name just a few of the skills required for smooth running of an efficient production line.

The sustainability of the hake resource has in recent years been a cause for concern. Low catch rates and small fish have impacted the economic performance of fishing companies, Several measures have recently been adopted by Government to improve catch rates in the longer term, such as a closed season and closed areas. It seems clear that the sustainability of the resource and hence the success of the fishing sector will determine the fate of the Namibian hake industry in future years.

The five largest hake companies in Namibia are integrated as regards fishing and processing. Three of them are commodity-based while two are brand owners and produce to specific customer demands. All five have a share in the annual TAC allocation, they process fish from that share and have their own sales operations. This makes them an attractive partner in supply management for retailers and foodservice in Europe. The importance of vertical integration is also underlined by the value structure in the industry. In 2005 the landed value of all catch was close to NZD 3 130 million while the export value was NZD 3 789 million (Kuntzsch 2007). The difference between landed value and export value, *i.e.* value addition by onshore fish processing, was therefore only about 21%, showing that the performance of the fishing sector is the most decisive factor in the financial performance of the industry.

Industry sources state that there is an urgent need for consolidation of hake quotas and especially production facilities. There are too many players for the limited amount of fish. This process could be initiated by introducing a system of individually transferable quotas, which would soon attain a price, enabling the companies to include quotas on their balance sheet. The improved equity base would give banks the incentives for further financial support and the improved financing would immediately be utilized for consolidation and much needed vessel renewal (Kuntzsch 2007).

The conclusion for this case study is that the Namibian hake industry is set to benefit greatly from a valuable resource, which is well known globally. The industry players seem ready to take control of their fate but Government must encourage consolidation and improved financing by tools in their power.

Alaska Pollock from the United States

Alaska pollock (*Theragra chalcogramma*) is the second most harvested species from capture fisheries with landings close to 3 million mt. It is the dominant groundfish species on world markets and is harvested primarily by the United States, Russia and Japan, all of which are superpowers in seafood production (FAO 2007; Ianelli, J. *et al* 2006; AFSC 2007; NOAA 2007).

The following figure for catches in the past 20 years shows that total landings have been relatively stable in the past few years, but in the late 1990s there was a significant decline in supply on both sides of the Bering Sea.

Figure 7.8. Alaska Pollock Landings 1987-2006

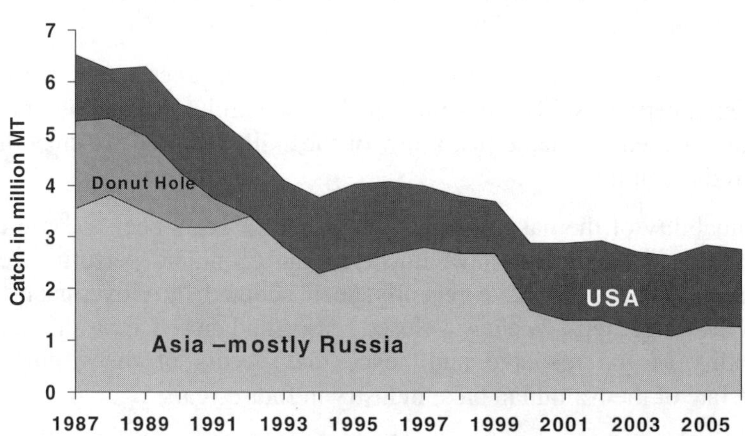

Processing of Alaska pollock has changed dramatically in the past decade and in many ways the changes have been a dominant factor on the traditional whitefish markets in Europe and North America. The rapidly developing re-processing industry in China has played no small part in these changes. Conversely, it can also be stated that changes on the major whitefish markets have served as a very powerful driver for changes in the pollock industry, with repercussions for the whitefish industry worldwide.

The following case study for pollock from the US, Russia and China serves to describe these changes within a business that is one of the most globalised of all. It also attempts to analyse the drivers for these changes and to make forecasts for the near future.

US pollock industry and markets

The Alaska pollock landings are presently close to 1.5 million mt from the Eastern Bering Sea and the Gulf of Alaska. The resource is conservatively managed and the industry is thriving after years of stagnation in the 1990s. Processing takes place in large freezer vessels, on-board motherships and in on-shore production plants. Chief products are fillet blocks and surimi, with approx. 53% of the resource used for fillet blocks and 44% for surimi in 2005, while the remainder was frozen whole. (Groundfish Forum 2006).

The US pollock industry was focused on surimi production for Southeast Asia and block production for the domestic market. Export of fillet blocks was non-existent prior to 2000. At that time the industry needed restructuring and required an injection of capital. It was apparent that surimi production for the declining Japanese market could not remain the mainstay of the industry for long. The key drivers for the positive turnaround in the US pollock industry in the past decade have been the following (Halhjem 2007).

- Changes in fisheries management
- Demand shift
- Product flow from China
- Freight costs
- Currency moves
- New technology in surimi and fillet processing.

The US pollock fishery was rationalized in 1999. Fishery management changed from an Olympic-style race for fish, where the fishing seasons took a few weeks, and into a system based on quota allocation to the operators, allowing fishing effort year-round, although there are basically still two seasons. The most significant operators are vertically integrated companies, fishing and processing at-sea and/or on-shore, and running their own sales and marketing functions.

The change in fisheries management also brought about rationalization in processing for all the operators. Market needs and profitability became the drivers and production planning became the tool to achieve this. Imminent spoilage of the catch was no longer an issue in production. Very soon the yield in fillet processing increased from 14-18% up to 28-30%, and new by-products were marketed, as opposed to previous usage of that material for fishmeal. Yield in surimi processing also increased, from an estimated 15% to 25%, mostly through use of new technology in processing.

Great changes were also taking place on the markets around the year 2000. Demand for pollock surimi in Japan was decreasing, and surimi made from tropical species was becoming significant on the Asian markets. At the same time a void was developing on the block markets in Europe, partly due to decreases in whitefish production in the EU as landings declined, but also due to a rapidly decreasing supply of pollock blocks from the Russian Far East, as their production of blocks decreased rapidly. All this meant that there was a huge driver pulling fillets and blocks from the US into Europe, mostly for the secondary-processing plants of Germany, France and UK. Added to this was the new drive announced by brand leaders to source fish only from fisheries that were certified as sustainable. Pre-assessment of US pollock was started in 2000 with the view of supplying products for Europe. The US Alaska pollock fishery was finally certified by the Marine Stewardship Council in 2004.

All of the above market changes were soon apparent in US pollock production as shown in the following figure. Surimi production remained at 180-200 000 mt, while fillet block production increased from 60 000 mt in 2000 to 170 000 mt in 2006 (NMFS 2007).

Figure 7.9. Trends in US Alaska Pollock Production 1997-2006 - main products

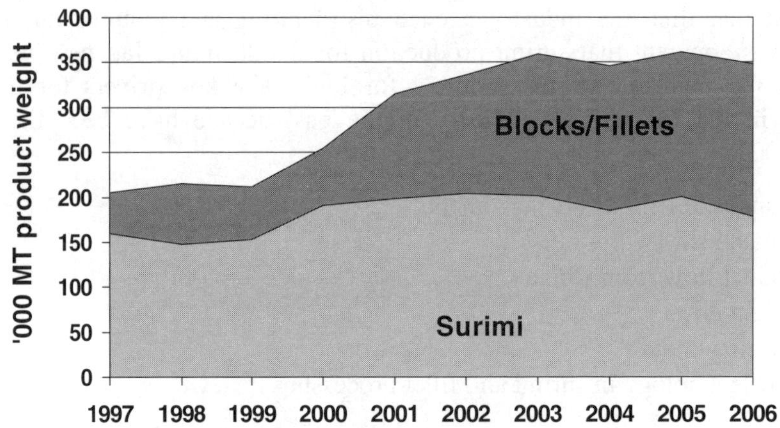

In 2000, the US market for pollock blocks was already supplied by a large volume (60 000 mt) of double-frozen blocks from China. The influx of Chinese blocks put pressure on the domestic market and underlined the need to look actively towards the markets of Europe.

Freight cost had long been a hindrance for US exports of pollock blocks to Europe. Prior to 2000, the cost of transport by containers was prohibitive but gradually tramper shipments became available at much more competitive prices. (Halhjem 2007).

Last but not least, the changes in currency values have had major effects in the marketing of pollock blocks. In the four-year period 2003-2006, the value of the euro increased by 25% relative to the US dollar[4].

All of the above factors have resulted in a dramatic increase in export of US pollock blocks and fillets to Europe. EU statistics show that 80 000 mt of pollock fillets were imported from the US last year, but import prior to 2000 was nil as shown in the next graph. The fillets are primarily blocks and to a small degree also interleaved fillets, while IQF fillets that carry a 15% duty into the EU are only a minor component (Eurostat 2006; Halhjem 2007). The graph also shows the US domestic market use in the same period by subtracting EU imports from the US total fillets/blocks production[5].

Figure 7.10. US Pollock Fillet Flow 1997-2006

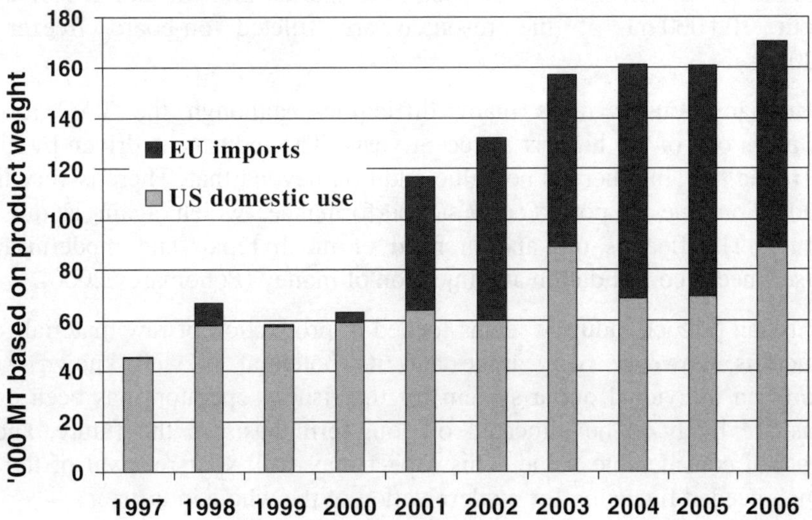

Since 2000 many advances in processing technology have taken place but especially in surimi tecnology. As a result, the US industry is now producing both the traditional A-grade surimi for Japan but increasingly also lower grades for Southeast Asia and Europe made from new mince sources, produced by recent technology. Recovery is now about 30% in surimi processing, but was closer to 18-20% prior to 2000 (Davis 2003).

In the years to come it is fully expected that the US pollock industry will continue to focus on the European market and expand the use of Alaska pollock in new markets worldwide. While the pollock block is a prime example of a commodity product, there are clear signs that value addition is taking place within the pollock market, as noted by

4 According to the *Wall Street Journal*, EUR 1 was worth USD 1.064 at the end of 2003 but USD 1.333 at year-end 2006.

5 It should be noted that export statistics from one country (US) do not usually agree with import statistics from others (EU), even if both are considered reliable. There is a time lag and product may be held for considerable time in cold stores in the importing country before being cleared for imports and declared as such.

increased emphasis on deep-skinned fillets for several large customers, such as McDonald's and Young's Bluecrest in products that used to be made from cod fillets. Interleaved fillets and even IQF fillets are gaining markets, although the 15% import duty on IQF fillets is a drawback. If this was to be lifted, a very large market for pollock fillets is likely to open up, which would most likely use MSC certification as leverage, especially in UK and Germany. Some private-brand pollock products are starting to sell at a noticeable price advantage based on ecolabelling (Halhjem 2007).

Russian Pollock Industry

Landings of pollock in Russia exceed one million mt in most years and the industry is a very significant force in international trade. The main character of the industry is, however, export of raw material to China for reprocessing. It is estimated that about 650 000 mt of the resource are wholefrozen (H&G) and exported to China every year. The remainder is wholefrozen for the domestic market and the South Korean market. Only about 100 000 mt of the resource are filleted on-board freezer trawlers (Radvan 2006).

The Russian industry faces many difficulties, although the TAC for 2007 at 1 220 000 mt is one of the highest in recent years. The industry is driven by high prices for wholefrozen fish, but there is no value-addition beyond that. There is very little land-based production due to poor economic performance, which again is due to weak infrastructure. The fleet is old and in need of much repair and modernisation. The industry itself needs consolidation and injection of money (Ponomarev 2006).

The Russian pollock industry seems locked in production of raw material for export. The resouce is, however, very large and its potential is well known worldwide. Furthermore, an individual quota system for the fishery operators has been established and quotas are likely to be allocated on long-term basis in the future, and thereby probably attain considerable value. This aspect may well spur renewal of the fleet and initiate much needed financing for modernization of the whole industry.

Chinese Re-Processing Industry

The Chinese reprocessing industry is a phenomenon, which has had significant impact on whitefish markets in Europe and the US in the past decade. Its mainstay is wholefrozen (H&G) pollock from Russia, but cod, haddock and hoki are also processed from sources as wide apart as New Zealand and Norway. Overviews for volumes and prognosis for the years ahead have not been widely available so far, but based on various industry information this now seems possible (Groundfish Forum 2006).

It is estimated that 700 – 800 000 mt of frozen whitefish (H&G) were processed in China in 2005, including more than 500 000 mt of Alaska pollock, and 175 000 mt of cod. Most of the products enter the markets of Europe and the US, while some remain in China or are re-exported to the domestic market of Russia.

The Chinese reprocessing industry is labour intensive and traditionally works on low margins, which have lately tended to get even narrower with escalating costs for raw materials. The success of the industry in past years is, however, noteworthy and it has influenced all the traditional whitefish markets by being able to offer low cost fillets, portions and blocks to the traditional further processors in Europe and the US. It is estimated that China is currently producing about 30% of the world market supply of whitefish fillets and fillet blocks from Alaska pollock, cod, haddock and hoki (Groundfish Forum 2006). Interestingly, there is little processing of hake in China, most likely because of the strong consumer markets for wholefrozen hake in South Europe.

Table 7.6. Global Supply of Major Whitefish Blocks and Fillets 2005

	Blocks mt	Fillets mt
Single frozen at source	300 000	425 000
Double frozen from China	172 000	126 000
Total blocks and fillets	**470 000**	**550 000**
% Chinese-produced	36%	23%
% Chinese overall	29%	

Further analysis of the whitefish products coming out of China (next table) shows the overwhelming importance of Alaska pollock for the Chinese industry, but also the importance of China for the world supply of interleaved pollock fillets, as opposed to block, (which is the principal product entering the market from the US fisheries). Pacific cod has also become more important in international trade after processing in China and most of this resource from both US and Russia is in fact whole-frozen at source and reprocessed in China. As regards Atlantic cod, it seems that close to 30% of the total supply of fillets and blocks is double-frozen product from China (Groundfish Forum 2006).

Table 7.7. World Market - Whitefish Blocks and Fillets from China (2005)

	Blocks mt	Fillets mt
Alaska pollock	130 000	71 000
Pacific cod	17 000	24 000
Atlantic cod	19 000	20 000
Hoki	3 000	4 000
Haddock	3 000	7 000
Total	172 000	126 000

The following figure from EU and US statistical databases show the split between the EU and US as markets for pollock fillets and blocks from China. In 2006, imports to the EU exceeded 150 000 mt and were on a steep increase, while US imports were lower than in previous years at 70 000 mt. The downturn in imports to the EU in 2002 was due to a temporary import ban imposed by the EU on grounds of product contamination (NOAA trade 2007; Eurostat 2007).

In 2005, EU imports of double-frozen pollock products from China equalled imports of single-frozen products directly from the two chief producers, US and Russia. Imports of blocks and fillets produced at-source were 91 000 mt from the US and 18 000 mt from Russia, while imports from China were recorded at 111 000 mt.

Figure 7.11. EU and US Imports of Pollock Fillets/Blocks from China

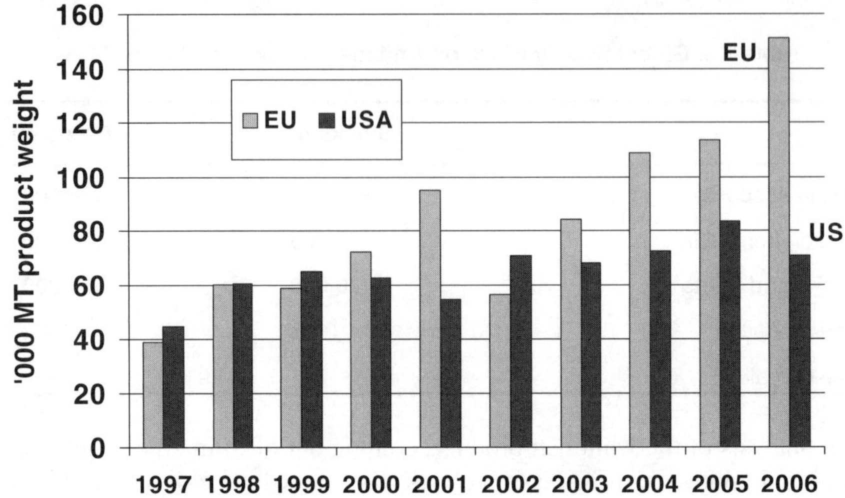

The steep increase in imports of cod products from China is also quite apparent in the next figure. The Chinese industry sources raw material chiefly from Russia (Atlantic cod and Pacific cod), the United States (Pacific cod) and from Norway. Imports were insignificant prior to 2000, but have increased steadily since, with the exception of 2002 in the EU. In 2006, cod imports to the EU exceeded 52 000 mt and to the US 36 000 mt. (NOAA trade (2007); Eurostat (2007)).

In 2005, EU imports of frozen cod fillet products from China also equalled imports of single-frozen products directly from the two chief producers of Atlantic cod, i.e. Iceland and Norway. Imports of frozen fillets produced at source were 32 000 mt from Iceland and 16 000 mt from Norway, while imports from China were recorded at 47 000 mt. The products appear to compete in similar market segments chiefly in the UK, but the indications are that the Chinese products are mostly sold coated in retail packs, while the single-frozen products have a more varied presentation on the retail market, including the chilled market.

Figure 7.12. EU and US Imports of Cod Fillets/Blocks from China (Pacific and Atlantic Cod)

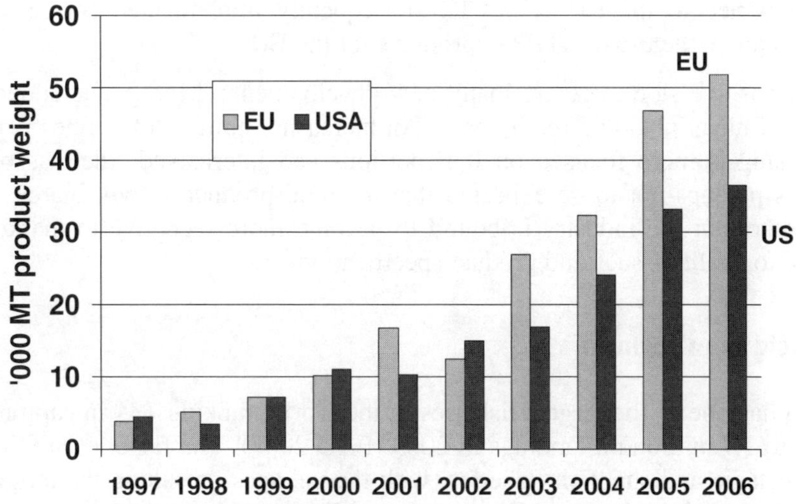

Evidently, such a large influx of frozen products onto the EU market has impacted the primary producers of whitefish in the North Atlantic. They have reacted in different ways to the competition from China. Some are exporting raw material to China or producing more saltfish, but others have increased fresh fillet production considerably.

As noted before, the Chinese processing sector works on tight margins and it has been difficult to see how any profit was to be made, given the raw material prices and product prices. High yield in processing is without doubt a large contributing factor, and in fact has raised questions about excessive use of additives, such as sodiumtripolyphosphate in some plants.

Issues in the Chinese processing industry that create concerns for its future include the following (Ng 2007):

- Strength of the RMB.
- Increased overhead and labour costs.
- Labour shortage.
- Quality issues.

The industry is currently impacted by strong economic factors with the Chinese currency having become stronger. In early 2006 there was RMB 8.27 to the USD 1 while early this year the exchange rate was at RMB 7.74 to USD 1.[6] This represents an increase of 6.4% in the value of the currency against the dollar. Inflation has also increased overheads and labour costs - the latter by 12% in the past year.

Labour shortage is becoming apparent. Companies have had to improve salaries, social benefits and working conditions in order to retain workers, and they are now expanding labour sourcing to the inland provinces in China.

6 Reported on: http://www.xe.com/ucc/convert.cgi.

Excessive use of water holding additives such as sodium tripolyphosphate (STPP) and selling short weight is still a problem in the processing industry, but quality has improved in the past year due to customer pressure. Last year the EU issued warnings and gave the China Inspection and Quarantine (CIQ) Bureau a list of plants using excessive STPP in products exported to the EU. The CIQ subsequently implemented corrective actions, including a ban on the use of STPP in products for the EU.

On the positive side there are many new developments. More value-added products are made in China, including retail packs. An increasing number of species is processed, and processing is more focused on IQF portions and interleaved fillets as opposed to blocks. It is perhaps not to be expected that the total production will increase much in years to come, but the industry is bound to become more receptive to market demand with regard to quality issues and product specifications.

Cod and Haddock from Iceland

Iceland has one of the largest fisheries in the world, ranking 13th in capture fisheries with close to 1.7 million mt landings in 2005 (FAO 2007). The fisheries are divided into demersal, pelagic and shellfish fisheries, with the pelagics providing the largest volume and the demersal fisheries the highest value - 70% of the total. Atlantic cod (*Gadus morhua*) is the most important of all the marine resources in Iceland. In 2005 it represented 38% of the total seafood export value, while the second most valuable stock - haddock (*Melanogrammus aeglefinus*), provided 11%.

The Icelandic fisheries are among the most global of seafood businesses in the world. They are also technologically advanced, and draw on one hundred years of landing data and 60 years of organised research on marine resources. Present day fisheries operate under a system of individual transferable quotas (ITQs), that include allocations for cod close to 200 000 mt and for haddock nearly 100 000 mt.

Running a modern business in the fisheries, however, continues to be a challenge and a learning process, ranging from unexpected findings in fisheries research to the constantly changing market environment.

The following case study for cod and haddock from Iceland describes recent changes in the processing and markets for these two important species. The study describes the rapidly expanding production of fresh fillets and portions which has been Iceland's response to consumer trends on its most important markets. At the same time, the processing of fresh products has also been Iceland's strongest response to new competition on the market, *i.e.* the influx of lower-priced frozen whitefish fillets from China. Previously it has been shown that import of frozen cod fillets from China to the EU now greatly outweighs import from Iceland.

The Icelandic producers do not provide wholefrozen cod and haddock for re-processing in China and have no capability to compete on price-levels with China on the fillet markets. Their response has been to decrease production of frozen commodity products, such as whole fillets and blocks and specialise in production of fillet portions cut by computer technology. A large range of products is now cut to customer specifications for marketing as fresh or frozen portions.

The fresh fillet industry in Iceland has matured into a stable and market oriented industry and is the only industry sector that has shown continuous growth in the past decade in terms of catch utilisation and export revenue. It is based on the concept of value addition, in which producers, transporters and the marketplace have all had a share. A further characteristic of this industry is the fact that all sales have been concluded with customers before the time of processing and export. This is in contrast to the image which most people have of fresh fish being sold at auctions, where prices are unknown beforehand and quality may be uncertain.

The largest markets for fresh cod products are within the EU, chiefly in the UK, France, Belgium and Germany. Only about 10% are exported to the United States. Fresh haddock fillets, on the other hand, have a more even market spread, with a third of export value to the US and two thirds to Europe.

Fresh fish marketing depends on an efficient supply chain management, including reliable supply, communication and transport to the market. Most of the export of fresh fillets is by airfreight on daily flights to a number of destinations in Europe and the US. In UK, which is the largest market, the fresh products are packed by large processors, supplying retail chains on a daily basis and usually under store-brands.

The following figure shows that sales of fresh cod and haddock fillets and portions from Iceland have tripled in the past ten years. In 2006 they amounted to 18,000 MT worth EUR 150 million (FOB[7]). In volume terms they were 26% of total cod and haddock fillet exports, but in value terms 35%. Prices for fresh products were generally 10-30% higher than for frozen portions, having taken the packaging cost and air-freight cost into consideration (Statistics Iceland 2007).

Figure 7.13. Export of Fresh Cod and Haddock Fillet Products from Iceland (1997-2006)

(total landings also shown)

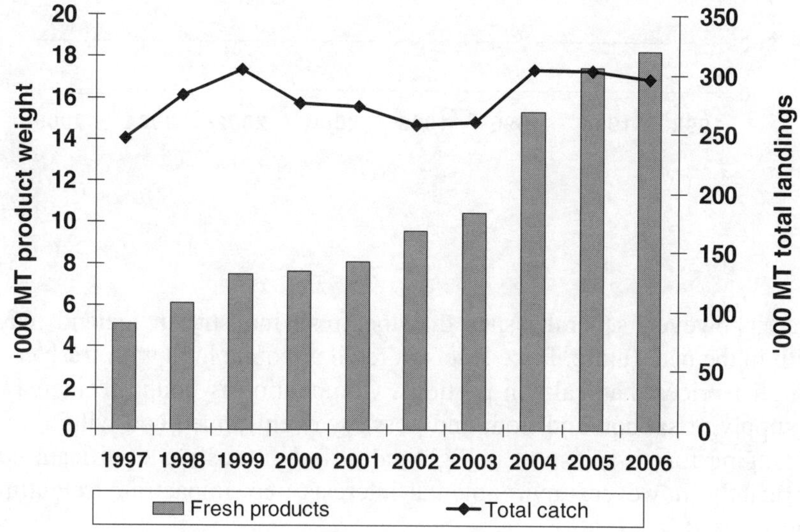

7 Free on Board: the seller pays for the transportation of the goods to the port of shipment, plus loading costs.

The modern fresh food market for highly perishable goods is driven by a variety of forces that not only originate in increased consumer affluence and lifestyle changes, but also in advances in food technology and logistics.

The image for fresh food is that it is inherently healthier and "fresher" than other foods, more natural and less processed. None of this stands to reason when looking at the attributes for many fresh foods vs. frozen food, including seafood, but the image is that of fresh and healthy - and image is among the chief factors to be considered in a sophisticated market.

Nowhere is the positive image of fresh (chilled) seafood better documented than in the UK retail market (Young's 2007). The following figure shows the marked and constant growth of the fresh market for seafood, which ten years ago surpassed the frozen market in value sales. All the indications are that this trend will continue, supported by an increasingly health-conscious market, and possibly underpinned by growing supply of fresh, farmed products such as salmon.

Figure 7.14. UK Retail Sales of Seafood (1992-2006)

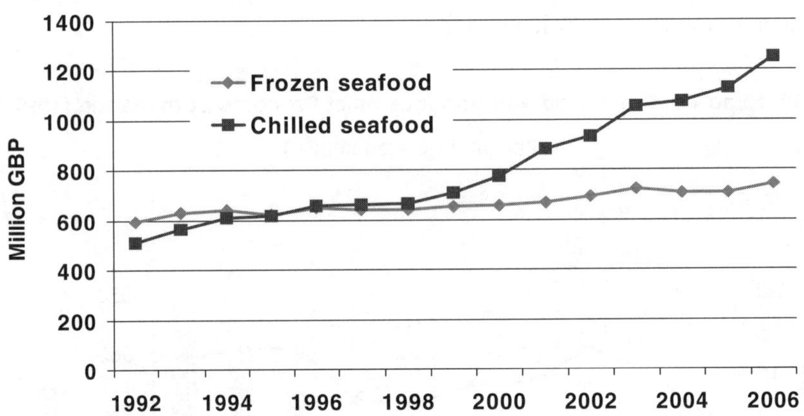

There are, however, several issues that the fresh industry in Iceland may have to contend with in the near future. Price levels in retail are quite high with fresh cod portions selling at higher prices than salmon portions. Competition is bound to increase with an increasing supply of salmon and trout and possibly even farmed cod within a few years. Cost of packaging for air-transport is high and airfreight costs a significant cost factor. More significantly, however, environmental interests may impact air freighting of food within a short period.

Recent reports (DEFRA[8] 2005) state that airfreight is the most damaging mode of transport in terms of climate change. It has the highest emission of carbon dioxide per tonne of product and is the fastest growing mode. Although, airfreight of food accounts

8 Department for Environment, Food and Rural Affairs, United Kingdom

for only 1% of "food tonne-kilometres" it causes 11% of carbon dioxide emission in food transport. Or phrasing it by another new concept – air transport leaves a much "deeper carbon footprint" than transport by sea or road, and is one of the key-indicators for reducing the environmental impact of food transport.

The way ahead for the fresh industry in Iceland is therefore likely to be development of more efficient transport by sea with super-chilling of products in transit.

Drivers for Changes in Seafood Processing

The case studies and product categories previously discussed in this report have shown that new developments in seafood processing are driven by many factors. Most of the drivers are in line with market trends and consumer preferences, but others stem from decisions made in production for increased supply, productivity, profitability and food safety, to name just a few.

The following is an attempt to focus on present and foreseeable market trends for seafood and how they are likely to affect seafood processing.

Consumer interests and needs

The following trends continue to be among the major drivers for the food market at large (Möller 2003).

Demographic and lifestyle trends, but especially increased participation in the work market and in further education, have led to several changes. Among them are the need for speed and convenience in food preparation, increased food consumption away from the home, smaller households in general, and an increased number of single-person households. Meals are less structured and less planned than before. Disposable incomes have increased in many countries and it seems that the proportion of food costs in total income is the lowest on record in many markets.

Consumer sophistication and knowledge mean that the consumer knows he has real choices and makes demands for quality and good value. The consumer is a traveler and likes to experience international foods, but also appreciates the comforts of traditional and local food. The consumer knows that diet is important for future health and takes steps to ensure that he makes healthy choices. The informed consumer will also make some food choices on the basis of the moral issues and safety concerns that are promoted by the media.

How can these very general drivers be translated into seafood preferences? The following seem generally agreed on (Möller 2003).

Simple preparation

People will strive to spend more time at home and eating at home, but cooking is nonetheless a low-priority task for most. Consumers are cutting down on side dishes and frequently increasing the amount of main-dish food prepared. The trend towards frozen or fresh prepared casserole dishes or one-dish meals is viewed as quite significant. For home entertainment there will be a growing selection of festive dishes sold in supermarkets.

Street foods

Today's busy society has created a new generation of food that may be called street food; a category of satisfying, portable, hand-held snacks that are often served hot, eaten on the street or in cars as sandwiches, *burritos*, burgers, snack rolls and beverages. Retailers and foodservice operators are now catering to "impulsive and mobile stomachs".

Fresh, healthy and functional

It is expected that products carrying fresh or natural labels will attract considerable consumer attention, and transparent date-markings will be the rule. Linked to this is the continued interest in organic foods.

A survey has indicated that the majority of American consumers (54%) believe that omega-3 fish oils have positive health-promoting benefits, such as reducing the risk of coronary heart disease and lowering cholesterol levels. There is a great deal of research in support of the claim that seafood consumption has health benefits. Seafood in general is promoted as a healthy food, although it is more logical to think of it as a necessary part of a healthy diet. This has doubtless been one of its strongholds in seafood marketing for many years.

Diversification of flavours

There seems to be a growing consumer affinity for spicy - even intensely flavoured - foods, but there is also a large need for the marketing of traditional flavours, often described as "home-style". Popular ethnic flavours include Mexican, Caribbean and Indian, but milder tastes are marketed as Vietnamese, Chinese or Californian. This trend for seafood has been noticed in areas as far apart as the UK, South Africa and the US.

Ethical behaviour in food choices

Purchase drivers for consumers in a number of countries are now based less on income groups than they used to be and more on the various moral issues and safety concerns that are actively promoted by the media, non-governmental organizations and action groups. Animal welfare is a highlighted issue, and so are chemical residues in farmed fish, together with over-fishing and sustainability issues for wild fish. Climate change is becoming a key political issue and the terms "food miles" and "carbon footprints" are increasingly heard as indicators of the environmental impact of food transport. A product with many food miles behind it may be found less suitable than a more locally produced food.

To sum up, the traditional drivers of demand for seafood are the quest for convenience in preparation, product quality, and the healthy image that seafood has gained in terms of nutritive value and even special health benefits. New drivers may have negative effects on transport of seafood over large distances.

Market response

Seafood marketing companies take a keen interest in customer trends, not least the quest for convenience, variety and good nutrition. Processing of ready-to-eat products and other prepared products often takes place in the market, in cooperation with the retailer or foodservice operator. Not often are the products made at source.

The market has often responded effectively to loss of supply in one category by increasing similar products from other categories. A clear example of this is the demand for farmed finfish on the US and European markets, which has increased in line with declining supply of traditional whitefish and flatfish. The market has exerted a pull on various farmed resources to fill the supply gap left by wild fish. Farmed production and trade in farmed products have grown because the market was receptive and possibly also because the marketing companies added farmed products to an existing range of wild products.

The following figure shows an example of the complementary nature of wild and farmed fish supply. The capture species are Alaska pollock, Pacific cod, various hakes and hoki, Atlantic cod, haddock, saithe and redfish. Flatfish is all the supply recorded by FAO. The farmed finfish is Atlantic salmon, coho salmon, rainbow trout, Nile tilapia, channel catfish and pangasius catfish, European seabass, and lastly gilthead and silver seabream (FAO 2006a, 2006b Groundfish Forum 2006).

Figure 7.15. Supply of Traditional Whitefish, Flatfish and Farmed Finfish

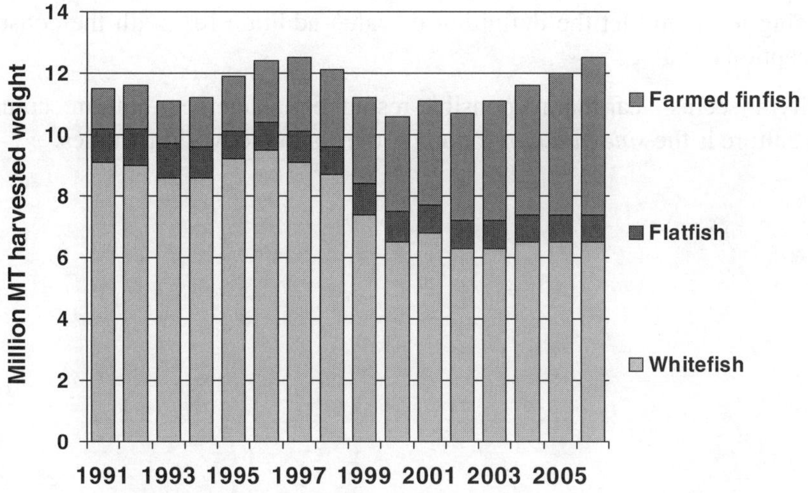

The response to customer interests in ethical food choices may take many forms in the future. Ecolabels are already established in some countries, and UK retailers have now pledged to combat carbon-dioxide emission or to become "carbon neutral" within a few years. Tesco, UK's biggest retailer, has recently stated that it will put new labels on every one of the 70 000 products it sells so that shoppers can compare carbon costs in the same way they can compare salt content and calorie counts. Marks & Spencer have also declared their commitments to become carbon neutral by 2012 (BBC news 2007).

Concern about climate change may soon affect the seafood business, especially transport from very distant suppliers and transport by air. In effect, this concern may work against international trade and create more interest in domestic supplies.

Terminology is still confused, but it is not unlikely that standards will be developed that will work against global trade as it is now known.

Processing sector adaptation

Requirements for food safety and hygiene are not negotiable in the food business, as many operators have experienced. Demands for safe foods, and especially regarding trace contaminants, are bound to increase in the global trade, partly because of more sophisticated detection methods.

Consolidation is very likely to take place in the fishing and processing industries worldwide in order to create leverage against large corporations and retail chains on the market, and for better financial performance. Similarly, it is clear that modern processing technology will be installed and in most instances it will increase productivity and output for many companies in developing countries.

Value addition has long been a keyword in seafood processing. Most often it conjures up the image of coated products, recipe dishes and ultimately of cooked ready-to-eat products. It seems that many primary producers around the world have achieved certain value addition in recent years, usually in cooperation with their customers, and often in preparation for further processing in the marketing country. As noted before, however, it also seems that there is not much processing of recipe products at source.

Perhaps it is time to consider value addition of seafood both as a marketing term and a processing term, and let the definition of value addition rest with the consumer in his own perception of value.

Finally, it seems clear that responsible resource management both in capture fisheries and aquaculture is the *sine qua non* for any progress in the world fisheries.

Bibliography

AFSC (2007), *www.afsc.noaa.gov/refm/cbs* . Report of the First Annual Conference of the Parties to the Convention on the Conservation and Management of the Pollock Resources in the Central Bering Sea (1996), Annual meeting from then onwards.

DEFRA (2005), The Validity of Food Miles as an Indicator of Sustainable Development. Prepared by AEA Technology for DEFRA, *http://statistics.defra.gov.uk/esg/reports/foodmiles/final.pdf*

FAO (2006a), FAO Fishery Statistics, Capture production, volume 98/1.

FAO (2006b), FAO Fishery Statistics, Aquaculture production, volume 98/2.

FAO (2006c), FAO Fishery Statistics, Commodities, volume 99.

FAO Fisheries Report 779 (2006), Report of the Third Meeting of the ad hoc GFCM/ICCAT Working Group on Sustainable Bluefin Tuna Farming/Fattening Practices in the Mediterranean, *ftp://ftp.fao.org/docrep/fao/008/y8870e/y8870e00.pdf*

Ianelli, J. et al (2006), Assessment of Alaska pollock stock in the Eastern Bering Sea. Report on: *www.afsc.noaa.gov/refm/docs/2006/EBSpollock.pdf*

Kurien, J. (2005), Responsible Fish Trade and Food Security, *FAO Fisheries Technical Paper 456.*

Möller, A. (2003), Studies on Seafood Value Addition, *FAO/Globefish, Special Market Study.*

Sloan, A.E. (1999), The New Market, Foods for the Not-so-healthy. *Food Technology, vol. 53/2.*

Vietfish International (2007), Issue (1), Publisher: VASEP, original source: Vietnam General Department of Customs.

Websites

BBC news (2007), Jan 17, *www.bbc.co.uk/2/hi/business/6276351.stm*

Eurostat (2007), EU trade statistics, *http://fd.comext.eurostat.cec.eu.int/xtweb/*

NFI (2007), Website of the National Fisheries Institute, USA, *www.aboutseafood.com/media/top_10.cfm*

FAO (2007) Fishstat Plus, *www.fao.org/fi/website/FIRetrieveAction.do?dom=topic&fid=16073*

FnBnews (2007), India's first food and beverage news, Feb 17, *www.fnbnews.com/article/detnews.asp?articleid=20034&sectionid=5*

NMFS (2007), National Marine Fisheries Service, *www.fakr.noaa.gov*

NOAA fisheries (2007), catch statistics, *www.fakr.noaa.gov/sustainablefisheries/catchstats.htm*

NOAA trade (2007), *www.st.nmfs.noaa.gov/st1/trade/index.html*

Statistics Iceland (2007), *www.statice.is/*

VASEP (2007), *www.vasep.com.vn/vasep/eDailynews.nsf/Homepage*

Vietnam industry website (2007), *www.worldofpangasius.com.vn/*

Personal communication

Bush, M. (2007), Partner, Endeavor Seafood, United States.

Davis, J. (2003), CEO, American Seafoods International, United States.

Groundfish Forum (2006), information collected by the groundfish industry worldwide.

Halhjem, T.K. (2007), Manager European Sales, Trident Seafoods Corporation, United States.

Hoe, T.D., (2007), VASEP representative, Vietnam.

Infante, R. (2006), General Manager, SalmonChile, Chile.

Ingolfsson, H. (2007), Marketing Manager, Marel, Iceland.

Kaptijn, J., (2006), Managing Director, Seafood Connection BV, the Netherlands.

Kuntzsch, V., (2007), Managing Director Hangana Seafood, Namibia, and Chairman of the Confederation of Namibian Fishing Associations.

Leroy, O., (2006), CEO, Leroy Seafood Group, Norway.

Namibia (2006), Based on information collected from the Namibian industry by Groundfish Forum in 1997-2006.

Neubacher, H., (2007), Correspondent for Seafood International, Seafood Processor and Visaktua magazine.

Ng, J. (2007), Executive Director, Pacific Andes Int. Holdings, Hong Kong.

Ponomarev, V. (2006), CFO, Okeanrybflot J.S.C., Russia.

Radvan, M. (2006), President E. Kratzenstein & Co GmbH, Germany.

Young's Bluecrest Ltd. Original ref: Taylor Nelson Sofres and Seafood Industry Authority.

Chapter 8

From Local Production to Global Trade: The Danish Seafood Industry in Transition

Poul Melgaard Jensen, Danish Seafood Association/Confederation of Danish Industries, Denmark

Poul Melgaard Jensen described the changing role of Denmark in fisheries processing and trade as a result of globalisation. The Danish seafood industry is in transition and increasingly acts as a broker in international fisheries trade. Complex supply chains are developing in which fish is sourced from a multitude of places, processed in different locations and then moved onto the market. In such a complex seafood world, the key to success is logistics.

Tariffs and tariff unpredictability, customs procedures, different food safety standards etc. create a complex and uncertain environment that increases risks for traders. For example, quotas for EU imports for 2007 had not been adopted by April of the same year, resulting in uncertainty for importers as to when quotas will enter into force, what amount and on what terms. To many Danish producers, the rate of the tariff is secondary to the need for predictability. In this regard, a Danish study focussing on especially low tariff rates suggests that the costs of administering the tariffs are higher than the revenue stemming from the tariffs.

The outlook for the Danish fish processing industry is for further outsourcing of the less complex processing and a continuation of further specialisation towards high value added products tailored for specific market segments. A reinforcement of the innovative capacities of the Danish processing industry is required to achieve this goal.

An Industry in Transition: The Global Seafood Company

The Danish seafood industry has undergone a radical transition during the last 20 years: From wage-production in many small and medium sized companies based on fresh fish primarily caught in local waters to a new type of company integrated in a global network of production, trade and ownership.

This transition has been brought about mainly due to steadily decreasing catches of fish from local waters surrounding Denmark and to a differentiation of costs of wage-production between Denmark and Asia and Eastern Europe. Decreasing international freight costs have accelerated the process and an increasing consumer-awareness and demand for high quality and new exotic products has further contributed to this transition.

Due to the unsteady supply of fresh fish attached to the current Danish quota system for regulating fisheries, the switch to global sourcing also carries with it the possibility to gain from a more steady supply, especially for farmed fish and fish products, so as to meet the supermarkets demand for constant deliveries.

The drivers on the market have been the outward-looking companies with an early "eye" for the global marketplace and it's potential.

Seen from a company-perspective the critical success-criteria for this new type of seafood company - call it the global seafood company - is *logistics*: To be able to effectively manage a complex, international supply chain from farm/catch to consumer and guard against inherent risks by relying on more exporters, more species and more countries.

A type of global seafood company manages from their Danish domicile their global trade of products never reaching or passing Danish territory. Other trading companies have specialised in efficient and timely delivery from own or remote cold stores confining their handling of the goods only for labelling and/or repacking.

The global seafood company is being integrated with foreign capital and buyers, for instance from Japan, Iceland and New Zealand and this international consolidation makes it easier to manage and oversee market developments and manages the global supply chain.

Basic processing like filleting is being outsourced to China and Asia, and for example Barent Sea cod are shipped to China, filleted here and shipped back to the final consumer in the EU. The industry basing its activity on transforming and processing in Denmark is generally squeezed by relatively high wage-costs and has become more and more specialized with a focus on high value-added, convenient products such as sushi, ready-to-make fillets in controlled atmosphere packaging and high-quality smoked products. The competition on raw material supply and with imported consumer-packed products has becoming intense.

Nearly 75 % of fish and fish products for consumption and processing are being imported into Denmark and 87 % is sold in countries belonging to the European Union. To put it short: "Danish" seafood is imported from all around the world and sold to the European consumer.

EU trade policy needs revision

As the trade and production of seafood becomes global barriers or restrictions to trade of any kind becomes more visible and costly - including tariffs, customs procedures, differences in food safety standards, anti-dumping measures, and administrative burdens of tariff arrangements such as *end use* or *inward processing*. A Danish trading company has lately hired a full time expert only to guide the company in matters on tariffs and customs procedures.

Import has surpassed the amount of landings in the EU, and for the global seafood company the trade policy of the European Union is not adjusted to this development and is generally regarded as protectionist, constituting barriers for the access to realise the full potential of globalisation.

The EU market regime for fish and fish products were founded when protecting the income of EU fishermen was the main motive, and this motive is prevalent even today. One sign of this is that if you want to import fish without duty for further processing the burden of proof lies with the importer: You have to provide the data showing that sourcing the fish from outside the EU has really become necessary and that sourcing from inside the EU is not sufficient.

Due to an increasing complexity of trade in the global seafood company custom procedures takes up time and money: According to a Danish study of especially low tariff rates the costs of administrating the tariffs are higher than the value of the revenues of the tariffs going to the EU budget.

When asking the global seafood company of their priorities the answer is clear: The uncertainty and unpredictability in tariff and customs arrangements is the first and most important problem. The *level* of the tariff rates is secondary. So in other words: Importers are prepared to trade for a more predictable system - even if higher tariffs would be the price.

Two examples illustrate the point:

- A Danish importer concludes a 6 month contract on hoki from New Zealand. The price is set taking into account the zero duty rate applicable to a limited EU quota - provided that the fish undergo further processing after being imported. However, because the quota was used before the consignment arrives in Bremerhaven the tariff rate of 15 % is applied. The import is loss-making to the importer; however to fulfil the contract the trade is nonetheless carried out.
- I would add that the quotas for EU import in 2007 have not been adopted yet, and importers had paid normal tariff rates as from January 1st. It is impossible for the importers to now for sure when the quotas will enter into force; the amounts and on what terms.

The EU trade defence instruments are also considered as a barrier.

Being revised at the moment the EU trade defence instruments has a heavy bias towards primary production. The restrictions on the Norwegian farmed salmon are an example: For more than 18 years this import has been subject to various restrictions and the reasons are now being investigated by the WTO.

The current minimum price being imposed on salmon from Norway is the result of a complaint from Irish and Scottish farmers representing around 260 employers. On the other hand more than 6 500 people are employed in the EU processing industry dependant on the free access to fresh, farmed Norwegian salmon.

According to the Danish National Agency for Enterprise and Construction the EU measures applied from 1993-2002 incurred a total net loss of EUR 22 million for the Danish society in general (=loss for importers, processors and consumers minus the protective benefit to the farmers).

Safe import conditions from less developed countries (LDC)

The global seafood company is importing more and more from less developed countries which are enjoying a preferential access to the EU market so as to facilitate their integration in the world economy.

However, this trade is risky to the importer: The documents underpinning this trade is not safe and reliable, leading to many corrective duties presented to importers in Denmark and the European Union. The proof of origin is issued by competent authorities in the country of origin at the time of export but often declared invalid when being subject to ex-post verification and control.

The EU Commission concluded in 2002 that this system clearly needed to be reviewed: The risk you encounter as an importer from LDC's is far beyond normal, commercial risk you could insure and cover.

There is a need for a reliable system underpinning the increasing import from the 154 countries - especially LDCs - with preferential access to the EU to help their integration and to help our sourcing in a win-win scenario.

Food safety as the main challenge

When tariffs are reduced the issue of food safety tends to become the main challenge for the free trade. The global seafood company experienced that in 2001 when EU border inspection posts found traces of antibiotic residues in crustaceans from South East Asia and farmed molluscs and crustaceans from China were banned for 18 months.

The figure shows how the otherwise rapidly growing import of crustaceans dived during this period.

Part of the problem was due to the development of increasingly refined analytical test methods in EU making it possible to detect traces on still lower levels.

Exporters from LDCs risk ending up in a situation where their products are being rejected or even destroyed at the EU border - other producers are not being able to export at all because they are not able to meet EU standards. Capacity-building (=assistance and education *on location*) projects would be the key to a solution with a potential significant trade promoting effect. It is important to note that reducing the tariffs into the EU does not promote trade when the problem is food safety.

A scenario for the future: 10 years ahead

If we look forward for the next 10 years a scenario for the seafood industry would point to a stronger and further consolidation in Denmark and with foreign companies realising investment opportunities in the Danish seafood industry.

Processing will be further outsourced - also for processing with higher value-adding like slicing, smoking and even ready meals. The processing industry in Denmark would be forced to further specialisation towards high value-added products tailored towards specific consumer segments and reinforce the innovative competences to be able to introduce and bring on the market new products more frequently.

The global seafood company would be more dependent on external factors for success. If the tariffs are successfully reduced on imports into the EU and the customs procedures could be simplified then food safety issues would definitively become the number one challenge.

Even though veterinary residues have become a lesser problem, other contaminants such as heavy metals could become a more serious problem. Being able to effectively cope with these risks would be critical for success. For the global seafood company the first priority for the global agenda would be development of a set of international recognized food safety standards based on objective, scientific criteria.

The strategic management of a more global supply chain and being able to spot and react quickly on new potential sources of supply becomes more important. Developing the raw material base to more and new species from a broader range of suppliers - including farmed species and importing species which can replace cod (i.e. tilapia and catfish) is important.

The stocks of wild fish in Danish local waters could become more stable or even turn into a sustainable development. Contributing factors would be the new management system for fisheries introduced in 2007 with a fixed quota for each vessel and with an option for pooling or trading the quotas between vessels. Also a more widespread demand for certified eco-labelled fisheries (as Marine Stewardship Council) would point in the same direction.

The new management system for fisheries would pave the way for bigger and more efficient vessels which could increase the horizontal integration between fisheries and industry directly by contracting on terms of delivery or by the industry acquiring vessels. It would improve the quality of the catch (freezing the fish on board after filleting), reduce discards, and give Danish fresh fish an advantage against cheaper imports - eventually supported by on a new "super quality" standard.

The Danish aquaculture *i.e.* freshwater and marine aquaculture would have increased production substantially within a sustainable framework due to new technologies to manage aquaculture for environmental purposes. Danish aquaculture could become an important new source for the Danish seafood industry.

As the markets with highest growth in demand are foreseen outside the EU it is critical that the Danish seafood industry becomes as global on the market side as on the sourcing side.

Conclusion

The Danish seafood industry has successfully embraced a global strategy exemplified by the global seafood company. If the global vision should be fully realised the EU should support it by opening up for trade with third countries: We need free and open import to develop and compete with the rest of the world.

Chapter 9

Fish Safety and Quality Challenges by Developing Countries: The East African Nile Perch Case

Nancy Gitonga, Fisheries Consultant, FishAfrica, Kenya

Ms. Gitonga spoke about the challenges facing the fishing industry with respect to globalisation in developing countries, particularly in light of international agreements such as the TBT and SPS Agreements. Developing countries need capacity building assistance to meet the safety and quality standards measures set by importing markets. As part of the two Agreements, developed countries are expected to provide assistance to developing countries to ensure such measures are not barriers to trade. Ms. Gitonga suggested that ways to overcome trade barriers include; building fish quality and safety capacity through training of health and safety trainers, establishment of effective quality and safety assurance systems, capacity building in negotiation skills, improvement of infrastructure, development of bankable projects to mobilise resources, and participation in standard setting.

Ms. Gitonga highlighted various ways that can assist developing countries to effectively participate in global market systems, some of which include proper and sustainable utilisation of national resources, building national and regional capacities to address emerging challenges such as risk analysis, traceability and eco-labelling and sourcing for technical assistance to enhance quality and safety standards, and making a deliberate effort to participate in international standard setting processes. The developing countries' inability to participate in standard setting fora or dispute settlement procedures, results in their weak position to dispute unjustifiably applied SPS measures. Other challenges in the global trading system that need to be overcome by developing countries include coping with many importing countries' requirements, which are all different. Ms Gitonga emphasized the need for global harmonisation of standards.

Introduction

Developing countries contribute over 50% of the world fish trade. Net earnings of foreign exchange from fish trade, in developing countries have continued to rise contributing more than meat, tea, bananas and coffee put together. About 58% of fish consumed in the EU come from non-EU waters, mainly from developing countries

The increasing demand for fish and fish products compounded by consumer demand for safe and high quality fish have compelled many countries including, East African countries, Kenya, Tanzania and Uganda, to undertake a comprehensive evaluation and re-organisation of their fish inspection and quality control systems in order to access and sustain global market share. The preventive fish safety and quality approach is the principle behind the current globally practiced and advocated risk analysis and Hazard Analysis Critical Control Point (HACCP) fish inspection systems. This has created enormous challenge on developing countries as they strive to comply with the requirement set by the main export market. The investment needed to comply with the fish quality standards are high and therefore, put developing countries at a disadvantage, largely affecting the small-scale fishing communities.

The current implementation of the EU hygiene standards in the fisheries sector, for example, threatens to severely weaken the market situation of small-scale artisanal fishing communities. Even where the artisanal fisheries sector has proved capable of modernising its operations and responding to market opportunities, their performance is threatened, due to considerable investment involved in meeting the EU and other international hygiene standards. This has easily resulted in monopoly of the export trade, where fishermen may have to accept the prices set for their fish by companies with approved EU hygiene standards. In the East African Community (EAC) Partner States, the fish processing establishments approved to export to the EU drive the exploitation and the prices of the Lake Victoria Nile Perch fishery.

It would be prudent for the EU, other importing countries and the processing establishments to support developing countries efforts to meet the EU and other international, standards with special attention to the situation of artisanal fishers. As one of the most important fishing powers of the world, the EU is expected to set high standards, but should also be prepared to strengthen research, evaluation of stocks, training, infrastructure, and sanitary installations and participate transparently in resource management. The EU has in the recent past approved a number of activities in the project aimed at upgrading fish quality standards in ACP countries in which Kenya, Uganda and Tanzania has benefited. This is a welcome step in the right direction.

Major Challenges of Fisheries Globalisation

Fish Safety and Quality Assurance – Global Perspective

Globalisation of the economy and development of regional economic groupings have highlighted the need for harmonisation of fish safety and quality assurance approaches to ensure fish safety and fair trade practices

It is important, therefore that at the regional and international levels regulation, agreements and guidelines are developed to enable countries and regional groupings develop harmonised fish quality and safety approaches that ensure consumer protection, but are not used as trade barriers. WTO, FAO and WHO are the main organisations that provide international regulatory framework for fish safety and quality.

International Regulatory Framework for Fish Safety and Quality

WTO Agreements

Agreement on the application of Sanitary and Phytosanitary (SPS) Measures and Agreement on Technical Barrier to Trade (TBT) are the most significant agreements for the food safety and quality in the WTO Agreements.

Sanitary and Phytosanitary (SPS) Agreement

The SPS Agreement confirms the right of WTO member country to apply measures necessary to protect human, animal and plant life and health. However, as tariffs and other restrictions to trade decline, there is growing recognition that SPS measures can be used to impede trade in agricultural and food products. The WTO Agreement on the Application of SPS measures has Special and Differential (S&D) provisions which provide among other things, for developed countries to take into account the special needs of developing countries in the preparation and application of sanitary and phytosanitary measures; to encourage and facilitate active participation of developing countries in relevant international standard setting organisations; to assist a developing country.

Under Article 3 of the SPS Agreement members are encouraged to participate in international standard setting organisations and are expected to base their SPS measures on the standards and guidelines or recommendations set by these organisations, where they exist. Under the Agreement members are also required to accept the measures of other members where they can be demonstrated to be equivalent and that they offer the same level of protection. The Agreement recognizes that SPS risks do not correspond to national boundaries.

These considerations, notwithstanding developing countries find it difficult to trade with developed countries due to the difference of quality requirements. A major problem is that there is lack of mutual recognition of inspection and standards. Some major importing countries are asking for sameness in the process rather than equivalence.

Impact of SPS measures on fisheries trade of the developing countries

Developing countries in particular, experience problems in meeting SPS requirements of developed importing countries and consider SPS requirements to be the greatest impediment to trade, especially in the case of the EU. In most instances in developing countries, the SPS requirements are incompatible with the prevailing systems of production and marketing. As a consequence, the structural and organisational change and associated costs necessary to achieve compliance can act to restrict trade in a similar manner to tariffs. In many countries, including East African countries, knowledge of SPS is poor and access to appropriate scientific and technical expertise is elusive.

The most significant constraint with respect to SPS Agreement has been the inability to participate effectively in the dispute settlement procedures and to demonstrate that domestic measures are equivalent to developed countries requirements. This is a reflection of poor scientific and technical infrastructure in many developing countries. Lack of financial or technical resources to implement stringent requirements or even to take a significant role in the standard setting process has exacerbated the weak position of developing countries in as far as mitigating the application of unjustified SPS measures is concerned.

Article 9 of the SPS Agreement requires (in theory) that developing countries be provided with technical assistance to assist them to comply with health and safety standards, but it is evident that much of the technical assistance is usually reactionary. Technical and financial assistance is normally provided to the developing countries, once problems of compliance to SPS requirements have been identified rather than as a part of a strategy for general capacity building.

Concerns on development of SPS Agreement

There are concerns about how SPS Agreement has been developed and implemented especially with respect to the developing countries capabilities and needs. These concerns include the following:

- Many developing countries have not actively participated in SPS Agreements. Many of them are not represented at SPS Committee meetings.
- Developed countries do not take sufficient account of the needs of developing countries when setting SPS requirements.
- There is insufficient technical assistance to developing countries to enable them build capacity to effectively participate in the agreement and dispute settlement in case SPS measures are unjustifiably used to impede trade.
- Insufficient capacity by developing countries to assess implications of the importing country SPS requirements and also to demonstrate that domestic SPS measures are equivalent to developed country requirements.

The SPS Agreement requires that SPS measures be based on internationally accepted risk assessment techniques. In this regard, it would be important for developed countries to take an active role in assisting developing countries to build capacity to facilitate effective participation in Agreements that involve fish processing and trade. This would consequently make implementation much easier. Developing countries must also find ways of sponsoring themselves to participate in these important agreements processes, through collaborative approach with resource users and other stakeholders such as processing and exporting establishments.

Coordination and harmonsation with other International Agencies

In many circumstances the harmonisation of SPS standards can act to reduce regulatory trade barriers. This would also protect members from arbitrary or unjustifiable discrimination due to different SPS standards. It would also reduce cost, as developing countries do not have to meet different standards.

The procedures for instance, of proving that some areas, such as fishing grounds are free from food safety risks or have low risk are usually long and burdensome and often include the need to provide complex scientific evidence and guarantees, which is problematic for poor countries. Indeed, given the complexity of SPS issues, harmonisation or equivalent standards would be the best option.

Effects of application of SPS measures on Kenyan fish trade

Embargo of fish trade with Kenya by Spain:

An example of an unfair application of SPS measures is the rejection of Kenya's fish product by Spain from 1996 to 2004, when Kenya was placed on List 1 of third countries harmonised to export to the EU. Spain declined to import fish from Kenya on allegations that Kenya's fish was not safe due to presence of *salmonella* in her fish products. It is suspected that Spain was using SPS measures to protect her fishers from Nile perch competition. Kenya did not then and still does not have the capacity to contest such unfair trade barriers.

Lengthy and unfair EU fish bans and harmonisation process

EU bans on East Africa's fish exports of the 1990s, are other forms of unfair trade barrier in pretext of consumer protection. East African countries had no capacity to contest the unfair trade barrier, and the bans were lifted only when the importing countries decided to do so. Now that the three EAC countries are harmonised with EU Hygiene and Sanitary Regulations, they are able to market their fish easily and freely in the EU including Spain, one of their biggest outlets for fish products. With the harmonisation of the EU safety standards, Spain would find it difficult to unilaterally put an embargo on EAC fish products.

Application of SPS measures to bar trade without sufficient scientific evidence

One of the bans (Cholera ban) imposed in 1998 was due to cholera outbreak in three EAC countries and Mozambique. The World Health Organization (WHO), however, came to the rescue of these countries and trade resumed. The organization, in a Note Verbale' from its Director General explained that despite the fact that at least 50 countries have been affected by cholera epidemic since 1961, there has been no documentation of any outbreak of cholera from commercially imported food. The foods that are of greatest concern to importing countries are seafood, freshwater fish and vegetables that may be consumed raw. FAO also confirmed that the cholera bacteria does not survive proper cooking or drying, and, therefore, cooked, dried or canned products are considered safe from cholera transmission. Epidemiological data suggest that the risk of transmission of cholera from contaminated imported fish is negligible.

The reaction of the European Union to the Cholera outbreak in 1998, with the consequential damage to EAC fish export earnings as well as the States' inability to rapidly and effectively challenge the imposition of the ban on scientific or technical grounds, is proof enough that much needs to be done to protect exporting developing countries which are at the mercy of importing developed countries.

Also according to Article 2(2) of the WTO SPS Agreement, members are expected to ensure that any sanitary or phytosanitary measures are applied only to the extent necessary to protect human, animal or plant life or health. These measures must be based on scientific principles and should not be maintained without sufficient scientific evidence. Furthermore, in another SPS Agreement Article, members are encouraged to take the least trade distorting measures. The EU did not obviously take all this into consideration when it imposed the ban on East African countries fish products in 1998.

Technical Barrier to Trade (TBT) Agreement

The TBT Agreement basically includes measures designed to protect consumer against deception and economic fraud. The Agreement, therefore, covers standards relating to all types of products including industrial products and quality requirement for foods, except those covered by SPS measures.

The objective of the TBT Agreement, therefore, is to prevent use of national and regional technical requirements or standards in general, as unjustified technical barriers to trade. The TBT Agreement provides that all technical standards and regulations must have a legitimate purpose. It also states that if there are two or more ways of achieving the same objective, the least trade restrictive alternative should be followed. WTO members are obliged to use international standards or parts of them except where these standards would be ineffective or inappropriate in the national situation.

Impact of TBT Agreement on Developing Countries

Due sometimes, to poor understanding of international agreements which countries are signatory to, even developing countries occasionally impose restrictive measures that prevent free trade, contrary to these agreements. The restriction to free fish trade and fishers' movement across Kenyan and Ugandan borders that occurs sometimes can be regarded as an example of TBT. Lack of capacity and expertise in dispute settlement in developing countries hinders quick solution to such situations.

Interventions necessary for fair trade in Fish and Fish Products

Both SPS and TBT call on member countries to facilitate the provision of technical assistance, especially to developing countries, either bilaterally or through the appropriate international organisations. Special and differential treatment provisions call for consideration of needs of developing and especially the least developed countries, when preparing and implementing SPS measures. Such consideration should include providing longer time frames for compliance on products of interest to developing countries. This consideration in the agreements is however, hardly ever applied by the developed importing countries even though the fish industry in most African countries play an important role in the economies of developing countries.

Way forward to overcome the unfair trade barriers

There is need for developing countries to pursue the following from their governments, development partners or through private/public partnerships:

i) Capacity building:

- Training of Trainers (TOT) on various fields relevant to trade and food safety.
- Building capacity (human resources and equipment) and maintaining an effective quality and safety assurance systems.
- Negotiations skills.
- Improvement of infrastructure, especially the cold chain.
- Establishment of testing and referral laboratories.
- Building of good and credible scientific database, through regular sampling and analysis of samples, to counter any false claims.
- Development of bankable project proposals to help in accessing financial resources.
- Participation in standard setting committees.

ii) Financial resources:

- Establish a sustainable funding mechanism.
- Mobilisation of funds from governments and development partners.

iii) Policies:

- Development of sound policies and strategies for food safety and quality.
- Development of an effective and relevant legal framework to facilitate good hygiene practices to ensure compliance with importers requirements.

iv) Co-ordination and harmonisation of important requirements:

- Given the complexity of SPS issues and other requirements by importing countries, harmonisation or equivalent standards would be the best option. The developing countries therefore need to lobby through active participation in international standard setting meetings for harmonisation of standards and legal requirements.
- The developing countries ought to request for adequate time to implement any new regulations and standards developed by the importing countries without consideration of the developing countries needs.
- Developing countries should strive to demonstrate the equivalency of their own standards with those of the importing countries.

Fish inspection and quality control systems

The major challenges the region faces in the development of Fish Inspection and Quality Control Systems include:

- Insufficient financial resources.
- Poor infrastructure.
- Inadequate technical and scientific expertise.
- Poor understanding of regulations and therefore poor application, insufficient awareness of fish inspection and quality Assurance by policy makers.
- Inadequate harmonisation at the regional and sub-regional levels.
- Inadequate capacity for Fish Inspection Quality Assurance.
- Poor communication and networking within and between countries.

In order to address these challenges, it will be necessary to:

- Explore possibilities to mobilize national resources, e.g. Fish Levy for Fisheries development and biodiversity conservation.
- Build national and regional capacities to address emerging challenges such as risk analysis, HACCP etc. Technical assistance could be solicited from FAO and the EU.
- Build fish inspection and quality assurance capacity at national level through training of trainers.
- Upgrade analytical capabilities at national and regional level to include training equipment, and analytical skills.
- Develop and strengthen co-operation and exchange of information with other regional and international institutions. Create databases on fish inspection and Quality Assurance *e.g.* Standard Operating Procedures, Codes of Practice etc. This would assist in the harmonisation of fishing inspection and Quality Assurance systems.

Frequent changes in the importing countries requirements

The exporting developing countries often find themselves in a dilemma while attempting to comply with new requirements from importing countries, such as new Regulations. A good example is the EU food law, which has undergone a recent major revision with the aim to:

- Harmonise and simplify the complex hygiene requirements currently scattered over 17 EU Regulations and three Directives.
- Create a single and transparent hygiene policy applicable to all food and all food operators.
- Develop effective instruments to manage food safety and potential future food crisis, throughout the food chain.

The EU has therefore created Regulation 178/2002, "…laying down the general principles and requirements of food law, establishing the European Food Safety Authority and laying down procedures in matters of food safety", which was published on 28th January 2002 and came into force on 15th January 2005.

Product Certification

Origin

Certification of origin in fish and fish products is tricky, because the origin of processed products would be the country that processed the product irrespective of where the material originated from. The marketing of fish requires that the country of origin of the marketed fish guarantee its quality and safety. The current global concept of HACCP ensures this is done, and any importing country satisfied that the exporter is Competent to apply HACCP principles, will easily accept the products.

Eco-labelling

This phenomenon though relatively new, will soon catch up with the developing countries and create a disadvantage in the marketing of fish and fish products. Though voluntary, once the consumers become aware of the eco-labelling principles, they will purchase eco-labelled foods by choice and this will create an unintended trade impediment to developing countries' fish commodities. It is important, therefore that the developed countries assist developing countries identify and present candidates (fishery or ecosystem) for eco-labelling before it becomes an issue. Some development partners such as GTZ and WWF have started to show interest through an initiative to sensitize stakeholders and to provide technical assistance for East African fisheries. Developing countries lack capacity, financial resources and ecosystems data and information to participate in eco-labelling certification process.

Traceability

The EU food law, Regulation 178/2002, Article 18 on Traceability makes it compulsory for all substances intended or expected to be incorporated into food to be traced back to the supplier and customers; and the systems are operated in order to provide this information to the competent authority upon request. It is also necessary that all food or animal feed placed in the market be adequately labelled to facilitate its identification and traceability. This is intended to enable product that may be a risk to consumer health to be identified and removed from the supply chain and to identify the likely source of un-safe food in order that appropriate actions can be implemented. This is another phenomenon that the developing countries have to comply with, though ill equipped.

Rapid Alert System for Food and Feed (RASFF)

The European Commission publishes weekly alert and information notification. Though this is meant to protect consumers and advise them on choices it also impacts negatively on the developing countries' fish exports. There have also been complaints that the Commission is not as quick to remove from the list, once the establishment have complied, as it is to publish. This problem is also experienced by developing countries and EU member states.

Precautionary Principle

Regulation EC/178/2002 formally establishes the Precautionary Principle as an option to risk managers when decisions have to be made to protect health, but scientific information concerning the risk is inconclusive or incomplete in some way. Although this principle is meant to guide the decision makers and or risk managers and though the regulation stipulates that such a decision must comply with the principles on non-discrimination, it opens avenues for application of unfair trade barriers.

Effects of Trade Liberalisation

Advantages

- Trade liberalisation has affected Fisheries sub-sector positively. The East African fish products can compete with any other fish products globally for international markets with respect to quality and safety.

- Economic empowerment for fishers. Better prices are realised by fishers due to high demand for fish as raw material.

- Development of auxiliary activities, such as net making, boat building, fish transportation.

Disadvantages

- Uncertainty and vulnerability of the export market and therefore income uncertainty for small scale fishers.

- Shortage of fish for domestic consumption.

- Overexploitation of natural fishery resources.

The East African Community (EAC) States Perspective

Preamble

The East African Community (EAC) consists of three countries; Kenya, Uganda and Tanzania. The three countries are linked together by Lake Victoria. Tanzania has the largest share of the lake; 51%, Uganda 43% and Kenya 6%. Kenya and Tanzania have a coastline and the two countries have a rich marine fisheries resource at their disposal. Uganda is landlocked, but is endowed with many inland lakes, besides L. Victoria and rivers, which are also rich fisheries resources.

Fisheries resources in East Africa

Lake Victoria is the major economic fisheries resource of the three countries, from which the world traded fish commodity (Nile perch) is harvested. Other commercially important fish from Lake Victoria include the Nile tilapia and the fresh water sardines, *Rastrineobola argentea,* which enjoy both the regional and international markets.

The marine fisheries are under-exploited by both Tanzania and Kenya. The marine fish trade commodity in these two countries includes prawns, lobsters, octopus and assorted fin fish. Kenya processes and exports tuna loins. There is however no other established land-based industry for tuna, despite its abundance in both Kenyan and Tanzanian Exclusive Economic Zones (EEZs). Most of these resources are exploited and processed by Distant Waters Fishing Nations.

Figure 9.1. Map of East Africa

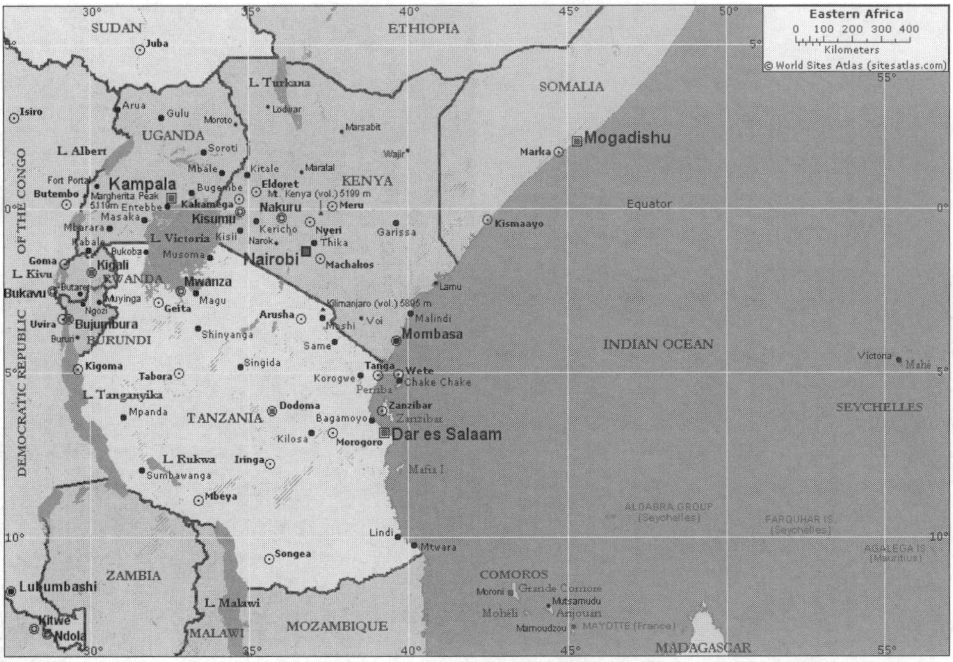

The Fish Industry in East Africa

The Fish industry in the three East African States, which include Kenya, Uganda and Tanzania, plays a very important role in the economies of these countries, through employment creation, income generation, foreign exchange earnings and contribution to food security and poverty reduction strategies. Fish production, fish trade, industrial processing/export and related enterprises such as packaging, boat building and net making, supports about 4 million East Africans directly. The fisheries sub-sector contributes about 3% to the GDPs of Uganda and Tanzania and about 0.4% to Kenya's GDP. Fish is also an important source of Government revenue and as an export commodity, fish earns the three countries substantial foreign exchange; USD 55 million for Kenya, USD 85 million for Uganda and over USD 100 million for Tanzania.

Most of the fish landed in Uganda, Kenya and Tanzania are fished in Lake Victoria. Lake Victoria, therefore, contributes approximately 49% of the total fish landed in Uganda, 92% of the total fish landed in Kenya and 61% of the total fish landed in Tanzania.

The Nile Perch industry

The Nile perch is an exotic fish introduced into the L. Victoria in the 1960s and again early 1970s. The Nile perch fishery started to gain its economic importance in the late 1980s, when Kenya ventured and introduced frozen fillets as export commodity in international markets. The Nile perch fishery has gained so much economic importance in the past 10 years that it is now the most important fish export commodity in the East African region.

Nile perch accounts for about 53% of the total fish production in Kenya and over 80% of Kenyan fish exports. It provides direct employment to over 200 000 Kenyans.

Figure 9.2. The Nile Perch (*Lates niloticus*)

Fish exports

The Nile perch constitutes over 80% of fish exports from the East African region. Over 70% of this fish was exported to the EU member States prior to the series of EU fish bans of 1996, 1997 and 1999. The market is now much more diversified since the fish ban experience, but the EU remains the most important export market for the Nile perch for the three East African States and is still the destination of more than 50% of total fish exports from East African countries.

The EU market was, therefore, the driving force behind the three States' effort to ensure the achievement of the highest safety and quality standards of their fish and fish product to sustain the world market share.

Fish marketing

Though the three governments have endeavoured to comply with international fish quality and safety standards, the need to build capacity in trade agreements skills, in order to broaden and maintain international market share for fisheries products cannot be overstated. The marketing of fishery products as mentioned earlier is vulnerable to unfair application of non-tariff trade barriers such as Sanitary and Phytosanitary (SPS) measures by importing countries. It is, therefore, important that the developing countries actively participates in WTO Agreement protocols and builds capacity in trade issues to help deter unfair trade barriers for EAC fish products.

The East African countries have been able to penetrate the international market for fishery products and compete well in a liberalised economy, because of their advancement in implementation of fish quality and safety standards. Fish exports which are mainly Nile-perch-based products, earn the three countries approximately

USD 350 million in Foreign exchange. There however exist enormous fisheries potential in the Kenyan and Tanzanian Exclusive Economic Zones (EEZs) whose resources are currently exploited by Distant Waters Fishing Nations (DWFN), without commensurate benefits to the countries from which the resource is harvested. There is need, therefore, to put in place an effective Monitoring Control and Surveillance System to ensure the DWFN operating in Kenya's and Tanzania's EEZs pay the governments appropriate dues either through negotiated fishing rights and agreements or payment of fishing licence fees. Kenyans are also being encouraged to invest in the EEZ through acquisition of fishing fleets, and establishment of fish processing plants along the Kenyan Coast. In order to achieve sound investments in marine fish industry, an elaborate quality and safety assurance systems, including testing laboratories is a pre-requisite.

The Competent Authority (CA) for fish and fish products

One of the requirements of the EU market is the establishment of fisheries Competent Authority by a third country wishing to export fish to the EU. The CA would ascertain that the handling of targeted fish for the EU market by third countries complies with the EU standards and the fish processing establishments adhere to their HACCP plans.

To ensure sustainability of fish safety, quality and trade, the three States have well organised Competent Authorities. In Uganda, Fisheries Resources Department of the Ministry of Agriculture, Animal Industry and Fisheries (MAAIF) is the agency responsible for statutory inspection, certification and control of fish and fishery products. In Kenya, the Competent Authority for all matters relating to fish safety and quality assurance, in the Ministry responsible for the Fisheries Department, which is the Ministry of Livestock and Fisheries, while in Tanzania, it is the Department of Fisheries in the Ministry of Natural Resources and Tourism.

All the three CAs have well trained and experienced inspectors, who are also adequately trained in HACCP - based inspections. These inspectors are charged mainly with inspection of the fish production areas, fish processing establishments and markets, certification of the products and the fishing environment.

With respect to requirements of EU Directive 91/493/EEC and also desiring to streamline their operations to achieve efficiency, the East African Governments have done the following:

- Appointment and establishment of CAs in each country, to create a one stop for licensing, management, quality control and development aspects of fisheries.
- Enactment of Fisheries legislation in all three countries on health and safety issues, which are harmonised with EU and other international fish safety legislation.
- Provided technical assistance to export processing establishments to meet required international processing standards, and to enable them develop and adhere strictly to HACCP plans, instituted in all their operations.
- Facilitation of formation of fisheries sector associations for fish traders, especially fish processors and exporters associations with enforceable code of practice as a means of self-regulation.

The EU bans on fish exports from East Africa

The EU fish bans on fish exports from East Africa into its markets, between 1998 and 2000, though unfortunate, because of the enormous economic losses incurred by fishers, traders and governments in the region, helped the industry focus more on the fish and fishery products safety issues and make necessary improvement. The governments, before the bans lacked comprehensive legislation to enforce necessary safety and quality issues. Certification systems were not well defined causing processors and traders to deal with several government agencies on fisheries matters.

The poor definition of CAs and their roles was partly responsible for the three EU bans on Lake Victoria fish exports. The three bans experienced by the East African countries were as follows:

- 1997: on claim of *Salmonella* contamination. This was limited to only Spain and Italy.
- 1998: following cholera outbreak in East Africa and Mozambique.
- 1999: on suspicion of presence of pesticides residues in Lake Victoria. This 3rd and last ban was lifted in April 2000 for Tanzania, August 2000 for Uganda and November 2000 for Kenya.

The Effect of Fish bans on Kenya Fisheries 1996-1999

The Nile perch export business in Kenya has suffered three (3) fish export bans by the European Union (EU) since 1996, which resulted in decline in fish prices as other markets were sought. (Figure 9.3). The Department of Fisheries embarked on corrective measures to remedy the situation as well as to ensure that the country is harmonised with EU safety and quality standards in order to fully access the EU market. The marketing of fish to the EU member states even after the lifting the last ban, and before the country was placed on list 1 in 2004, was through bilateral agreement with individual member States.

The Government Intervention

The Kenya government is giving priority to the improvement of infrastructure at selected landing sites and other fish quality and safety facilities in order to enhance fish marketing and reduce post harvest losses. Modernisation of fish depots, upgrading of fish roads, provision of electricity, telecommunication access, availability of clean water, establishment of chill rooms and ice making plants are some of the developments planned to be accomplished in the medium term. Other planned developments include establishment of three accredited fish quality laboratories and one referral laboratory in the country. The improvement of fish standards for quality and safety will guarantee fish markets which in turn guarantees revenues for the government and earnings for fishers.

Figure 9.3. Effect of Fish Bans on Kenya Fisheries

(1996-1999)

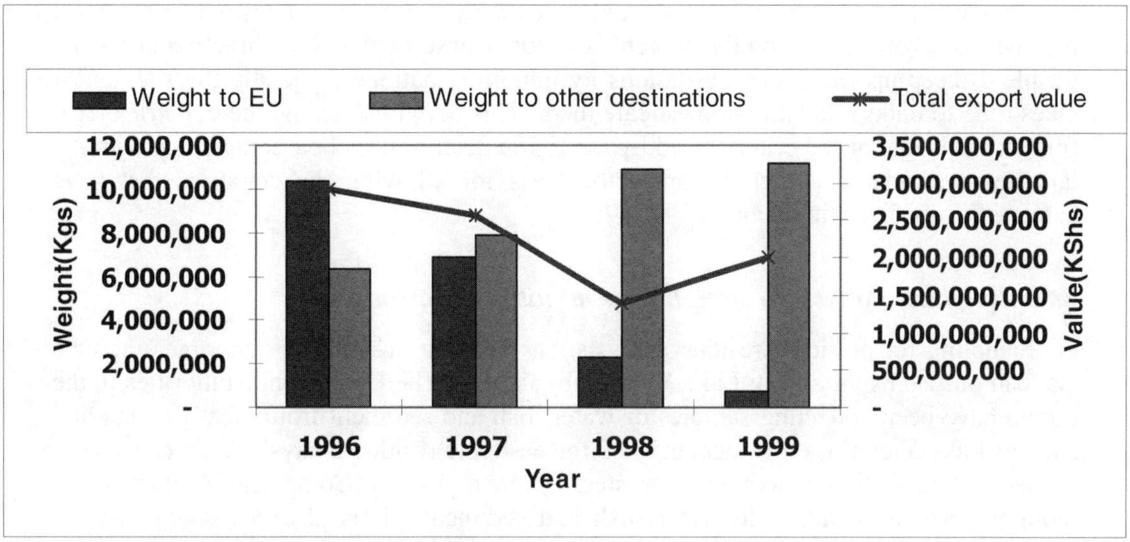

Harmonisation with EU regulations

The Council Directive 91/493/EEC of 22nd July 1991 lays down the health conditions for marketing of fish and fishery products in the EU. Third countries fulfilling the conditions of this Directive are listed in the Commission Decision 97/296/EEC of 22nd April 1997, which is regularly revised and updated, and in the last few years many third countries have been included. The EC classifies third countries into two categories (lists) for import of fishery products for human consumption.

The first category (List I) includes countries whose processing systems and health standards are at least equivalent to the EU systems and whose Competent Authorities have been audited by the EU inspection team. The countries in the second category (List II) are those that have not been audited by the EU inspectors, but have supplied the Commission with written guarantees that fulfils import conditions. Fish products from List II may enter the EU market under bi-lateral agreements with individual EU member States and may be subjected to additional national legislation of importing State.

The three East African States have had their CAs, fish processing establishments and landing sites inspected by teams from EU DG SANCO. The three countries have thus been harmonised with EU standards after complying with requirements and are all now on List I. Tanzania was the first to be upgraded to List I in 1997. The country enjoyed this status until March 1999 when EU imposed a ban on fish products from Lake Victoria. The ban on the exports was lifted in Tanzania in April 2000, after which the EU inspectors visited the country twice to ascertain that safety measures were in place, before reinstating the country back to list I. Uganda was placed in List I in 2003 and Kenya in 2004. The major impediments for Kenya and Uganda to achieve List I status were the lack of clarity on the definition and the roles of their Competent Authorities. The EU inspection teams were concerned that inspection and certification roles were not clearly defined and were handled by various government agents.

Fish quality and safety assurance legislative framework

All the three countries have developed Fish Quality and Safety Legislation that ensure fish and fishery products placed on any market are safe for human consumption. The penalties for not complying are stipulated in the three countries statutes, some of which include cancellation of compliance certificate or license denial. The three countries are unable to keep up with new legislations by importing nations especially the EU, and it takes time to understand and domesticate them. This notwithstanding, the countries legal framework is one of the criteria used by the EU in determining their compliance with the laid down standards, which becomes the basis for allowing the countries that have achieved equivalency to export.

Sampling and analysis of samples for monitoring purposes

Sampling for pesticide residue analysis, has been an on-going programme since the last ban on fish exports from Lake Victoria by the EU. The Competent Authorities in the region have been collecting samples of water, fish and sediment from identified landing sites in Lake Victoria, on a regular basis for pesticide residue analysis. This exercise is carried out to fulfil the written guarantees given to EU by Competent Authorities to monitor pesticide residues in water, fish and sediments. It is also envisaged that the programme will assist the CAs to build a database on the status of the fish and the fishing environment that would in turn assure consumers of the safety and quality of fish in Lake Victoria. The results of the analysis have so far been satisfactory, as they have shown no detection of pesticides at the level of Detection (LOD).

The microbiological analysis of fish samples and swabs collected from fish contact surfaces is regularly conducted by CAs in fish processing establishments.

Public participation in safety and management of fisheries resources

The East African Community countries have realised the need to harmonise the upgrading of safety and quality systems. This realisation was brought about by the bans imposed on fish exports from Lake Victoria, which for Kenya lasted for 20 months. The last ban, which was shorter for Tanzania and Uganda, was as result of suspicion of use of chemicals for fishing in Lake Victoria, and though the suspected report originated from one country, all the three countries were affected. The three countries also noted that they are all at different levels of Fish Inspection and Quality Assurance development and decided to work together to harmonise fish quality control and inspection systems co-coordinated by Lake Victoria Fisheries Organization (LVFO) secretariat. A task force was, therefore, established to look at fish quality issues.

Beach management units popularly known as BMUs, have now been adopted by the three East African countries as the primary grass-root institutions that would ensure compliance with fisheries management measures and fish quality and safety by community.

The role of the BMUs includes the following:

- Ensuring that fishers carry out fishing and fish handling responsibly and hygienically.
- Dissemination of information on responsible fishing and handling methods to the community and stakeholders.

- Ensuring that hygiene standards at the landing site are maintained.
- Management of environmental issues.

Upstream control of fish safety and quality

To address some of the severe constraints on the fisheries post -harvest issues, the three East African States have endeavoured to improve fish handling facilities at fish landing sites. Funding for improvement of these facilities is usually done on an integrated approach, where the Government, the fish processors, the local communities and donors make contributions.

To ensure traceability of fish from fishing grounds to the consumer, the three East African States have introduced local health certificates at landing sites. The certificate accompanies raw fish from the landing sites to fish processing establishments. This ensures that fish for export is collected at designated areas that have been certified by CAs.

Private sector fish processing and export associations

One of the most useful outcomes of the EU harmonisation process was the silently implied requirement for an organised fish industry private sector, capable of self regulation. The Fish Processors and Exporters of the three countries organised themselves into strong member driven associations. These associations include Association of Fish Processors and Exporters of Kenya (AFIPEK), Uganda Fish Processors and Exporters Association (UFPEA) and Lake Victoria Fish Processors and Exporters Association of Tanzania (LVFPAT). The three associations, who form the East African Fish Processors Association, have been instrumental in ensuring standards are sustained by establishments through development and implementation of the Code of Conduct for the members. These Associations have also acted as useful interfaces with governments on fisheries co-management and trade issues.

The fish processors and exporters associations are basically focused on exports only, a sector which takes less than 50% of total fish landings. The need for the business linkages between large and small-scale players in fisheries cannot, therefore, be overemphasised. The Associations are expected to play a key role towards achieving this linkage and inclusion, as a means of expediting development of the small scale fish trade and industry. Kenya with the assistance of GTZ and in collaboration with AFIPEK has embarked on the development of a small scale –fish traders programme to improve market linkage.

Emerging Fish Trade Developments and Challenges within the Region

Trade-Mark

The Nile perch is now registered as “Lake Victoria perch” through Lake Victoria Fisheries Organization (LVFO). This is expected to give the popular product a niche in the global fish market.

Value addition

The three States ought to look into value addition to maximise revenues derived from the Nile perch products and also to enhance sustainable management of the natural resource, as less would be harvested for the same amount of earnings.

Value-Chain reduction

Fish reaching the EU market from this region often goes through 4 or more steps before it is retailed for the consumer. The result of this is a substantial difference between retail and export price. The Nile perch fillet FOB value to the East African exporters is USD 3-4, but retails at between USD 15-17 in the EU supermarkets. This area, if properly explored and developed would enhance revenues from Nile perch products for the three EAC States.

Slot size

Limitations on exports of whole Nile perch and fillets below certain sizes and weights if implemented would discourage fishing of juveniles and encourage stock recruitment and hence contribute to increased sustainability of fishing in Lake Victoria. The region is implementing slot-size harvesting with very encouraging results of success.

SPS issues

Procedures and durations required to implement SPS issues in fish and transparent verification mechanisms sometimes presents problems in the fish industry. Unilateral demands, such as sudden introduction of new standards, will only continue to hurt the region's fishery economy. A case in point is the current practice whereby an exporter whose consignment has suffered a rapid alert (marked as having a pathogen or unauthorized substance) has no access to the sample in question for independent verification, and even when he does, the ensuing results may not be considered, when the verifying laboratory is not EU based. It also takes much longer time to remove the rapid alert from internet after mandatory compliance, than it takes to place it there.

Harmonisation of fish inspection and quality assurance standards and certification in EAC

This was initiated by a regional task force (RTF) composed of 6 members; two from each of the East African States was formed to undertake the activity in 2000.

The RTF has developed three draft documents:

- Code of Practice (COP) for fishing, fish handling and processing in Lake Victoria.
- Inspectors Guide.
- A manual of Standard operating procedures for fish inspection and quality Assurance.

The East African Community (EAC), made a decision in 2003 to harmonise sanitary and phytosanitary measures in East Africa. A joint regional harmonisation team consisting of the members of RTF and Regional Working Group on Fish Quality and

Assurance members was established to produce Fish and Fishery Products sanitary standards, measures and procedures. The joint team through consultative meetings has thus developed the following documents:

- A Code of Practice for Fish and Fishery products: Part I Capture Fisheries, Part II, Aquaculture,
- Inspectors' Guide, Part 1 and II
- Manual of Standard Operating Procedures (MSOP) Part I to IV

The drafts were presented to stakeholders at a workshop and adopted in 2006.

Conclusion

For developing countries SPS measures are considered to be the most important impediment to food exports to the developed world. To a large extent this reflects lack of scientific and technical expertise, information and finance. In addition, the incompatibility of production and/or marketing methods in developing countries is also another factor affecting access to developed country markets. The major problems associated with the operation of the SPS Agreement, however, is that developed countries do not take sufficient account of the developing countries needs in setting standards. The length of time allowed between notification and implementation of the SPS requirements and the level of technical assistance provided by developed countries are also considered to be big challenges to the developing countries export market.

The developing countries must therefore be in a position to challenge measures taken by developed countries and ensure active participation as stakeholders in trade negotiations. The private sector also ought to actively partner with governments so that their concerns are represented through their governments.

References

CFFA-CAPE (2000), "Who benefits at what costs from EU-ACP Fisheries Relations".

Codex Alimentarius (1997), Report on: "Guidelines for the application of the Hazard Analysis Critical Control Point (HACCP) system" (ALINORM 93/13A, Appendix II). FAO, Rome. 58 pages.

EEC (1991), Council Directive 91/493/EEC of 22nd July laying down the health conditions for the production and placing on the market of fishery products Official Journal of the European Communities. No. L268. p.15.

Europa - Food safety- General food law: Regulation (EC) 178/2002; http://ec.europa.eu/food/food/foodlaw/index_en.htm

European Commission (2001), First Report on the Harmonisation of Risk Assessment Procedures. Brussels, Belgium. 173 pages.

European Union, (2001) Commission Decision of 12 February 2001 amending Decision 97/296/EC drawing up the list of third countries from which the import of fishery products is authorized for human consumption. Official Journal of the European Communities. L42: 6-8

European Commission Communication (2000), *Fisheries and Poverty Alleviation,* November.

FAO. (1995) "Code of Conduct for Responsible Fisheries", Rome, 41 pages.

Henson, S.J., R.J. Loader, A. Swinbank, M. Bredahl, and N. Lux (2000), *Impact of Sanitary and Phytosanitary Measures on Developing Countries.* University of Reading: Department of Agricultural and Food Economics, Reading, United Kingdom.

Hillman, J. (1997), "Non-Tariff Agricultural Trade Barriers Revisited." in *Understanding Technical Barriers to Agricultural Trade.* D. Orden and D. Roberts, eds., pp. 1-32. St. Paul, Minnesota, United States: University of Minnesota.

Johnson N, (1995) "The Economics of EU Fisheries Agreements; the Case of Senegal".

PART 4

Identifying the Parameters with which We Are Working in the Harvesting Sector

Chapter 10

Are Present International High Seas Governance Structures Sufficient to Reap the Benefits of Globalisation?

Michael Lodge, Counselor, Round Table on Sustainable Development, OECD

Michael Lodge presented the legal framework governing access to high seas fish stocks. He explained that sub-optimal utilization of high seas fish stocks is an economic problem, caused by the common property nature of the fisheries and the open access regime on the high seas. Inefficient utilisation of shared stocks could be costing the global economy an estimated USD 50 billion per annum.

The United Nations Law of the Sea (LOS) and the UN Fish Stocks Agreement form the basis for a legal framework governing access to high seas resources. A fundamental condition arising from LOS is the duty for states to co-operate in managing shared fish stocks and high seas resources. The paradigm for co-operation is the Regional Fisheries Management Organisation (RFMO). There is no possibility for unilateral action and members have a duty to cooperate even when they fail to agree. Non-members who do not agree with RFMO members are not discharged from the duty to cooperate. Where no RFMO exists, states are under a duty to cooperate to establish one or, at least, apply conservation measures consistent with the LOS.

Michael Lodge highlighted the major threats to the stability of RFMOs. They include free riding; IUU fishing; the failure to heed the special requirements of developing countries; failure to deal with new members; and failure to agree on allocation. This last point is made more difficult by the dual role of States who act in a capacity of both allocating and managing. In addition, developing countries form the bulk of RFMO participation, but not the bulk of allocation rights. Other challenges include a greater role for industry and overlapping areas with national management.

Conditions for success include effective conservation and management practices (such as adopting ecosystem and precautionary approaches), emphasis on better implementation, RFMO performance assessment and review, guidelines on best practice, and global standards for RFMOs accompanied by more transparent accountability.

Introduction

For the fisheries sector, the process of globalisation may be considered essentially as the closer economic integration of fisheries markets throughout the entire value chain. Globalisation creates significant opportunities to improve economic and social welfare for all involved in the fisheries sector. These come about through increased trade opportunities as a result of increased demand for fish and more efficient supply networks, strengthening of the value chain and international capital movements. On the other hand, the processes of globalisation also exert increased pressure on the fishing industry and fish stocks by demand for lower input costs and increased raw material.

Alistair Macfarlane (Macfarlane 2007) argues powerfully that vertical integration between the catch sector and the rest of the value chain is only taking place in fisheries that can assure stock abundance (for which, read long-term sustainability) and secure access rights. In other words, that the inabilities of current governance structures to provide secure access rights to fishers discourages globalised investment in fisheries. If that is the case for fisheries within areas of national jurisdiction, it is even more so for high seas fisheries.

The question this paper aims to address, therefore, is whether present international high seas fisheries governance structures are sufficient to reap the benefits of globalisation. The paper aims to examine the issue primarily from an international legal perspective. It begins with a brief review of the current legal regime for high seas fisheries, concluding that the so-called "freedom" to fish on the high seas is in reality a conditional freedom, to be exercised in accordance with international law. The paper notes that current international law demands cooperative management of high seas resources. The paradigm for such cooperation is the Regional Fisheries Management Organisation or Arrangement (RFMO). The paper goes on to consider the conditions for successful management through RFMOs, identifying the conditions that must exist if the two banes of cooperative management arrangements, namely non-compliance and "free-riding" are to be avoided. The paper concludes that, although the situation is far from perfect, efforts by the international community over the past decade have brought about significant improvements, particularly in terms of more effective enforcement and an enhanced understanding of the need for conservation measures that will promote long-term sustainability. Furthermore, whilst the conditions surrounding the freedom to fish do not yet equate to secure access rights, the regime provides a sound platform on which future rights-based access regimes could be superimposed. The paper concludes, however, by conceding that in most regions, security of access to high seas resources has been dealt with imperfectly, if at all. More effort will be needed if these problems are to be addressed. Some key aspects include more imaginative and constructive engagement by and with the fisheries industry and improving governance and sustainable capacity in developing countries.

The Legal Regime for High Seas Fisheries[1]

It has long been recognized that the root cause of the economic problem in capture fisheries management lies in the traditional common property, or "common pool," nature of fisheries resources, and the resulting open access, or near open access, fisheries. In

1 This section of the paper is summarised from Lodge *et al.*, 2007.

open access type of fisheries, it is irrational for any individual fisher to invest in the resource. By refraining from harvesting the fisher can expect to do no more than increase the harvests of his competitors. In these circumstances, the individual fisher has every incentive to regard fishery resources as if they were non-renewable, in other words, as resources to be mined.

The magnitude of the threat to the stability of fishery resources of the consequences of the "common pool" nature of capture fishery resources will depend upon the potential profitability of the fisheries based on these resources. If the potential profitability is minimal, the threat may be minor. Conversely, where the potential profitability is high, the objectives of long-term sustainability and optimum utilization of fishery resources become extremely difficult, if not impossible, to achieve without an effective cooperative management regime. The more likely scenario is severe overexploitation of the resources and their sub-optimal utilization. That there has been widespread overexploitation of world capture fishery resources (excessive resource disinvestment) and that the utilisation of the resources has been decidedly sub-optimal has become increasingly evident. A recent study, undertaken for the FAO and the World Bank, estimates that the net contribution of world capture fisheries to the world economy is negligible. If, on the other hand, these fisheries were being optimally managed, their net contribution to the world economy would be in the order of USD 50 billion per annum (Arnason 2006). Serious management of world ocean fisheries is a relatively recent phenomenon, dating back only about 60 years. Until comparatively recently, the legal regime for high seas fisheries was based around two fundamental premises, namely (a) the impossibility of the high seas being subject of effective occupation and, (b) the inexhaustible nature of marine fishery resources (Orrego Vicuña 1999, pp.4 – 5; but see as well: Lodge 2002). In the early part of the twentieth century it became increasingly apparent that these assumptions were no longer tenable. This led eventually to States taking measures to conserve living resources through voluntary curbs upon the exercise of their freedom to fish on the high seas. Of necessity, because of the nature of international law, such arrangements were entered into voluntarily and on a regional basis through international agreements.

With encouragement from the FAO starting in 1945, several regional commissions were established. An immediate post-World War II example is that of the International Commission for the Northwest Atlantic Fisheries (ICNAF), established by Canada, the United States and a few European countries in 1949. Other early examples included the Indo-Pacific Fisheries Commission and the General Fisheries Council for the Mediterranean (GCFM). This model for international cooperation for conservation and management of high seas resources through regional conventions based on the geographical and biological distribution of the marine populations concerned was reflected in the provisions of the Convention on Fishing and Conservation of the Living Resources of the High Seas of 1958 which in turn became the source of articles 116 to 120 of the 1982 United Nations Convention on the Law of the Sea ("LOS Convention").

In the mid 1970s, when the new 200 nautical mile exclusive economic zone ("EEZ") regime was clearly on the horizon, it was estimated that over 90 per cent of the world commercial marine capture fishery harvests (by volume) were taken within 200 nautical miles from shore. It was assumed, therefore, that conferring sovereign rights on coastal States for the purpose of exploring and exploiting, conserving and managing these EEZ resources would lead to better management. That proved not to be the case. Coastal States soon found that many of their EEZ resources were shared with other coastal States or with distant water States now operating on the high seas (straddling and highly migratory fish stocks). Indeed, the reality is that over 60% of world capture fisheries are based on straddling and highly migratory fish stocks. Over-exploitation of the high seas segments of straddling and highly migratory fish stocks (or, conversely, mismanagement of the EEZ portion) could and did undermine conservation and management action.

Eventually, a global crisis in the management of certain types of transboundary fish stocks in the 1990s (well described elsewhere) led to the adoption of a binding global agreement – the United Nations Fish Stocks Agreement ("UNFSA") of 1995 – which significantly strengthened the position of RFMOs as the paradigm through which States are to cooperate to achieve and enforce conservation objectives both on the high seas and in areas under national jurisdiction. The implications of that Agreement will now be considered.

RFMOs post-UNFSA

UNFSA significantly strengthened the position of RFMOs as the paradigm for the adoption of fisheries conservation and management measures. In this sense it implements the call for cooperation over high seas conservation in the LOS Convention and so represents a progressive development of concepts of cooperation, compatibility and responsibility that are set out in the LOS Convention.

Under UNFSA, RFMOs are to be seen as the primary mechanism for achieving the cooperation between, and among, coastal States and fishing States that is deemed to be necessary if the objectives of the UNFSA are to be met. Those objectives are set forth in articles 2 and 5 of the Agreement. Together, these make it clear that the primary objective of the Agreement is to seek effective and compatible conservation and management regimes both inside and outside areas of national jurisdiction. UNFSA is meant to apply, unless otherwise provided, to the conservation and management of straddling and highly migratory fish stocks beyond areas of national jurisdiction. There is, however, a significant proviso. Article 6, entitled application of the precautionary approach, and (importantly) article 7, on compatibility of conservation and management measures, "apply also to the conservation and management of such stocks within areas under national jurisdiction ..." Conservation and management measures should be established on the basis of a precautionary approach and should use reference points for establishing the level of utilization of stocks. They should be based on the best scientific information available. For this purpose an essential element in the management procedures is the requirement for the collection and exchange of data and information.

RMFOs (or arrangements) are designated as the appropriate medium through which States are to cooperate to achieve and enforce conservation objectives both on the high seas and in areas under national jurisdiction. Where no RFMO exists for an existing or emerging fishery, States must co-operate to establish one. Where an RFMO does exist, States that wish to fish for the resource are obliged to join the RFMO or, at the very least, conduct themselves in accordance with its rules. At the same time, UNFSA emphasises that States with a "real interest" in the fisheries concerned are entitled to become members of a relevant RFMO. This important and difficult provision is designed to ensure that, on the one hand, UNFSA cannot be used to protect the position of States currently fishing on the high seas by freezing out potential new participants, whilst, on the other hand, RFMOs should not be open to all States unless they have a recognisable interest in the fisheries concerned. The implication is that only those States which are members of the relevant RFMO, or which agree to apply the conservation and management measures established by the RFMO, may have access to the fishery resources to which those measures apply.

In addition, UNFSA specifies in more detail than the LOS Convention the duties of flag States with respect to their vessels fishing on the high seas and introduces innovative provisions on enforcement by non-flag states, as well as providing for port-state jurisdiction in respect of fishing vessels.

Freedom of fishing on the high seas after UNFSA

It is still sometimes argued that, in so far as the high seas is concerned, the basic principle is freedom of fishing; and that this principle is what applies among the members of an RFMO whenever no agreement has been reached within an RFMO on the allocation of fishing opportunities or what applies between its members and a non-member State. Today, these arguments are seriously misleading.

It is true, of course, that article 87 of the LOS Convention, entitled freedom of the seas, includes the freedom of fishing. However, in the modern world where stocks have been shown to be exhaustible, this freedom is not absolute. Article 116 of the Convention subjects this right to other treaty obligations, the rights and duties as well as the interests of coastal States as provided for, *inter alia*, in article 63, paragraph 2, and articles 64 to 67 and to articles 117 to 120. Articles 117 and 118 provide for the duty to co-operate with other States whose nationals fish in the same area or for the same stocks in taking measures necessary for the conservation of the stocks. As the International Court of Justice has found,

> "It is one of the advances in maritime international law, resulting from the intensification of fishing, that the former laissez-faire treatment of the living resources of the sea in the high seas has been replaced by a recognition of a duty to have due regard to the rights of other States and the needs of conservation for the benefit of all." (Fisheries Jurisdiction Cases 1974).

The duty to co-operate, which is clearly of universal application as part of customary international law, can be discharged by direct co-operation or by establishing an RFMO.

Two other matters stand out from these provisions of the Convention. First, obligations under "other treaties" include obligations arising from the constitutive treaties of RFMOs. Secondly, the duty to cooperate in taking conservation measures is a legal duty carrying significant consequences. The International Tribunal for the Law of the Sea has applied the duty to cooperate in its Order prescribing provisional measures in the Southern Bluefin Tuna cases (1999).

The relevant provisions of the LOS Convention have been implemented and further elaborated by UNFSA. Part III of UNFSA contains detailed provisions concerning the duty to cooperate over conservation and management. This duty applies to members of RFMOs, as well as to non-members. The duty of members is not discharged by merely creating or joining an RFMO: they also have the continuing obligation to respect conservation measures adopted by the RFMO concerned. In cases where the members have failed to reach agreement on the allocation of quotas for a particular species or for a particular area, the members remain under a legal duty to cooperate. Non-members which do not agree to apply conservation measures of an RFMO do not have access to the stocks covered by measures. Unilateral action such as authorising nationals to fish in a manner that undermines an RFMO's conservation measures may well be found to amount to a failure to cooperate contrary to general international law and the LOS Convention, especially where prior warnings had been given by the member States through diplomatic channels. Such a failure would engage the responsibility of the State under international law. The same could go for failing to prevent nationals from undermining such measures despite warnings.

Thus, the conclusion that may be drawn is that the freedom of fishing exists generally, but it is now very much a conditional freedom. Especially far-reaching conditions exist in regard to the regulatory areas of RFMOs. There is no freedom to fish for straddling or highly migratory stocks contrary to applicable conditions. These conditions include measures laid down by RFMOs: there is no freedom to undermine any of these measures. On the contrary, fishing in such circumstances could violate general international law, the LOS Convention, UNFSA and the constitutive instruments of the relevant RFMO, depending upon the particular facts of a case.

In summary, as far as the right to fish on the high seas is concerned, the following conclusions may be stated with some confidence:

- The duty to cooperate inherent in the LOS Convention applies equally to members and non-members of RFMOs
- The duty to cooperate persists even where members of RFMOs fail to agree: there should be no unilateral action.
- Non-members which do not agree to apply RFMO measures are "not discharged" from the duty to cooperate.
- Where no RFMO exists, States are under an obligation to cooperate to establish an RFMO or apply appropriate (precautionary) conservation measures.

Regional Fisheries Management Organisations

Applying the FAO definition of an RFMO as an "intergovernmental fisheries organisations or arrangements, as appropriate, that ha[s] the competence to establish fisheries conservation and management measures", there are now some 18 RFMOs worldwide. There are some 20 other regional fishery bodies with an advisory capacity but which do not have the capacity to establish conservation and management measures. Some RFMOs, such as the International Whaling Commission (IWC) and the North Atlantic Salmon Conservation Organisation (NASCO) have very specific mandates or deal with single species. Others have broader mandates. Since 2003, new RFMOs have been established for the Western and Central Pacific Ocean (WCPFC), South-East Atlantic (SEAFO) and South Indian Ocean (SIOFA). Processes are also underway at present to establish RFMOs for the Southern Pacific Ocean and North West Pacific Ocean. Thus, while some important gaps remain, both in terms of species and area coverage, the majority of the world's marine fish resources are now under management by one or more RFMOs.

The conditions for success in RFMOs

If RFMOs are to be the key mechanism for achieving cooperation in the management of straddling and highly migratory stocks, it is necessary to consider what conditions must prevail, if a RFMO cooperative resource management arrangement is to prove to be stable over the long run. This is an important question because experience has demonstrated that the negative consequences of ineffective cooperation are severe. Furthermore, RFMO cooperative resource management regimes, involving, as they typically do, large numbers of participants, are inherently fragile. The more players that are involved, the more difficult it becomes. RFMOs lacking the resilience to cope with their inherent weaknesses and the threat of unpredictable shocks are likely to founder over time. The predictive power of the economist's model of non-cooperative management of such stocks has proven to be brutally strong. Breakdown in the cooperative management of shared resources, particularly straddling and highly migratory fish stocks, carries with it threat of results comparable to open access, unregulated fisheries, with all that that implies. Put simply, we know that non-cooperation leads not only to over-exploitation, but may also lead to near or total stock extinction. Conversely, cooperation always leads to higher rents and higher stock levels than non-cooperation.

Unsurprisingly, UNFSA, and the various instruments that have been developed since UNFSA such as the FAO Code of Conduct for Responsible Fisheries and the various FAO International Plans of Action (IPOAs), place great emphasis on elaborating the basic parameters for RFMOs to operate effectively. The main contribution of UNFSA in this regard is to define (in its article 10) the desirable institutional characteristics of an effective RFMO by listing, in a legally-binding form, the matters upon which States are expected to agree in order to achieve sustainable management of fisheries. These include agreement on conservation and management measures to ensure long-term sustainability, agreement on participatory rights such as allocations of allowable catch or levels of fishing effort, agreement on decision-making procedures which facilitate the adoption of conservation and management measures in a timely and effective manner and agreement on mechanisms for obtaining scientific advice and ensuring compliance with and enforcement of conservation and management measures.

It is suggested that fundamental preconditions for success if RFMOs are to achieve their overriding objective of long-term conservation and sustainable use of fish stocks (MacFarlane's first precondition for reaping benefits from globalisation) include:

- Effective conservation and management practices (including the application of ecosystem-based management principles and the precautionary approach)
- Effective decision-making procedures
- Effective dispute settlement procedures
- Transparency
- Effective participation by developing States
- Robust institutional structures

Furthermore, measures agreed by RFMOs have to be implemented by members of the RFMO and complied with by individual fishing vessels. To achieve this, effective compliance and enforcement of agreed conservation and management measures, supported by adequate monitoring, control and surveillance regimes is crucial.

The extent to which RFMOs actually meet these conditions for success in practice varies considerably. In part this is because, although UNFSA sets out, in broad terms, minimum requirements for RFMOs, many RFMOs were established prior to 1995 and do not necessarily possess the mandates to carry out all the functions ascribed to them. Further, there has to date been a lack of a systematic approach to the implementation of the Agreement by RFMOs, including the lack of effective means for cross-learning among RFMOs about best practices. Consideration of the degree of success experienced by RFMOs is beyond the scope of this paper. Suffice it to say that no single RFMO has a monopoly on best practice and there is ample scope for cross-learning between RFMOs. New practice (both good and bad) is constantly being generated by RFMOs.

A full analysis of the requirements of international fisheries instruments and best practice in their application has recently been carried out by an independent panel commissioned by a number of governments and international conservation organizations (Lodge *et. al.* 2007). The Panel's analysis indicates that there is clearly scope for more effective cooperation between members of RFMOs and between RFMOs themselves, particularly in the area of compliance and enforcement. Practical steps that could be taken without the need for any change to existing paradigms include, for example, standardizing and sharing or consolidating vessel registers and information from vessel monitoring systems. Implementation of the FAO port state model scheme on a regional basis, combined with a standardised approach to catch documentation schemes, would also reduce the opportunities for IUU fishing. The introduction of alternative dispute resolution procedures, such as technical panels of experts, would help to promote more effective decision-making. And a more systematic approach to the problem of non-members would help to reduce the scope for RFMO measures to be undermined.

Major threats to stability

The greatest single threat to the stability of management regimes introduced by RFMOs is probably the failure to allocate fishing opportunities on an equitable basis. An RFMO's response to the question of defining and allocating rights and the subsequent impact this has on the resource is an integral component of its overall effectiveness. It is essential to address the issue of allocation if a breakdown in the cooperative management of the resource is to be averted. Furthermore, it is apparent that many of the difficulties experienced by RFMOs in preventing overfishing of key stocks and other problems driven by overcapacity, including detrimental impacts on non-target, associated and dependent species are inextricably linked with failure to address the problem of allocation.

Experience in RFMOs to date shows that gaining agreement on the basis for allocations, acceptance of the outcomes of an allocation process and then controlling the resulting impacts on the fishery is extremely difficult, with potentially negative consequences for target species and non-target species as well as the broader marine environment. The key impediments to effective allocation of rights are so-called "free riding" and non-compliance by members and non-members. The latter is generally referred to now as "IUU" – Illegal, Unreported and Unregulated Fishing – although in fact, the boundary between free-riding and unregulated high seas fishing is blurred and the categories often overlap.

It should be obvious that the consequences of non-compliance, *i.e.* willful violation of the terms of a cooperative arrangement by a participant in that arrangement, can only lead to undermining of the arrangement. Equally damaging is "free riding", when a non-participant in the arrangement simply takes the benefits from the arrangement with sharing any of the burden.

Much effort has been expended in recent years to deal with IUU fishing. These measures include, for example, detailed measures adopted by all RFMOs (based on UNFSA provisions) that place compliance obligations on flag States, such as the recording and timely reporting of fisheries data, cooperation in MCS, including vessel monitoring systems (VMS) and observer programmes. Most RFMOs have also agreed on procedures to deal with infringements by their members, including required follow-up actions and reporting. Nearly all RFMOs emphasize the importance of registers of authorized vessels and have adopted a practice of blacklisting vessels which continue to fish illegally. Some RFMOs have also established specific measures concerning transhipment and landing of catches as well as other trade and market related measures. Some of these measures, particularly those relating to port States and trade, extend also to non-members of the RFMOs concerned. In an attempt to widen the scope of compliance and enforcement measures, in recent years there has developed the concept of the "cooperating non-member" of an RFMO. This is a status that increasingly carries with it certain rights and obligations.

Active use of port State jurisdiction has proved to be a particularly effective weapon against IUU fishing operations, both because all harvested fish must be landed at some point and because the use of such measures does not require substantial resources compared to other measures such as, for instance, inspection at sea. Enhanced port State controls can act as a disincentive to IUU operators by increasing the cost of their operations (*e.g.* by forcing them to seek out more remote and hence more costly ports). The key is to ensure that port State controls are applied widely and consistently in order to avoid the development of so-called "ports of convenience". Thus, in recent years the importance of coordinated port State measures has become increasingly recognized by RFMOs and by the FAO (which recently adopted a model scheme for port State measures to be applied on a regional basis). Trade measures, such as catch documentation schemes, have also proved effective in eliminating illegal catches from the value chain and are becoming widespread in their application.

The difficulty is that these advances have nearly all targeted only the "illegal and unreported" elements of IUU fishing. Much less has been done to deal with "unregulated" fishing. The legal situation is clear. UNFSA requires that States give effect to their duty to cooperate with each other by joining the relevant RFMO or agreeing to apply its conservation measures. As long as the newcomer is prepared, by doing either of these things, to accept the same conditions as apply to existing members, those members cannot, consistently with article 8(3) of UNFSA, deny the newcomer's right to participate. Conversely, the right under article 87 of the LOS Convention for any State to fish on the high seas must be regarded as conditional and includes the duty to cooperate with and not undermine the conservation measures determined by an RFMO.

To date, RFMOs have adopted a range of responses to the problem of accommodating new members, with differing degrees of success. Arguably, no RFMO has yet fully succeeded in balancing the rights of new members with the rights of its existing membership while maintaining the integrity of its conservation and management measures. When faced with a choice to either reduce the allocations of existing members or to increase the overall catch of a stock to accommodate new members, a number of RFMOs have routinely chosen the latter approach, regardless of the status of the resource and whether access is already fully subscribed. For example, CCSBT wrestled with issues relating to new members over a number of years before agreeing to allow the Republic of Korea and Chinese Taipei to join the Extended Commission with national allocations that were in addition to the existing TAC, rather than reducing those of the three charter members. This decision was taken despite scientific advice that the spawning biomass of Southern Bluefin Tuna has been reduced to a low fraction of its original biomass. The pragmatic argument in favour of this approach is that by accommodating all current interests (both existing and new) a more stable environment is created that then allows for future decisions to be taken to effectively address stock status.

The approach used by both NAFO and NEAFC is to welcome new members, but to warn them that stocks currently managed are fully allocated, and that fishing opportunities are likely to be limited to new fisheries (i.e., stocks not currently allocated by TAC/quota or effort control). Whilst this approach of simply closing the door to new members at the regional level may be effective for NAFO, since its membership already includes all relevant coastal States and so there is little potential for new members, it is unlikely to prove effective on a global scale. This is demonstrated by the experiences of ICCAT, IATTC and CCSBT, where not all relevant coastal States were members of the RFMO when quotas were originally allocated. The process of bringing such States into the regime, in the absence of pre-agreed mechanisms to accommodate their interests, resulted in short to medium term lack of stability. The NAFO approach may also come under future scrutiny if applied in such a way that it fails to recognise legitimate expectations of developing countries (see below).

More flexible allocation systems

If these problems are to be overcome it is clear that RFMOs need to consider more innovative approaches to the allocation problem. This requires that they are empowered to consider the use of a wide range of mechanisms for achieving acceptable economic benefits to all parties from cooperation and compliance, including access arrangements, quota trading and leasing.

One approach which is receiving increasing attention is for fishing opportunities within RFMOs to be traded. The idea of tradeable allocations to facilitate new entrants to a fishery is not new, having been a fundamental element of many national fisheries management schemes for some decades. Trading in national allocations between contracting parties has already occurred in a number of RFMOs including ICCAT, NAFO, NEAFC and the former IBSFC. The concept has also received preliminary consideration within both CCSBT and WCPFC. Serdy (in press) provides a detailed summary of the maturity of RFMOs in moving towards a form of tradability of national allocations and concludes that transferability may provide opportunity for new entrants to "discharge their duty of cooperation."

There would appear to be no obstacle in international law to trading of quota within RFMOs. The treaty regimes upon which most RFMOs are built would not require significant amendment. Some sort of administrative arrangement would be necessary to support the trading system, including rigorous systems of accounting for catches as well as a reliable register of rights and transactions. Such systems are far from novel, variants of the same having been applied successfully since 1987 in the context of the multilateral treaty between the United States and certain Pacific Island Countries, as well as in the regional management arrangements within the South Pacific Forum Fisheries Agency. The key to the success of any such system, however, remains the question of the initial allocation upon which all subsequent trades would be based.

Clearly, there would be significant challenges involved in introducing any system for allocation. It is apparent that consideration of any such mechanism would need to be interdisciplinary in nature, involving scientists, lawyers and economists as well as the RFMOs themselves. Responsible industry should also be given a greater role. In a globalised world, where the fisheries value chain is multinational in nature, it is far from clear that the "shareholders" in RFMOs should be governments alone.

The special requirements of developing countries

Developing countries form a specific subset in any allocation regime. One of the major contributions of UNFSA is its specific articulation of the special requirements of developing States and the nature of the obligation on States to recognize and take these special requirements into account. UNFSA sets out a range of obligations on States with respect to developing States including assisting them to develop their own fisheries for highly migratory and straddling stocks and to access those stocks on the high seas. Article 11 of UNFSA also establishes that the "…interests of developing States from the subregion or region in whose areas of national jurisdiction the stocks also occur" should be considered as a criterion in determining participatory rights for new members.

One of the impediments to progress in RFMOs has been a failure to properly recognize or account for the development aspirations of developing countries, especially those that are coastal States in the relevant region. Allocation of fishing opportunities or quotas among members has generally been based on historical catch which has meant that States not actively fishing over the relevant period did not receive an allocation. In many cases this approach failed to recognize either the nature of the sovereign rights over resources within a country's EEZ or the dynamic nature of flag State activities on the high seas.

It is interesting to note that this approach to determining access rights preceded the articulation of not only the rights of developing States but also those of coastal States. For example, both IATTC (1949) and ICCAT (1969) pre-date the conclusion of the LOS Convention and the acceptance of the EEZ regime. In ICCAT, the lack of provision for the interests of developing coastal States in the initial allocation process resulted in a long period of disagreement in respect of overall catch levels and led to significant over-harvest of already depleted stocks. Both ICCAT and IATTC have since taken steps to address this situation in their allocation regimes with efforts mainly directed towards recognition of the rights of coastal States, which in both cases are comprised mostly of developing States. More recent RFMOs, including SEAFO and WCPFC, contain extensive reference to the special requirements of developing States and the factors to be taken into account in the allocation of participatory rights. In both cases these include specific criteria relating to developing States. The operation of these criteria, including how such concepts might be quantified and weighted, is yet to be tested.

The unsustainability of the traditional approach to allocation can be demonstrated by considering the balance between traditional harvesting States and developing country participants in selected RFMOs and in UNFSA as a whole (Figure 10.1). This makes it clear that developing countries represent more than half of the participants in RFMOs; yet the use of historical catch as a basis for allocation would inevitably result in the largest share of the resources going to developed, distant-water States.

Figure 10.1. Developing Country Participation in RFMOs and UNFSA

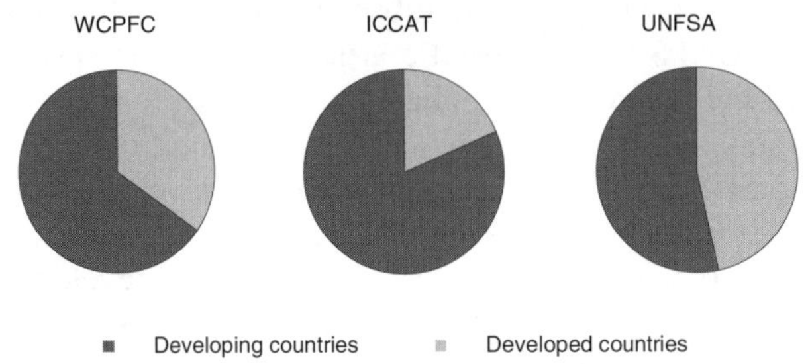

Moves by developing countries to, for example, increase their national allocations or insulate these from a need to reduce overall catches, is likely to destroy any equilibrium reached in an initial allocation process between parties. Developing countries want access to high seas resources (*e.g.* tuna), but existing fishing countries are loath to reduce their holdings in already fully subscribed fisheries. The result is that allowed catch levels are effectively pushed higher to accommodate both, in the hope of a later mutual phase down being agreed, and stocks are put at risk. There will be a point at which individual States may reconsider whether they are better off complying with their allocation in the light of redistributions of rights driven by developing States' claims.

The analysis suggests that, if the problems of international fisheries governance are to be resolved, greater and more decisive, coordinated and coherent engagement with developing countries is essential. This includes moving more rapidly towards an operational basis for introducing developing countries to the economic benefits from high seas resources. While the UNFSA provides for the special requirements of developing States to be taken into account, the manner in which these are accounted for needs to be balanced not only with the conservation status of the stocks but with other participants' shares. Novel operational solutions need to be found to accommodate the legitimate fishing aspirations of developing countries (general principles are inadequate): for example attrition, whereby a small percentage of all existing holdings revert to a central pool each year for redistribution should be considered.

Conclusions

Considerable effort has been expended over the past decade in strengthening RFMOs, reviewing their mandates and modernizing their approaches to fisheries management. For example, the most recent UN General Assembly resolution on sustainable fisheries (61/103 of 3 December 2006) urged RFMOs to strengthen their mandates and modernize their measures and approaches to fisheries management and called upon States to make further efforts to strengthen and enhance cooperation among existing and developing RFMOs. The same resolution also called upon States to develop and apply best practice guidelines for RFMOs and to undertake performance reviews of RFMOs, based on transparent criteria.

At the same time, the expectations placed on RFMOs have also grown exponentially in recent years. The last twenty years in particular have seen a plethora of hard and soft law instruments aimed at addressing the problems of international fisheries governance. However, despite the proliferation of RFMOs and the development and evolution of instruments aimed at empowering them, it is often suggested that RFMOs have generally failed to prevent over-exploitation of straddling and highly migratory fish stocks and degradation of the marine ecosystems in which they occur.

This is probably unfair criticism. The efforts made by the international community over the past decade have brought about significant improvements, particularly in terms of more effective enforcement and an enhanced understanding of the need for conservation measures that will promote long-term sustainability, such as the application of the precautionary approach and ecosystem-based management. Great advances have also been made in the field of developing market-based incentives to fish in compliance with management regimes, such as catch documentation schemes and port State schemes. The so-called freedom to fish unrestricted by conservation concerns is a chimera.

Nevertheless, whilst the current regime provides a sound platform on which future rights-based access regimes could be superimposed, it has to be conceded that the conditionalities surrounding the freedom to fish do not yet equate to what may be regarded as secure access rights. There remain some issues that are likely to prove very difficult for RFMOs to deal with. Such issues include dealing with excess capacity in the world's fishing fleets and the allocation of high seas fishing opportunities on an equitable and sustainable basis. Key areas where progress could be made include more imaginative and constructive engagement by and with the fisheries industry and improving governance and sustainable capacity in developing countries. Models of more flexible allocation systems, developed through a multidisciplinary approach, for example, through the OECD Fisheries Committee, could no doubt better inform RFMOs and governments of the likely merits and demerits of different approaches to the problem.

References

Arnason, R. (2006): "The 'Rent Drain': Towards an estimate of the loss of resource rents in the world's fisheries", paper prepared for the FAO/World Bank workshop on "The Rent Drain," Washington, D.C., January 17-18, 2006.

Fisheries Jurisdiction Cases, United Kingdom v. Iceland, [1974] *ICJ Rep.* 3.

Lodge, M. (2002): "Review of factors of unsustainability in fisheries: Relationship to international fisheries instruments", paper prepared for the FAO Workshop on Factors of Unsustainability and Overexploitation in Fisheries, Bangkok, Thailand, 4-8 February 2002.

Orrego Vicuña, F. (1999): *The Changing International Law of High Seas Fisheries*, Cambridge University Press, Cambridge.

Serdy, A. (in press). *Fishery Commission Quota Trading under International Law*, Ocean Yearbook, vol. 21.

Southern Bluefin Tuna cases, New Zealand v. Japan, Australia v. Japan, [1999] ITLOS Rep.

Chapter 11

Fisheries Agreements and their Impact on the Globalisation of the Fisheries Sector in Mauritania

Chérif Ould Toueilib, International Consultant, Fisheries, Environment and Development, Mauritania [1]

Fisheries are vitally important to Mauritania and constitute an important export earning, Only marginal quantities of the fish landed are for domestic consumption. Chérif Ould Toueilib highlighted that Mauritania is highly dependent on access agreement revenues, although it does have its own national fleet and highly developed infrastructure.

The EU plays an important role in the EEZ of Mauritania through access agreements. Access agreements provide important income for Mauritania generating in excess of EUR 100 million in export revenue per year. However, access agreements have also resulted in the overexploitation of resources, fleet conflicts for resource access and losses to the national fleet. The high level of raw material exports means that the national processing sector is fairly weak. This makes the Mauritanian fishing sector vulnerable and exposed to market stocks.

According to Chérif Ould Toueilib, the future of the Mauritanian fisheries sector in a globalised world relies heavily on the development of a high quality and food safe domestic processing industry that can have access to international markets. This will require a major institutional drive to reduce reliance on fisheries access agreements. Concurrently, a diverse marketing system and a set of internationally recognised food safety and hygiene standards should be developed.

1. This document was originally written in French by Chérif Ould Toueilib,

Executive Summary

Mauritania, a country renowned for its abundant fishery resources, has relied on the revenue generated by the fisheries sector for the past 30 years. Since the emergence of this young country, the fisheries sector has been the mainstay of the national economy in terms of budgetary resources, employment and food security.

Today, despite declining yields and intense pressure on fishery resources in the Mauritanian Exclusive Economic Zone (EEZ), fishing continues to play a major role but its profile is declining with the emergence of other productive and revenue-generating activities.

It is highly likely that institutional, political, economic and social developments in Mauritania will bring about structural change in the fisheries industry. This may completely transform the sector's current vocation as a source of budgetary revenue to that of an industry providing jobs, value-added and food security.

It should be borne in mind that, over the past thirty years, the fisheries sector has been subject to a variety of regimes and approaches, ranging from a nationalisation policy focused on the domestic fleet and landings to a policy intent on striking a balance between the creation of value-added (small-scale fishing and the domestic fleet), rent optimisation (fisheries agreements) and the export of frozen products through the Mauritanian Fishery Product Marketing Company (SMCP).

Striking such a balance has certainly been neither easy nor successful; while rent-related targets have generally been met, liberalised access has marked the beginning of the end for the domestic fleet and proved disastrous for the creation of value-added by land-based industries.

This report assesses the strengths and weaknesses of fisheries agreements and their impact on liberalisation in the fisheries sector in Mauritania. It notes the positive impact of the agreements on budgetary revenue, debt reduction and institution-building in the sector, but also the intense pressure placed on stocks and its perverse effects on the sustainability of resources and their environment.

Despite the pressure from biological exploitation, meeting public spending needs will remain the top priority for at least the next six years. It will be logically impossible to maintain the pressure beyond that date for various reasons, some of which relate directly to Community policy as the EU will be unable to meet the funding requirements of its subsidised fleet system. Similarly Mauritania, increasingly required to implement good governance in line with the expectations of local communities, will be obliged to opt for policies that create jobs and value-added. This will mean smoothing out the inconsistencies between sustainable fisheries policies and fisheries agreements, as analysed in a later section of this chapter.

Clearly then, the conflicting aims of the different parties to commercial agreements represent a real barrier, which militates in favour of an in-depth policy review, both in

Mauritania and at EU level. This will show that integrated economic partnerships are the only viable option. But striking a balance in such partnerships is not self-evident. It will only be achieved with the economic and social sacrifices that may stem from an overhaul of the legal and institutional frameworks, required to adapt to the new situation.

The other aspect of this report on the globalisation of trade in fishery products and its effects on the fisheries sector in Mauritania is an analysis of behaviour in this sector, which is a growing – but not the sole – supplier of an ever more demanding market.

Section V highlights signs of liberalisation in this sector, such as the privatisation of the SMCP (1980s) and the IRM/EU agreements which marked the opening up of access to demersal species, formerly confined to the domestic fleet. The lifting of the SMCP's monopoly on wetfish certainly contributed to this liberalisation process, which started in the South (Nouakchott) where small-scale fisheries prevail. As the report emphasises, this situation is having a direct impact on supplies to domestic consumption markets.

With regard to the globalisation of trade, the report addresses Mauritania's ability to cope with competition for market access, as it loses the advantages it enjoyed as an ACP country. Also analysed in this report are the other technical and hygiene-related barriers, which can be viewed as either a "silver bullet" or a market veto, and their effects on a fisheries sector that is heavily dependent on exports and therefore on the dynamics of trade liberalisation, whether in relation to markets or to the terms of trade between the dollar, the euro and the yen in international transactions.

The implications of the fisheries agreements for the future of Mauritania's harvesting industries, covered in Section 6, are indicative of a difficult situation, requiring solutions that integrate the sector more closely in the national economy and promote equitable co-operation between Mauritania and the EU.

Finally, the report concludes that the fisheries sector does not seem prepared for the challenges of globalisation. The problem lies in its limited ability to confront the following risks:

- Shocks caused by sharp falls in prices attributable to international supply and demand mechanisms.
- Shocks caused by the technical barriers imposed by import markets.
- Shocks caused by competition between Asian and ACP products as preferential regimes come to an end.
- Shocks caused by resource scarcity and declining output by domestic vessels, which face unfair competition for access from subsidised European vessels.

On the basis of that evidence and those challenges, the report identifies some avenues to explore (subject to national priorities to be set with the relevant authorities) in giving Mauritanian fisheries the assistance they require. In particular, assistance should focus on the following priority areas:

- Promoting development to adjust the sector's health, hygiene and technical capacity (the EU's EDF programme).
- Building national capacity for processing and the creation of value-added.
- Enhancing the institutional and business environment to attract joint ventures and integrated industries exploiting both European and ACP assets, with a view to

achieving the transfer of know-how and the mutual trust that will provide sustainable benefits for all stakeholders.

- Supporting the promotion of a fisheries system based on diversified and selective techniques in order to relieve pressure on endangered stocks and conserve vulnerable cohorts as well as their environment.
- Fostering the emergence of private partnerships that place the focus on quality rather than quantity and encourage the voluntary withdrawal of part of the fishing effort (capacity adjustment).
- Promoting new landing points in the South and equipping them with storage facilities and the necessary services.
- Developing export certification for clams, a major local resource aimed solely at the export market.

Introduction

Covering an area of 1 030 000 km², the Islamic Republic of Mauritania (IRM) is bordered to the west by the Atlantic Ocean, to the north by the Western Sahara and Algeria, to the east by Mali and to the south by Mali and Senegal. This position ensures that it has a definite future in North/South trade.

In administrative terms, the country is divided into 13 *Wilayas*[2], which are in turn subdivided into 53 *Moughatas*[3] and 250 smaller units, or communes.

The Mauritanian population, which has undergone far-reaching economic, social and cultural change, was put at 2 548 157 by the official statistics published in 2000. Today it is reported to exceed 3 million, over half of them under 18 years of age.

The key feature of the Mauritanian economy is the divide between the traditional sector, which is labour- intensive, and the modern capital-intensive, export-dependent sector, consisting largely of mining industries, industrial fisheries, irrigated agriculture and, very recently, petroleum products.

Since gaining independence on 28 November 1960, Mauritania has relied on fisheries as an economic mainstay providing budgetary resources, foreign currency revenue, GDP growth and jobs.

In the 1960s, for instance, the government introduced a policy whereby it sold fishing permits and developed infrastructure as a means of localising fishery rents in Mauritania and creating value-added. This initial attempt was eventually to fail as foreign fleets did not like landing their catches in the country's only port, Nouadhibou, so land-based firms were operating at less than 10% capacity, thereby jeopardising the entire future of the newly established fisheries industry.

The government drew on the lessons learned from this and shifted to a policy aimed at creating a domestic fishing fleet, strengthening its rent-localisation measures and creating value-added. It was this determination that led to the New Fisheries Policy, and the subsequent series of five-year plans that have focused for the past 20 years on the

2 Similar to a region.

3. Similar to a county.

sustainable development of fisheries resources in the EEZ and the expansion of small-scale fishing.

There have, accordingly, been seven new policies since then, the overall outcome being the emergence of a Mauritanian fisheries industry operating alongside foreign vessels fishing under bilateral agreements and conventions. The main policy achievements can be summarised as follows:

- Enhancing the stock-assessment capacity of fisheries research;
- Establishing the Mauritanian Fishery Product Marketing Company to promote exports and bring foreign currency back into the country;
- Strengthening the surveillance and inspection system that ensures compliance with management measures and protects water quality;
- Establishing recent new development plans as integrated management tools;
- Setting up the Advisory Council on Fisheries as a framework for dialogue and good governance;
- Promoting a regulatory and institutional framework of law to underpin management policy in a system based on the rule of law;
- Developing a national demersal and small-scale fishing fleet;
- Developing land-based and support facilities.

In spite of their achievements, these fisheries policies have admittedly been unable to prevent:

- The gradual emergence of overcapacity, leading to the excessive harvesting of species with a high market value, including cephalopods, crustaceans and demersals (Sparidae);
- Overinvestment in and undersupply of land-based industries;
- Increasing sectoral reliance on fisheries agreements, which are gradually taking over from national fisheries and creating such keen competition with the domestic fleet for both resources and markets that the sector is obliged to take a pro-active approach.

Share of Fisheries in the Mauritanian Economy

Fisheries, the mainstay of the domestic economy since the emergence of this young country, continue to play a vital economic and social role. 50% of its export revenue which, in value terms for fisheries alone, was as high as MRO 46 406 million for 114 810 tonnes in 2005, compared with MRO 26 807 million for 189 680 tonnes in 1998.

The sector's contribution to the national budget is estimated to have averaged between 25 and 27% of budget revenue in recent years. Revenue in 2006 was put at around MRO 41 889 million as against MRO 36 690 million in 2005. The sector is therefore still in the lead in terms of its contribution to the State Budget.

This is because the fisheries sector is increasingly focusing on the export of high-value species (demersals). The introduction, in recent years, of a policy to free up access to foreign vessels, largely the European fleet, has made the sector more dependent and increased agreements as a share of budget revenue. The domestic fishery sector's contribution to GDP, estimated at over 12% in the 1970s, barely exceeds 5%, excluding the agreements[4]. Its contribution to the creation of value-added is still lower than expected (just 3%), which shows how poorly integrated the sector is in the national economy.

In terms of the labour market the fisheries sector is a major employer (directly or indirectly providing 30 000 jobs), particularly in small-scale fishing, but has declined since the 1980s when fisheries provided over 36 000 jobs.

Fish consumption in Mauritania remains relatively low compared with other countries in the sub-region, at 4.3 kg per head per annum, compared with over 15 kg per head per annum in Senegal. But the figures in urban areas are around 7.5 kg per head per annum, owing to living standards and purchasing power in the towns. Catches are aimed more at the export market, and local consumption market is increasingly undersupplied.

In terms of foreign debt, the sector is helping indirectly, through export revenue and EU fisheries agreement revenue, to reduce the debt by participating in the funding of public investment.

The fisheries sector is also perceived as a sector with the potential to combat poverty, particularly in small-scale mainland fisheries which offer considerable opportunities in terms of jobs, income and food ratio. The sector's input can be seen from the following table:

Table 11.1. Fisheries Revenue in 2005/2006 (MRO million)

	2005	2006[5]	Growth factor
Direct access fees	2 022	1 158	0.57
Indirect access fees	54	910	16.85
EU fees	2 650	3 382	1.27
Contribution	27 606	29 326	1.062
Open cephalopod licences	3 007	3 424	1.13
Other open licences	214	166	0.77
Chartering	492	524	1.06
Fines	645	2 936	4.55
Total	36 690	41 889	1.14

Source: MPEM.

4. If revenue from the agreements were to be included, the figure would be 10%

5. Data for 2006 are not final

Table 11.2. Direct and Indirect Jobs Provided by the Fisheries Sector

	Direct jobs	Indirect jobs	Total jobs in fisheries
Nouadhibou	15 400	400	15 800
Nouakchott	7670	430	8100
South	270	30	300
South Cape Timiris	500	100	600
PNBA	470	130	600
Migrant fishermen	4000	600	4600
Total	28 340	1060	30 000

Source: MPEM.

Table 11.3. Export Trends

(Tonnes and MRO million)

Exports in volume and value terms	1998	1999	2000	2001	2002	2003	2004	2005
Fishery exports in volume (tonnes)	189 680	216 439	210 958	159 312	107 438	100 260	85 310	114 810
Fishery exports in value (MRO million)	26 807	32 609	35 442	36 338	32 903	35 365	44 620	46 406
Octopus exports in volume (tonnes)	18 744	27 772	26 499	30 972	20 424	19 432	26 471	27 378
Octopus exports In value (MRO million)	13 442	16 956	16 634	20 637	19 824	24 519	35 670	34 865

Source: Based on Customs and SMCP data.

Mauritania's Fisheries Management System

Physical harvesting environment

Mauritania's coastline stretches along the Atlantic from 16°04'N to 20°36'N for some 720 km. The Mauritanian continental shelf is fairly flat, with the exception of the seabed south-west of the *Banc d'Arguin* (PNBA) where ocean trenches 300 metres deep are commonly found alongside 10-metre shallows.

Influenced by the displacement of the intertropical front, to the north of which is a trade-winds system, Mauritanian waters are the source of a strong, steady upwelling which has made the name of these rich fishing grounds.

Fishery resources and allowable catch

The fairly abundant fishery resources in the Mauritanian EEZ or MEEZ (170 marketable species out of a total of some 350 identified species, excluding seaweed) consist basically of cephalopods, crustaceans, demersal fish, pelagic fish and clams.

Small cephalopods are estimated to account for over 80% of the allowable catch (excluding unharvested clams and seaweed) and only 10% in value terms, whereas demersals, cephalopods, crustaceans and fish are thought to account for over 90% in terms of value and less than 10% in terms of allowable catch.

Table 11.4. Updated Table of Leading Fishery Stocks in the MEEZ

Resources	MSY1998 tonnes	1998 Results tonnes	MSY2006 tonnes	2006 Results tonnes
Cephalopods • Octopus • Cuttlefish • Squid	51 000	Overfished by 25%. To be cut by 25%	45 to 50 000	Overfished by 25 to 45% To be cut by 30%
Crustaceans • Deepwater shrimp • Inshore shrimp	7 000	Fully harvested	Not known	Little or no surplus Fishing effort to be frozen
Demersal fish	12 000	Fully harvested	12 to 18 000	Fully harvested
Inshore fish	200 000	Estimate (JICA/PA)	200 000	
Small pelagics	980 000	Underfished	Over 1000 000	Underfished
Clams	301 000	Unharvested	Above the 1998 potential catch	Experimental harvesting
Total potential catch	**1 501 000**		**Over 1 501 000**	

Source: Updated IMROP figures, 2006.

Institutional and legal framework of fisheries management

The sector known as Fisheries and the Maritime Economy has had an institutional and legal framework since 1978: The Merchant Navy and Fisheries Code, followed in 1988 by The Fisheries Code, and in 2000 by the revision of The Fisheries Code to incorporate the principles of sustainability and international instruments. At the time of writing, the framework has just been partially revised. The main features are as follows:

Instruments

Since 1979, when the New Fisheries Policy (NFP) was drawn up and Mauritanian sovereignty was extended to a 200-mile EEZ (1978 Merchant Navy and Fisheries Code), the government has introduced a series of management measures and developed an institutional and legal framework to underpin them.

For instance Act 144/88, which gave fisheries a specific Code, was drawn up and implemented but replaced by Act No. 2000/025 establishing a Fisheries Code and an Act establishing a Merchant Navy Code, together with the relevant implementing legislation. In some respects, relating to the conservation and protection of the environment, these Acts have been supplemented by the Environment Act, based on the Kyoto Declaration, and Act No. 2000/024 on the PNBA (national park run according to the precautionary principle). The implementing legislation for these Acts lays down the rules for managing fishery-related vessels, gear, operations and activities and the marine environment.

There have been major developments in the management system, which has progressed from being a system based on open licences to the current approach that involves controlling fishing effort, setting harvesting levels (TAC), controlling the export of frozen products, and conducting inspections and surveillance at sea.

The latest developments in the system have been:

- The adoption of the first development plan for octopus, which is the country's main export resource but subject to overfishing for the past decade. This plan steps up the enforcement of technical management measures, in particular the extension of biological recovery periods from 2 to 4 months a year, restrictions on the number of vessels, zoning (coastal zone protection) and assessment forecasts to set annual TACs. The plan provides for an advanced phase of implementation for ITQs (individual transferable quotas) depending on how the current system evolves, particularly with the prospect of institutional development guaranteeing the cost-effectiveness and transparency of the system.
- The adoption of a management plan for small-scale fishing, aimed at controlling capacity.
- The introduction of an access regime based on annual licences to be issued on either a quarterly or six-monthly basis. Access is not open but subject to a fee based on the characteristics of the vessel and the commercial value of the species it will be fishing. This access fee, which has been subject to a series of reforms since 1995, the most recent being Decree 2006/019 of 1 January 2006, is divided into a direct access fee payable in local currency upon entry, per month and per unit of gross tonnage (around 30%) and an indirect access fee payable in local currency, per tonne exported according to the type of conservation on board or on land (around 70%).

Table 11.5. Direct and Indirect Access Fees

	DAF[1] (MRO/GT/month)	IAF[2] (MRO/ET)
Freezer trawlers	1 900	
Wetfish trawlers	1 400	
Selective wetfish vessels	900	
Cephalopods and crustaceans frozen on board	45 000	
Cephalopods and crustaceans frozen on land		34 000
Frozen demersals		30 000
Wet demersals		23 000
Frozen pelagics		15 000
Wet pelagics		12 000

1. Direct access fees expressed in ouguiya per GT and per month (Decree 2006/019/PM).
2. Indirect access fee in ouguiya per GT and per month (Decree 2006/019/PM).

Further steps have been taken to regulate the entry of new vessels into Mauritanian fisheries. They are set out in Circular 026/MPEM of 29 June 2006, which lists three regimes, namely:

- Acquisition;
- Chartering;
- Open licences.

The details are given in the tables below:

Table 11.6. Acquisition Regime

Vessels	Below 750hp	Equal to or over 750hp but below 1500	Equal to or over 500hp but below 3 000 hp	Equal to or over 3 000hp but below 7 500hp	7 500hp and above
Fees in MRO per category	200 000	400 000	600 000	800 000	1 000 000

Table 11.7. Chartering Regime

Fishery	Shrimp	Octopus.	Hake Merlus	Demersals	Tuna	Pelagicsp halopods	Comments
EUR	176	200	172	80	20	14.5	Chartered pelagic vessels pay back 25% of product value to their Mauritanian partner

Table 11.8. Open License Regime

Fishery	Shrimp	Octopus	Hake	Demersals	Tuna	Pelagics
Fees (EUR) per GT	2 530	2 875	1 035	1 150	288	207

Players

The Ministry of Fisheries and the Maritime Economy oversees the management and harvesting of fisheries resources on behalf of the State and for the benefit of the national community. As the institution in charge of management, it encompasses several government departments with a range of responsibilities, including the Mauritanian Institute for Ocean Research and Fisheries *(IMROP)*, in charge of stock assessment and monitoring.

Responsibility for fisheries enforcement lies with an autonomous institution, the DSPCM (Delegation for Fisheries Surveillance and Inspections at Sea), which has at its disposal some relatively satisfactory operational resources, compared with those in the sub-region as a whole. This paramilitary structure enforces fishery rules and reports to the Department of Fisheries.

The SMCP (80% privatised) still has a monopoly over the sale of frozen products and is a major player in the industry owing to its role in repatriating the revenue from market sales and to the bargaining power it gives producers.

The other leading fisheries players, both upstream and downstream, are the Central Bank, the Independent Port Authority, the Small-scale Port Authority, the Maritime Training School and the National Fisheries Federation.

Harvesting and processing industries

Harvesting in the MEEZ focuses on two types of fishing, one being small-scale and inshore fishing and the other industrial trawling and selective fishing.

Small-scale and inshore fisheries involve boats ranging from a maximum of 12 m in length (small-scale fishing) to 26 m (inshore fishing). They are under an obligation to use selective techniques and have an ice storage system.

Small-scale and inshore fisheries are booming. The engines used are 15 hp, 40 hp[6] and 250 hp. The techniques are fairly diverse, depending on the species, the fishing ground, the time of year and the fisherman's financial resources. The most common fishing gear is octopus pots, following by lines and nets.

Industrial fisheries involve factory vessels in excess of 26 m in length which use trawling or selective fishing practices. Any fishery using trawls or dredges is considered to be an industrial fishery, irrespective of the vessel's length or capacity. Trawling is the most common industrial fishing technique but others include purse seine netting, pots, long-lines and live-bait).

Mauritania is generally presented as an area where vessels flying every flag are found and, over the past five years, has had an average potential fishing effort of 513 vessels, over 337 of them operating on a regular basis in 2007. The industrial fishing effort over the past nine years is reported to stand at around 381 vessels, 165 of which are in the domestic fleet, in other words just over 43% of the nominal fishing effort. The overall standardised effort in 2005 is reportedly 3.5 million fishing hours, slightly higher than the 2001 figure of 3.2 million hours, whereas the 2006 figure appears to indicate a slight fall that will most probably continue in 2007. The percentage of Mauritanian vessels laid up on a temporary or prolonged basis is estimated to be over 25%. The small-scale and inshore fishing fleet consists of an average of 3 400 units.

6. 40 hp is the most widespread engine power among small local fishing boats. These are made from wood, plastic, glass fibre and sheet metal.

Table 11.9. Trends in the Number of Working Fishing Vessels

	1998	1999	2000	2001	2002	2003	2004	2005	2006
Demersal Fisheries	181	159	149	158	170	170	227	211	211
Mauritanian Vessels	136	110	100	103	115	109	173	161	161
Foreign Vessels	45	49	49	55	55	61	54	50	50
Pelagic Fisheries	64	80	54	53	69	70	74	71	71
Mauritanian Vessels	01	0	0	0	0	0	0	4	4
Foreign Vessels	63	80	54	53	69	70	74	67	67
Other Fishing Vessels	159	130	131	133	142	165	203	75	55
Mauritanian Vessels	11	18	13	18	17	28	25	0	0
Foreign Vessels	148	112	118	115	125	137	178	75	75
Total Ind. Fishing	404	369	334	344	381	405	504	357	337
S-Scale/Inshore[7]	2 430	2 431	2 790	3 890	3 700	3 850	3 600	3 400	3 400

Source: DPI/MPEM.

Catches fluctuate enormously and in many cases are not fully reported. The following table shows the changes, reflecting catches by foreign fleets.

Table 11.10. Trends in Catches in the MEEZ

Fisheries	2000	2001	2002	2003	2004	2005
Pelagics	56 319	630 871	786 117	512 945	727 811	580 312
Demersals	34 476	46 343	53 161	55 875	33 112	27 644
Cephalopods	34 970	33 816	26 138	38 515	27 358	30 837
Crustaceans	3 864	5 000	3 964	7 839	3 797	4 070
Total Ind. F. In T	636 503	716 030	869 030	615 174	792 078	642 863

Source: MPEM

Land-based processing and marketing industries, with the exception of the system introduced by SMCP, are still not very structured or organised. Currently they offer very limited scope for the sector's integration into the economy. 95% of catches in Mauritanian waters are exported.

Virtually all demersal fish is exported raw, in spite of some limited initiatives involving peeled shrimps, dried or frozen *poutargue* (roes), or filleted sole and croaker. Yet minor processing operations such as preparing cuttlefish or squid, processing octopus in some form, or slicing demersal fish can generate considerable value-added. However, only 10% of the products landed in Mauritania and intended for export undergo any form of processing.

7. Small-scale and inshore fisheries, expressed in numbers of boats, each boat representing one unit of GT.

With regard to approved factories in Mauritania, there are 20 in Nouakchott and 19 in Nouadhibou, giving a total capacity of 700 tonnes/day. Thirty per cent of the factories have had their approval temporarily suspended or, in around 5% of cases, permanently suspended.

Fisheries Agreements: Inputs to and Impacts on the Fisheries Industry in Mauritania

Background

Mauritania's recourse to fisheries agreements as a form of access to and means of procuring revenue from the fishing grounds in its Exclusive Economic Zone is based on the principle of international maritime law whereby coastal states are in charge of managing and harvesting the fishing grounds in their EEZ. In the event of a surplus, the coastal state has a duty to allocate some of those resources to other states (whose vessel-owners had previously fished in Mauritanian waters before the new maritime and EEZ arrangements came into force), who are in a position to harvest them sustainably in return for economic, social or political benefits.

In the early 1970s, Mauritania therefore drew on the provisions of international law (the work that was to become the United Nations Convention on the Law of the Sea 1982) to exploit these resources according to the options available which, at the time, were confined to open licences under bilateral agreements with numerous African, European and Asian countries.

In 1979, Mauritania adopted a policy of optimising fisheries rent based on localising landings and processing (on national soil), but the policy option was strongly opposed by some domestic and foreign vessel-owners who were directly responsible for its failure and eventual withdrawal. It was in this context of a crisis in the domestic fleet and the impact of a structural adjustment programme that the first agreement was signed with the European Union[8] as a means of ensuring the financial needs of the national economy.

The first three fisheries agreements were signed for a period of 3 years (87/90, 90/93, 93/96). The 93/96 Agreement marked the start of access by Community vessels to the cephalopod fishery and the landing of these products in foreign ports (contrary to the provisions of Acts 114/88 and 2000/025); this was a first step towards liberalising access to these resources.

Since then, fisheries agreements with the EU have made cephalopods a key target and this, in spite of its impact on fishing pressure, has allowed Mauritania to secure substantial revenue for the Treasury. It has also given the country back-up for its fisheries management tools, particularly in terms of institutional support for areas such as research, surveillance, training and the promotion of small-scale fisheries. The five-year 1996/2001 Agreement, representing a total of EUR 53 million, was followed by the 2001/2006 Agreement for a total of EUR 86 million (excluding fees) and the 2007/2012 Agreement for an annual total of EUR 108 million (including all revenue).

8. The other fishery agreements with Senegal (small-scale fishing), Japan (tuna), and Russia (pelagics) are comparable to non-commercial bilateral agreements that make little economic contribution.

Fishery agreement objectives

In spite of the perverse effects that fisheries agreements may have, in particular on resources and their physical, social and economic environment, they were signed to meet objectives that are relevant and strategic, for both Mauritania and the EU.

From the Mauritanian standpoint, the agreements are aimed at:

- Providing substantial resources (financial contribution plus fees) for the public purse;
- Ensuring permanent co-operation ties with a major partner involved in bilateral and multilateral funding for Mauritania.
- From the EU standpoint, the agreements are aimed at:
- Ensuring the strategic positioning of the industrial fleet, which has increasingly been experiencing declining output and problems of access since the amendments to the Law of the Sea and the emergence of EEZs;
- Supplying the European market, where demand has grown sharply in recent years (30% in 10 years) (see Failler P. *et al* 2003 and reports on IRM/EU negotiations, 2006);
- Providing jobs for seamen, officers and land-based personnel in the fisheries sectors of EU Member States.

Contribution of the agreements to the national budget

In the late 1980s, Mauritania was in serious need of budget revenue and foreign currency for its economy, which had had conditions imposed on it by international financial institutions (IMF and WB).

With the domestic fleet in crisis and national players lacking in creditworthiness, those urgent needs could no longer be met; this paved the way for a new approach to the allocation of fishery resources, namely fisheries agreements.

Thanks to those agreements, Mauritania found itself with a large financial windfall that was readily available. The 2001/2006 Agreement, which has just ended, provided total net financial contribution (excluding fees and knock-on effects) of EUR 430 million over five years, giving an annual average of EUR 86 million. To this should be added an annual average of EUR 9.5 million in fees paid by vessel-owners.

The agreement that has just been signed also provides EUR 86 million per annum for two years, with two two-year periods of tacit renewal, giving a total of six years. Fees will provide an additional EUR 22 million per annum, making an estimated total of EUR 108 million, and hence a total of EUR 648 million for the longest agreement period in the history of Mauritanian agreements. The table below shows how the revenue from the previous two agreements has been used.

Entirely fungible, the revenue from the agreements is used to constitute foreign currency reserves and contribute to the balance of payments, as well as covering part of the expenditure on public investment, thereby indirectly reducing the volume of debt.

The other part, *i.e.* some 11% of the latest agreement compared with 4.7 % of the 2001/2006 Agreement, is being used to support development in the fisheries sector, in particular management tools (development, research, surveillance and training), and the expansion of small-scale fisheries. The resources given over to those areas under the 2001/2006 Agreement came to EUR 4 million a year, compared with EUR 11 million a year under the current agreement, including EUR 1 million for the Banc d'Arguin National Park, a heritage site providing a base for the entire fisheries system.

Table 11.11. Use of the Fishery Agreement Financial Contribution

	1996/2001 agreement	2001/2006 agreement	2006/2008[9] agreement
Financial contribution	53 360 000	86 000 000	86 000 000 including 11 000 000 for development
Training	250 000	300 000	
Research	600 000	800 000	
Surveillance		1 500 000	
Small-scale fisheries		800 000	
Other support	200 000	600 000	1 000 000
Treasury	52 310 000	82 000 000	

Source: IRM / EU fisheries agreements.

Other revenue comes from fines, payrolls and services, and some of the resources from fines are used for surveillance, scientific observation at sea and other government expenditure.

The EUR 11 million given over to development support under the new agreement will be used in the following areas:

- Enhancing fisheries governance;
- Integrating the fisheries sector into the national economy;
- Increasing capacity in the sector;
- Enhancing governance.

To summarise, the revenue from the agreements has not only contributed to the national economy but has also, in the fisheries sector, increased institutional capacity for fisheries research, fisheries surveillance, training for Mauritanian seamen, the development of small-scale fishing and the inspection of fishery products for export purposes.

9. Of a total of EUR 108 million, EUR 11 million is for development initiatives targeting the fisheries sector, EUR 1 million for conservation in the PNBA (National Park). This includes fees and financial contribution. Taking into account other indirect forms of revenue flowing from the agreement, the overall amount may be as high as EUR 116 million.

Impact of the agreement on the sustainability and development of the fisheries sector in Mauritania

Impact on overall fishing capacity

For some years now the status of demersal stocks in general has been deemed critical, in particular for octopus and demersal fish. Since 1998 IMROP[10] working parties have been recommending an average reduction of 25% in fishing effort for those stocks.

Recent studies at the national and regional levels do not appear to offer better data or greater hope. Almost all of them confirm the increasingly critical status of demersal stocks and the ineffectiveness of the mechanisms to reduce fishing pressure in general.

The arrival of European vessels in Mauritania's EEZ in 1993, and more specifically of vessels fishing for cephalopods, triggered a bitter race to fish these species and gave rise to heavier fishing pressure and keener competition. The repercussions were felt immediately, in the form of declining yields and gradually the prolonged laying-up of the country's own fishing vessels, which are the least well-equipped in terms of technology and know-how and are thus less competitive.

For instance, between 25-30% of the country's industrial fishing vessels are now immobilised, while almost all of the remainder are laid up temporarily and are experiencing financial difficulties (e.g. *financing trips, maintaining production equipment).*

On the other hand the EU fishing effort, which is over 80%-subsidised, has increased on average over the period concerned. The fishing opportunities granted to Community vessels in Mauritania have risen from 18 vessels in 1993 to 55 under the 2001/2006 agreement, *i.e.* a threefold increase. In the 1996/2001 Agreement, the tonnage allocated to demersal species was 36 600 GRT as well as 57 boats for tuna fishing and 22 boats for small pelagics. Under the 2001/2006 Agreement, the European fleet's authorised capacity for demersal fisheries was 38 500 GRT (largely cephalopods) as well as 67 boats for tuna and 15 for pelagics.

The 2006/2008 Protocol to the 2006/2012 Agreement provides for a theoretical reduction of 20% on average (187 vessels in the 2007 Agreement rather than 250 in the previous one). This decrease, combined with technical measures and increased surveillance could be beneficial and, in all probability, release some of the pressure on these species. But fisheries do not always respond as expected to these theoretical rules on nominal effort reduction, because a single vessel may well land twice as much as two vessels in the same category, depending on fishing time, harvesting techniques, experience and strategies.

---

10 Mauritanian Institute for Ocean Research (IMROP), formerly known as the National Centre for Ocean Research and Fisheries (CNROP).

Table 11.12. Fishing Opportunities as Set Out in the 2001/2006 and 2006/2008 Agreements

Fisheries[1]	2001/2006	2006/2008	1996/2001(for reference)
Crustaceans (Total)	6 200 GRT or	10 040 GT	5 800 GRT
Cat.1	6 000 GRT (35 vessels)	9 440 GT	5 500 (31 vessels)
Cat.10	200 GRT (2 vessels)	300 GT	300 (2 vessels)
Cat.6		300 GT	
Demersal fish (Total)	15 800 GRT	6674 GT	9 700 GRT
Cat2	8 500 GRT (27 vessels)	3600 GT	4 200GRT (32 vessels)
Cat.3	3 300 GRT (25 vessels)	2324 GT	5 500 GRT (14 vessels)
Cat.4	4 000 GRT (10 vessels)	750 GT	
Cephalopods	16 500 GRT	18 600 GT	12 600 GRT
	55 vessels	43 vessels	42 vessels
Pelagics			
Cat.7	15 vessels	36 vessels	22 vessels
Cat.8	36 vessels	31 vessels	40 vessels
Cat.9	31 vessels	22 vessels	17 vessels
Cat.11		1 500 GT	

1. The ratio of GT to GRT is estimated to be 1.4

Source: Protocols to the IRM/EU Fishery Agreements

Pelagic fishing which, in any case, is said to pose fewer problems in terms of fishing effort, increased noticeably but is still below capacity for MEEZ waters.

To summarise, the agreements have in general created excess capacity, the effects of which can be seen from:

- The rise in the proportion of small sizes in the fishery;
- The steady decline in yield per unit of effort;
- The fact that target species are moving further away;
- The decline in biomass and hence stock abundance;
- The intensity of user conflicts;
- Discards and economic losses;
- Laying-up of vessels.

Table 11.13. Vessels Operating under the 2001/2006 Agreement

	2001	2002	2003	2004	2005	2006
EU	196	223	207	178	160	162
IRM	156	163	170	177	168	165
Others	159	168	156	160	124	10
Total	511	554	533	515	452	337
EU Share	38.4	40.2	38.8	34.6	35.4	48.07

Source: DPI/MPEM

Impact on catches by the domestic fleet

The catches reported by Community vessels are deemed to be incorrect by the Mauritanian public authorities, who put under-reporting in excess of 40%. Reported EU catches range from 190 000 tonnes to 337 336 tonnes at the most, or an annual average over the five years of over 244 000 tonnes. These figures differ very little from those announced by the EU, which puts under-reporting at 14 to 15%.

On the basis of the under-reporting coefficient put forward by the Mauritanian authorities, actual catches landed by the EU fleet in the MEEZ would appear to range from 350 000 tonnes to 475 000 tonnes out of a total of 650 000 tonnes, equal to the total average catch each year in Mauritanian waters.

In terms of effort, it has been estimated that the European fishing effort therefore accounts for 60% of the MSY. The greatest pressure (63%) is on crustaceans, followed by small pelagics (60%) and then cephalopods and hake at around 52 to 54% of MSY.

In any event, catches by the EU fleet have led to a steady decline in output by the Mauritanian fleet, a marked decrease in landings in Mauritanian ports and an increasingly preoccupying break in supplies to land-based factories.

To summarise, then, the domestic fleet, whose technical and economic performance comes nowhere near that of the EU fleet (in terms of technical status, know-how and funding, for instance), has been left far behind in this competition. The reasons stem from the following factors, all of which are to the advantage to the EU fleet:

- Subsidised access to resources;
- Ready access to world markets and appropriate prices;
- Comparative cost of inputs;
- Funding facilities for the purchase of production facilities, equipment and trips;
- Know-how and technical skills.

Impact on land-based industries

Prior to the signature of the fishery agreements, all Mauritanian products were subject to a landing requirement. Landings ranged from 550 to 650 000 tonnes, a portion of which was transhipped in the harbour area but nevertheless recorded.

Fishery agreements have gradually shifted away from their original objective of localising fishery rent and creating value-added by dispatching catches directly to foreign ports, thereby noticeably reducing the volume of landings on Mauritanian soil and supplies to land-based factories.

Landings in Mauritania are estimated at 100 000 tonnes a year, or just over one-seventh of total catches in the Exclusive Economic Zone. This is bound to have major repercussions for the sector's integration into the national economy and constitutes a serious handicap for the expansion of Mauritania's fishery industries.

At another, strictly commercial level, domestic products are facing direct and unfair competition on world markets from the products deriving from EU catches, which benefit from advantages such as:

- Rapid access to global markets thanks to logistics and marketing facilities;
- Difference in quality on board and on land compared with the situation on board Mauritanian vessels;
- Procedures, formalities and non-tariff barriers (relating to hygiene);
- Storage capacity.

All these aspects are in some way detrimental to the value and sale of Mauritanian fishery products, and this may have enormous implications for both producers and the entire Mauritanian economy (see *Section VI)*.

In terms of local consumption, it must be said that the best species are not sold on the local market but exported for their high commercial value. Demersal and inshore species are rare and, where available, excessively expensive and not within the means of the majority of the population. So the agreements thus contribute strongly to shortages of species on the local market and therefore to the decline in consumption of fishery-based animal protein.

Impact on employment

The fact that Mauritanian vessels are lying idle and factories on land are awaiting supplies pushes up unemployment on the maritime labour market, and the employment situation on board European vessels operating in the MEEZ does nothing to offset this.

The Mauritanian regulations provide for 100% Mauritanian crews on board vessels operating in the MEEZ. This clause applies solely to domestic vessels. Vessels operating under the agreement take only an agreed percentage, which is well below 100%.

Data from the latest agreement (2001/2006) show that the percentage of Mauritanian seamen on board Community vessels stood at around 65% in 2003, 76% in 2004 and 72% in 2005. The figures vary from one category to another. In 2003, for instance, tuna boats which very seldom enter Mauritanian ports took on **35** local seamen rather than the regulatory 105, *i.e.* 31% of the agreed percentage. In the same category, the figure reached 33% in 2005 with the hiring of 57 local seamen instead of the 171 that should have been hired. Rather than the agreed total of over 1 020 local seaman planned for 2005, only 625 were hired, *i.e.* 61% of the scheduled number.

The situation is far more critical in the case of officers, as Community vessels in 2003, 2004 and 2005 took on board only 20, 38 and 18 Mauritanians, respectively.

The latest agreement (2006/2008) provides for 1 100 Mauritanian seamen to be taken on, rather than 700 as previously, plus an increase in officers, but the impact will not be measurable for at least one year into the agreement. The importance of this change, if it does actually take place, lies in the fact that it is fully convergent with fishery policy objectives and promotes coherence between the agreements and domestic policy.

On the other hand, the agreement considerably restricts scope for employment on land, inasmuch as all catches are landed abroad (with the exception of the wet fish catches by seine netters which must be landed in Mauritania) and vessels only enter Mauritanian ports to pick up their crews and licensing paperwork.

To conclude, the Community vessels – reported to account in real terms for some 60% of the pressure on MEEZ resources – employed fewer than 4% of Mauritanian seamen under the previous agreement and, under the new one, may well hire 10% of the country's seamen - an estimated total workforce of 15 000. This means that the agreement's input in terms of jobs is negligible and constitutes one of the basic weaknesses that will have to be tackled in future agreements.

Impact on the environment

Responsible fishing is not always the strong point of fleets that operate in the MEEZ, whether or not under agreements.

Discards (hake and continental shelf crustacean fisheries), industrial waste, illegal, unreported and unregulated fishing (IUU), are all perverse effects that can be engendered by fisheries operated at a distance and with little or no control.

So the impact of the agreements on the environment is far from certain and will grow with the increase in overcapacity and the harvesting of new components such as clams and seaweed, which require a more careful approach because of their potential impact on the resource and the environment.

Coherence between agreements and fisheries policy

To sum up, then, the report so far has shown that agreements in general, and more specifically the past three agreements, have had some positive and some negative effects, all of which should be reviewed in an objective light. The most negative impacts are those relating to fishing capacity, in particular for cephalopods, to landings abroad, to the employment of Mauritanian seamen, to competition with the country's own demersal fleet and land-based industries and to specific forms of irresponsible fishing (environment and operations that harm marine wildlife). Similarly, the reliance of the Mauritanian budget on revenue from agreements, like the reliance of its research and surveillance structures on that same revenue, accentuates the vulnerability of the whole management system and indeed of economic policy as a whole.

The positive aspects of these agreements lie in the substantial budgetary revenue they provide, their indirect contribution to debt reduction when part of that revenue is ploughed back into investment programmes, the support they represent for research, fishery surveillance, training and pelagic fishing (for which Mauritania does not yet have a fleet), and the spillover effects they can have on partnerships even if these are still not very common.

Table 11.14. Summary of the Strengths and Weaknesses of Fishery Agreements in Mauritania

Indicators	Positive impacts	Negative impacts
Revenue	• Major resources • Institutional support	• Poor integration • Heavy reliance
Employment	• On-board employment • Knowledge transfer	• Job losses on land • Unfair working conditions
Fishing effort	• Readily redeployable • Harvesting of surpluses	• On fishing • Irresponsible fishing
Market supply	• Covering marketing costs	• Market shortages • Local
Land-based supply		• Inadequate supplies • Little scope for partnerships
Domestic fleet	• Spillover effects on competitiveness and know-how	• Feeling of unfairness

Source: Based on agreement data

Future Role of Fishery Agreements in Mauritania's Management and Sustainable Development Policies

The major challenge facing Mauritanian fishery policies in years to come is to strike a balance between fishery resource sustainability, rent optimisation and the economic and social development of the sector.

To date Mauritania, like several other coastal states subject to financial constraints, has found that its options are revenue-driven. In spite of the focus on creating value-added, jobs and food security, the fishery sector's main task is to act as a lifeline for the national economy, particularly in terms of budget revenue and the inflow of foreign currency for imports of goods and services. In terms of support for sectoral policy, *as this report has already pointed out,* agreements are the most vital input for fisheries research, fisheries surveillance, training and support for the expansion of small-scale fishing.

The country's growing need for expenditure on investment and economic and social development suggests that fishery agreements are still most useful. This is because the agreements generate in excess of EUR 100 million in foreign currency revenue. That generated by petroleum products, at least over the first five years, is reported to amount at most to some USD 100 million, whereas the lowest figure put on public requirements (excluding development assistance) ranges from EUR 250 to 300 million a year.

However, given Mauritania's economic prospects, particularly for petroleum, mining, services and tourism, the government could shift the sector's focus towards value-added, jobs and poverty reduction by making budget revenue a secondary objective.

The issue that will remain central to the debate is to find a solution that is optimal and sustainable for both Mauritania and the European Union. That solution must necessarily protect fishery resources and their environment, while at the same time ensuring that the benefits to both parties will live up to expectations.

Types of agreement in line with sustainable fishing policies in Mauritania

The first-generation agreements, based on a financial contribution (commercial agreements category) enabled Mauritania to mobilise the financial resources it needed to meet its policy requirements, not without perverse effects on resources, structural development in the fisheries sector and social policy

The future, however, calls for new forms of economic co-operation in the field of fishery agreements. To the EU, the costs of vessels operating in Mauritanian waters may seem high, although it hardly exceeds 18 to 20% of the first-sale value of reported catches. Maintaining this level of contribution would not be possible without the subsidies that appear to pose serious problems for the EU, in both the European and global context.

From Mauritania's standpoint, in spite of the great value of the agreements as a source of government revenue, it is becoming clear that resource harvesting should comply with stock renewal requirements, ensure better integration, and create both jobs and value-added; this can only be achieved by reducing the demersal fishing effort, localising landings and expanding land-based industries.

The commercial agreements whose objectives we have reviewed, from both the Mauritanian and EU standpoints, cannot meet the medium- and long-term needs deriving from the issues set out above. The solutions lie in a form of equitable co-operation, gradually integrated through beneficial Economic Partnership Agreements geared to the country's situation.

Economic Partnership Agreements (EPA)

Here the challenge, both for Mauritania and the EU, is how to reconcile the rationale behind fishery agreements for the EU with the three main recurring objectives of Mauritanian fishery policy, namely

- Striking a balance between fishing effort and MSY for demersal stocks;
- Optimising rent;
- Integrating the sector into the national economy.

Future agreements will no longer be able to escape the demands of globalisation and will have inevitably to shift towards a new framework of economic partnership agreements. These will have to promote the sector's integration into the global economy by fostering the creation of an environment that is conducive to private investment, dynamic and transparent, bringing together natural assets, know-how and financial assets from the North.

It is therefore clear that a far-reaching review is required of the objectives of both Mauritanian fishery policy and the CFP on fishing in Mauritanian waters. This review, aimed at promoting equitable economic partnership agreements [11] should focus more specifically on the following:

11. These agreements should not be confused with disguised transfers of overcapacity which have serious biological and economic implications for Mauritania and do not further transparent, sustainable, partnership-based solutions.

- EPAs should be discussed and conducted with a view to sustainable management, in terms of both fishing and investment. The allocation should therefore be based on scientific data from joint programmes and receive financial and technical support from EU Member States;
- EPAs should be confined to fishery resources for which there is a surplus. Redeployment may be envisaged between Mauritanian and EU operators, while maintaining acquired economic and social rights.

These fair agreements should necessarily guarantee the protection of native fisheries (Imraguen fishery[12], small-scale fishing), promote local landings, encourage selective fishing practices and ease the pressure on cephalopods. Furthermore, they should:

- Promote know-how in the field of hygiene inspections for fishery products and environmental quality, in particular for clam resources which hold great potential in Mauritania;
- Promote private partnerships in the fields of investment, on-board technology transfer and vessel sales. The latter should be offset by the permanent withdrawal of less competitive vessels of equal capacity, as part of a buy-back programme. Another possibility might be the development of alternatives based on the country's existing domestic production capacity, supported by EU capital and by access to European markets;
- Develop a selective inshore fishery for wet fish, landed in Mauritania to supply European and related markets via land-based processing industries in Mauritania. This should be feasible through private IRM/EU partnerships backed by a government development scheme;
- Develop a joint programme involving the industry and mobilising the resources required to combat illegal, unreported and unregulated fishing and flags of convenience. This multilateral programme should reflect the new realities of illegal immigration, a scourge that is putting coastguards on both sides of the Atlantic to the test;
- Promote an institutional and legal environment conducive to foreign investment and partnerships in the fisheries sector and related activities;
- Develop and rehabilitate landing facilities used to preserve, process and market fishery resources.

Effects of the Agreements on the Mauritanian Fisheries Sector

Since the late 1980s, which were marked by the crisis in the Mauritanian fleet, the end of the monopoly for access to living marine resources in Mauritania and the marketing monopoly, the government has made a firm move towards liberalisation, with trade policies aimed at maintaining economic growth by focusing on fishery exports produced by the domestic fleet and the sale of access rights to foreign vessels, the main feature being fisheries agreements.

12. Community of small fishers who have subsisted on fishing for centuries, on the coast between Nouadhibou and Nouakchott, famous for fishing mullet with shoulder nets.

But this liberalisation, which was also a reaction to the growth in international demand for fishery products, was initiated in a context marked by:

- The overfishing of most of the high-value resources as a result of overcapacity;
- Conflicts of interest owing to declining output and less competitive tools;
- Institutional weaknesses and a poor understanding of the challenges and strategies of the international market;
- Inadequate local capacity in the fishing industry.

Faced with a situation requiring performance and quality, how has the fisheries sector reacted to competition from fishery agreements and how has it been affected by them? This is the set of issues we shall be addressing below, by analysing the impact of fishery agreements in terms of access to resources and to markets, in the context of globalisation. It is worth noting, however, that this section of the report, which is in fact the most relevant to this Workshop in Paris suffers from a lack of reliable information and visibility on trade policies in Mauritania, which are still disparate and somewhat inconsistent (fisheries policy, trade policy, agricultural policy, financial policy and policies to combat poverty, all of which have different objectives for the same issues).

Context of the agreements

Fishery sector relations between Mauritania and the EU are governed by the Cotonou Agreement and the bilateral fisheries agreement, which is the benchmark for current agreements. These two complementary frameworks have enabled Mauritania and the European Community to develop co-operation which, although not devoid of perverse effects, has been beneficial and useful to both parties.

Admittedly, Nature reacts to external effects far more rapidly than Mankind, who is very slow to regulate effects that may become irreversible.

That said, the impact of fishery agreements in Mauritania on the fisheries sector – be it the harvesting or the marketing system – in terms of compliance with international regulations and tools for the sustainable management of fishery resources and international trade (*including the 1982 United Nations Convention on the Law of the Sea and the 1995 Agreements on Straddling Stocks, the FAO Code of Conduct, the 1993 FAO Agreement, CRFB and the WTO rules*), are prompting some major comments on the lack of coherence that could increase with the scarcity of natural resources, the increase in overcapacity, the rise in global demand for fishery products and Mauritania's growing financial needs.

This is because the complementarity provided for by international legal instruments and used as a benchmark for new agreements only applies to demersal species, subject to 25% overfishing for several years now. Hence this chapter will try to highlight in the following sections.

Effects of fishery agreements on the sustainability of the sector against a background of globalisation

EU vessel access - compatibility with international legal instruments

The United Nations Convention on the Law of the Sea (UNCLOS) authorises coastal States to give fleets from other States access to surplus fishery resources in its EEZ within the framework of international and regional co-operation. It is therefore on this basis that EU deep-sea fishing vessels have been allowed access to resources in Mauritania. The decision reportedly took place at a time when the crisis affecting the Mauritanian fishing fleet made it technically and financially incapable of harvesting fishery resources optimally, and when the country was subject to structural adjustment programmes and under an obligation to find financial resources for its economy and the welfare of its population.

Contrary to the second of these reasons, which is the basic rationale behind the agreements, the crisis affecting the domestic fleet cannot justify European access to Mauritania's demersal resources, in that all demersal stocks, according to the latest assessments by IMROP's scientific working parties, are fully harvested or overfished (see Fishery resources and allowable catch).

Regardless of their value to both Mauritania and the EU, these fishery agreements thus pose a problem in that they are incompatible with the international provisions cited above, granting access to demersal stocks that are already overfished (in particular cephalopods and crustaceans); this is of concern not only to the EU States whose vessels are fishing but also to the coastal states, in this case Mauritania, regarding their mutual responsibility to co-operate and prevent overfishing.

The same constraints are found in the implementation of the development plan for cephalopods – in particular octopus which is the main source of revenue for the fisheries sector – adopted in 2006 after a long process of dialogue and expert reports. It is the most significant expression of Mauritania's determination to ensure responsible fisheries, as recommended by the Code of Conduct and other international legal instruments aimed at sustainable management and the protection of the marine environment.

In another respect, again contrary to the international provisions and recommendations that have emerged from various meetings on fisheries and environmental management (including Rio, Cancun, Reykjavík and Johannesburg), fishery agreements are increasingly tending to influence scientific opinion and/or subject it to interpretation so as to allow vessels (from States that have signed an agreement) access to these resources. The most telling example has been the thorny issue of biological recovery periods and first-catch sizes/weights upstream and downstream from the fisheries system.

Effects on resources and on the equity principle

Fishery agreements are the major feature of the harvesting system in Mauritania, placing heavy pressure on stocks, and their repercussions have a direct impact on the domestic fleet fishing for the same species (see section on the Impact of agreements). Foreign vessels, with their high-tech equipment and valuable know-how in competing for space and resources, mean a loss of earnings and unfair conditions for the domestic fleet, compounded by the subsidies received by the European fleet under fishery agreements.

These subsidies skew competition, are unfavourable to the least affluent and lead to overfishing and depleted fishing grounds, as well as the trade distortions caused by subsidies.

According to studies on the impact of fishery agreements in Mauritania in 2005 and 2006, the European fleet is 80% subsidised. European owners pay only 18-20% of their access costs at the very most (in 2001/2006 they accounted for only 10% of the overall cost of agreements). This is said to have caused the withdrawal of Mauritanian operators for reasons of profitability (*FAO in 2001, WTO/Doha in 2001*). In addition to their subsidies, European vessels receive benefits in kind, which is even more detrimental to the domestic fleet. They are, for instance:

- Exempt from the obligation to land cephalopods (Act 2000/025);
- Exempt from the obligation to hire all-Mauritanian crews (Merchant Navy Act);
- Allowed to fish for shrimp in grounds where trawlers are banned (fishery agreements);
- Allowed in some cases to fish during biological recovery periods (this arrangement will not be renewed).

Effects on fisheries governance

This section addresses the concept of fishing effort and its measurement and responsible fisheries (compared with international standards), together with their impact on overcapacity and the loss of earnings for Mauritania.

Estimating fishing effort

When negotiating individual fishing agreements, Mauritania and the EU have reviewed fishing effort on the basis of the data available on catches, number of vessels, number of days at sea, capacity in GRT or in Gt, and by-catch.

Each time, the data disparities are huge and indicate a big information gap that might in several cases lead to underestimated fishing effort, which might in turn have serious implications for the sustainability of fishery resources, not only at national level but also in terms of stock assessments worldwide. Almost every European fishery in Mauritania is concerned by this, inasmuch as catch inspections are not carried out on landings and fishing logs are kept poorly, if at all.

This underestimation, in addition to its impact on fisheries management, causes a loss of revenue for the national economy. Using non-standardised vessel numbers and approximate measurement units (inconsistent capacity measurement unit; London and Oslo Conventions), and underestimating the impact of technology and the input of know-how, mean that the financial contribution awarded to Mauritania under the fishery agreements does not necessarily correspond to (and hence underestimates) the volume of catches and the cost of the perverse effects on both fishery resources and the economic/social environment in the fishing sector.

Fishing practices

Fishery agreements continue to give rise to a large amount of juvenile fishing, by-catch, discards and fishing in the vicinity of protected areas which play a major role in the regulation of the West African fisheries system (PNBA and PND)[13]. Catches of undersized and underweight species (350g for octopus) and the use of gear subject to national and international bans such as monofilament nets have been common, at least under earlier agreements. The impact of such practices is highly detrimental to the sector in general and encourages other fishermen to do the same.

Declining output due to the overfishing of demersal species is prompting foreign vessel-owners to request, via the EU, a review of Mauritanian regulations to allow the fishing of smaller octopus, under the pretext that such sizes are sold on the EU market, whence the need (according to them) to downsize to the realities of market demand with sizes and weights that suit consumers' culinary practices. So demand from the European market, one of the main destinations for Mauritanian fishery products, is encouraging the fishing of juveniles and thus jeopardising sustainability. It is worth noting that larger sizes are usually exported to the Japanese market, whereas smaller sizes are becoming a specialty in European menus.

Discards at sea and the processing of much of the by-catch into fishmeal on board pelagic vessels naturally affect fisheries governance, because fishing pressure is underestimated and data are lacking.

Effects of agreements on trade in fishery products in the context of globalisation

International markets

The fisheries sector is an export industry *par excellence*. In earlier years, before the 1996/2001 Agreement, fishery exports accounted for over 60% of all Mauritanian exports. The European market was not the main destination. The 2001/2006 Agreement has been a milestone in an era that had begun tentatively with the 1996 Agreement and went on to be characterised by the domination of the European market, the leading destination for demersal exports with the exception of octopus, much of which is now exported to the Japanese market.

But the export path is not always smooth. In some cases, tariff and non-tariff barriers hamper the promotion of fishery exports. In spite of WTO efforts to reduce tariff barriers and regulate non-tariff barriers, the Mauritanian economy – which reacts directly to external shocks owing to its reliance on fishery exports – is giving cause for concern as to its ability to tackle the globalisation of fishery products after 2008 and the end of the preferential regime it has enjoyed to date. That regime has enabled its unprocessed products to maintain their market share. After that date, however, the country's products will face insurmountable competition from Asian products, which will benefit from the removal of customs tariffs and be sold on the European market at more competitive prices than those from Mauritania.

For processed products, the current tariff structure is already hampering the promotion of this high-value-added activity in that tariffs are higher than for unprocessed products. Removing tariffs on unprocessed products will to some extent aggravate the situation for such products, which require skilled labour, industrial inputs and costly compliance with standards, and will therefore be less competitively priced than those

13. Banc d'Arguin National Park and Diawling National Park, two world heritage areas where motorised and commercial fishing are banned.

from Asian countries that benefit from the removal of such barriers and the advantage of low production costs.

With regard to non-tariff barriers, the WTO's SPS and TBT agreements should logically limit the risk, but one should bear in mind the famous embargo that was overcome by the fishing industry, and even the economy as a whole. These barriers may be applicable to Mauritanian exports, which would be neither fair nor transparent and constitute a genuine problem for the fisheries sector.

These barriers are sometimes used as a means of restricting access to resources, or under the pretext of human or animal health, all of which runs counter to the WTO principles of non-discrimination and transparency (WTO Agreement on Technical Barriers to Trade) and the FAO Code of Conduct for Responsible Fisheries (11.2.7).

Other HACCP requirements and initiatives such as traceability and eco-labelling (which may benefit Mauritania with its two national parks) might affect the export system.

Barriers linked to vessel origin may also hamper the country's export trade and oblige its fleet to acquire European technology or make room for foreign overcapacity. This is because the domestic fleet consists mainly of vessels flying the Mauritanian flag but built in Asia (*e.g.* China and Korea) and owned by Mauritanians, either fully or in joint ventures. Yet its catches are exported to European markets and have to compete with catches by EU vessels harvested in Mauritanian waters.

Local market

Fishery agreements target all species and compete directly with the local market, particularly for demersal species and small pelagics which are popular with Mauritanian consumers.

The large volume fished under the agreement and the substantial volume of exports covering virtually every species are creating an ongoing shortage of fish on the domestic market and making consumer products expensive. The problem is all the more urgent in that fishing vessels are now targeting inshore pelagic and semi-pelagic species, once viewed as having a low market value (sardine, sardinella, croaker and mullet).

By and large, Mauritanian fishery products are increasingly confined to the species targeted by the European market, and are virtually all directed at that market. This could have four major implications that would be detrimental to Mauritania and infringe WTO rules:

- Heavy pressure on high-value species, which may cause sudden breaks in the biological chain (stock structure);
- Heavy reliance by the fisheries sector on the European market, with possible implications for the sector's prices and commercial strategy in the event of barriers;
- Depleted local markets and a drop in local consumption, which would undermine the country's food security strategy (adverse social and nutritional impact, running counter to the objectives of international trade and the Code of Conduct for Responsible Fisheries, as well as the Millennium Goals);

- More conflict and greater social tension vis-à-vis the agreements and public policy, particularly in a context of democratisation.

Consequences of the Fishery Agreements on Industry

The initiation of commercial agreements marked the beginning of the end for the fisheries industry in Mauritania, which was still embryonic at the time and unable to withstand the keen competition that emerged with new, more efficient players. The outcome was a permanent fall in catches, followed by a break in supplies going to the land-based facilities that are the sector's gateway to integration in the national economy.

Under-supply of land-based factories and facilities

The land-based factories that sprang up under the new policy in 1979 and again in 1987 are now in an appalling state, from a technical standpoint. Currently used as storage and chilling facilities, these firms –originally designed in a global, integrated context – now number only 39 export-approved facilities with an overall storage capacity of 700 tonnes a day. The first problem they face relates to supply, owing to the declining output of the national fleet which is in strong competition with Community vessels operating under the agreement, while the second relates to phytosanitary barriers that restrict access by their products to EU export markets. These two constraints are increasing firms' production costs and making them less competitive.

The handful of factories that attempted to develop a processing/value-added industry, including Société MIP specialising in fillets and ready-cooked meals and the canning firm MEIPP S.A., suffered from supply shortages and were eventually forced to close or reduce their output.

Attempts at a solution to maintaining supplies to land-based factories by introducing clauses on compulsory landings, as set out under Article 17, Section 4 of the 2000/025 Act of 24 January 2000, to be amended by the new Act in 2007[14], have failed. None of the agreements between Mauritania and the EU have managed to solve the issue. In answer to Mauritania's request that catches be landed in the country's own ports, European officials cite technical conditions in land-based factories and ports there, and poor support services.

The principle of voluntary landings (Chapter VII of the Agreement), which replaces compulsory landings, has destroyed all hope of Community catches supplying land-based firms in Mauritania.

This is because the number of Community vessels landing their catches on a voluntary basis under the 2001/2006 agreement, has been estimated at 15. And even those vessels actually landed their catches in the country for exceptional reasons, solely of benefit to their owners.

The new agreement endeavours to resolve one aspect of the problem of factory supplies by making it compulsory for seiners (an annual total of 15 000 GT) to land wet fish catches in Mauritania, in exchange for lower fees. This fish is intended to be sold to land-based factories, processed locally or dispatched overland using national haulage

14. The new Act, as yet not published or numbered, improves certain aspects of Act No. 2000/025 which still remains the main reference.

services. To date, the vessels concerned have shown little enthusiasm for this arrangement.

In addition to all these problems, which cast a shadow over both the present and the future of the land-based fisheries industry, we should point out that the industry is not in any financial or technical state to overcome hygiene constraints, in spite of the government's efforts both to bring the regulations into line with international rules and standards and to bring the industry up to compliance. The industry needs to be afforded the technical and financial help it requires to cope with the perverse effects of globalisation and overcome non-tariff barriers, in particular compliance with European standards.

National harvesting industries

The impact that the agreements have had on the country's harvesting industry can be seen from the decline in vessels' catches and their gradual withdrawal from fishing. Landings by Mauritania's industrial fishing fleet have declined because of keen competition from the European fleet and because of the scarcity and technical status of the resource. On average, total domestic landings stand at around 100 000 tonnes a year, whereas European catches are 4 to 5 times that figure.

Landings, for instance, have dropped sharply since the signature of the agreements, and now represent only 1/5 of their historical volume and 1/7 of all catches at present, whereas just along the coast in Senegal, which has far fewer fishery resources than Mauritania, small-scale fishermen are landing over 300 000 tonnes. As highlighted throughout this report, the number of vessels being withdrawn from the domestic fleet today is said to be over 25% of those operating prior to the 2001 agreement, and vessel efficiency has never been lower.

Furthermore, fishery agreements have led to a shift to species destined for the European and Japanese markets, although those species (excluding pelagics) account for only a tiny fraction of the potential catch. That shift, coupled with the use of not very selective techniques, encourages discards and wastage which have implications for the volume of landings and are harmful to the environment and biodiversity.

Small-scale fisheries have not been spared here either, as they are increasingly turning to the export market and bearing the full brunt of competition from European vessels.

Conclusion

Notwithstanding their adverse effects, extensively described above, fisheries agreements have provided Mauritania with an inflow of foreign currency revenue that has been extremely useful, particularly during the lean years that preceded its current status as a petroleum-producing country which, in addition, has just seen the cancellation of its foreign debt, estimated at around USD 819 million.

These agreements have also been the first sign of liberalisation in the fisheries sector, in terms of both access to demersal resources and competition on export markets (Japan, Europe and Africa).

When the 2007/2012 agreement comes to an end, will Mauritania – now in a more comfortable financial position – be sufficiently prepared to develop a sustainable fisheries policy based on good governance, with special emphasis on the conservation of fishery resources and the marine environment, in line with international legal instruments such as the United Nations Law of the Sea Convention 1982 and its agreements linked to the Code of Conduct for Responsible Fisheries, as well as a more integrated role in the national economy?

Such a policy might not necessarily exclude the signature of fishery agreements but would seek to develop production, development and marketing systems that obey the rules of responsible management and meet the demands of liberalisation.

To conclude, then, the future of fisheries in Mauritania will depend on the country's ability to promote quality rather than quantity, which has reached its upper limit in such a poor harvesting environment. This goal would require a major institutional drive, developing approaches likely to diminish Mauritania's reliance on fishery agreements and lessen the potentially perverse effects of globalisation. That drive should necessarily focus on developing:

- A production system conducive to stock renewal;
- A diversified marketing system, targeting more than one market and entirely free of export constraints or barriers;
- A set of hygiene and technical standards to meet the requirements of the international market; and
- A small-scale fishing industry that targets inshore fishery products and keeps local markets well supplied and consumers satisfied.

Bibliography

Catanzano J. (2003), Régulation de l'accès aux ressources marines vivantes pour un développement durable des pêches, FAO, Saly, Senegal.

CNROP (1999), *Rapport du 4ième groupe du travail sur l'évaluation des stocks et des pêcheries mauritaniennes;* Nouadhibou, 7 to 13 December 1998.

CNROP (2002), *Rapport du 5ième groupe du travail sur l'évaluation des stocks et des pêcheries mauritaniennes*; Nouadhibou, 9 to 17 December 2002;

EU/IRM Fisheries Agreements (1987; 1990; 1996; 2001; 2006).

European Union (2006) *Evaluation of the EU/IRM fisheries agreement 2001-2006.*

Failler P., United Nations Environment Programme, *Environmental Impact of Trade Liberalisation: A case study on the fisheries sector of the Islamic Republic of Mauritania.*

Failler P, (2002), *Cogestion des Pêches, Aspects Institutionnels Economiques et Contractuels., Séminaire Aménagement,* IMROP, Nouadhibou, Mauritania.

FAO, Document Repository, *The effect of subsidies on sustainability, a selective history of fishery subsidies, the nature of subsidies*, from *Introducing Fisheries Subsidies.*

FAO, *Code of Conduct for Responsible Fisheries*, Rome, Italy.

FAO (2000), *International Fish Trade: Experiences with Provisions of the Technical Barriers to Trade Agreement (TBT) and Informal Trade Barriers.*

IMROP (2006), *Rapport du 6ième groupe du travail sur l'évaluation des stocks et des pêcheries mauritaniennes;* Nouadhibou, Mauritania.

IMROP/RIVO (Netherlands Institute for Fisheries Research) (2003), *The Demersal Fisheries of Mauritania in 2003, Description of Fleets and Assessment of Resources*, December.

IMROP, (2002) *Séminaire National sur l'Aménagement des Pêches en République Islamique de Mauritanie*, Bilan et Perspectives, Rapport Final.

IMROP (2002), *Rapport final " Bilan et perspectives",* Séminaire national sur l'Aménagement des Pêches en République Islamique de Mauritanie.

IMROP (2004), *Rapport du groupe de travail sur les crevettes*, Doc 01, Nouadhibou, Mauritania

IMROP (2006), *Evaluation des Ressources Démersales en Mauritanie*, Rapport du groupe de travail IMROP/RIVO, Doc 03, Nouadhibou, Mauritanie.

IMROP (2005), *Effort Development in the Industrial Fisheries of Mauritania from 1992 to 2004*, Doc 02; Nouadhibou.

IMROP (2006), Bulletin Scientifique, volume 30 January 2006, Nouadhibou, Mauritania

Lionel K. and Cherif Ould T. (2004), Indicateurs de performance du secteur des pêches, MPEM, Nouakchott, Mauritania.

Lionel K. (2004), Sociétés Mixtes et accords de deuxième génération UE/Argentine, MPEM, Nouakchott, Mauritania.

MAED (2004) (Ministère des Affaires Economiques et du Développement), *Plan Stratégique de lutte contre la pauvreté.*

Mahfoud, T. (2000), *Evaluation de l'activité des flottes industrielles étrangères ciblant les petits pélagiques dans la ZEEM.*

MPEM (1987, 1990, 1996, 2001, 2006) (Ministère de la Pêche et de l'Economie Maritime) *Accords et Protocoles de pêche RIM/UE.*

MPEM (2006), *Stratégie d'Aménagement et de Développement du Secteur des Pêches et de l'Economie Maritime*, Nouakchott, Mauritania.

MPEM (2006), *Evaluation de l'accord de pêche RIM/UE 2001-2006,* Nouakchott, Mauritania.

MPEM (2000), Loi portant code des pêches 2000.025, Nouakchott, Mauritania.

MPEM/DEARH, (2006), *Plan d'aménagement du poulpe*, Nouakchott, Mauritania.

MPEM (1979, 1987, 1994, 1998, 2004, 2006) *NPP, LPP, SADSPEM,* Nouakchott, Mauritania.

MPEM (2006), *Stratégie d'Aménagement et de Développement du Secteur des Pêches et de l'Economie Maritime* March.

MPEM (1998, 2004*) Stratégie d'Aménagement et de Développement du Secteur des Pêches et de l'Economie Maritime.*

Oceanic Developpement (2005), *Evaluation ex post du protocole d'accord de pêche entre la Mauritanie et la communauté européenne et analyse de l'impact du futur protocole sur la durabilité, incluant une évaluation ex ante*, FISH/200/02 project, Brussels, Belgium.

ONS, (2002; 2003, 2004, 2005, 2006), Agrégats de la comptabilité nationale et indicateurs socio-économique, Nouakchott, Mauritania.

J. Paul T. (2002), *Régulation de l'accès dans les pêcheries ouest Africaines, Séminaire national sur l'aménagement des pêches*, IMRP, Nouadhibou, Mauritania.

SGG (2000), Loi du PNBA, Nouakchott, Mauritania.

Toueilib, Chérif Ould (2005) et al, *Accords de pêche; outil de développement des activités de pêche de manière durable*, United Nations Conference on Trade and Development, Geneva, 24 to 25 October 2005.

Toueilib, Chérif Ould (2002), *Politiques des Pêches en Mauritanie, mécanismes de régulation, Séminaire Aménagement des pêche*, IMROP, Nouadhibou/ Mauritania.

Toueilib, Chérif Ould (2001), *Cohérence au niveau national entre accords de pêche et politique de développement*, DFID progress report, 18 pages.

PART 5

Identifying the Parameters With Which We Are Working in the Aquaculture Sector

Chapter 12

Impact of Globalisation – Challenges and Opportunities for Indian Aquaculture

G. Mohan Kumar, Chairman, Marine Products Export Development Authority (MPEDA), India

Mr. Kumar explained how India is coping with the pressures of globalisation and how frameworks have been adjusted to facilitate its response. India is the world's second largest aquaculture producer, with shrimp a major export item from aquaculture. Shrimp provides 58% of the total value and 28% of the total quantity of India's aquaculture production. Major markets for shrimp are the EU (one third of production), USA and Japan. However, India has a large number of small farms that practise extensive farming with limited stocking densities; 90% of farmers have ponds smaller than two hectares. This means that farming is highly fragmented and traceability can be challenging and certification prohibitively expensive.

Uncertainty in the international trading system has been a large problem for Indian shrimp farmers, particularly as a result of anti-dumping measures; for example the number of Indian exporters to the United States fell from 169 in 2002-03 to 77 in 2006. For Mr. Kumar other concerns include disease, lower prices on the international market, increasing cost of production, rising quality requirements, lack of insurance and financial support and exploitation by middlemen. Indian shrimp farmers also experience difficulties in international trade, including market access issues due to tariff escalation. The increased dominance of supermarkets that raise entry barriers through the use of brands creates difficulties in supply logistics and raises costs for shelf space and market promotions. Many producers are forced to pack for private labels, thereby heavily foregoing margins. Food safety issues and non-tariff barriers such as SPS measures and EU directives for residue levels also raise production costs.

To help mitigate some of these costs, aquaculture is now regulated by the Coastal Aquaculture Authority Act, which includes Code of Practices for shrimp hatcheries, farms and their registration. A comprehensive database is also under preparation that will cover 150 000 farms. A programme to introduce participatory farming through the formation of Aqua Clubs is also under way. Finally, clustering in groups of farms is likely to pool resources and allow synergies and economies of scale to be realised.

Introduction

Globalisation exerts its impact on fishery production and trade in the form of changes in production technology and pattern of production as well as increasing market interdependence. The international fish trade has soared from the level of about USD 15.5 billion per year during 1980 to a new high of over USD 71.5 billion during 2004. This accounted for 0.8 % of the total international export trade (USD 8 880 billion as per World Trade Report-2005).The share of developed countries decreased from about two thirds of total trade to about half during the period. The developing countries have benefited from this boom, with their net earnings rising from USD 3.4 billion to over USD 20 billion over the same period (George Kourous FAO 2006). Fish trade today ranks highest among the net foreign exchange earning food commodities, including coffee, tea etc. for the developing countries. Currently, about 77 % of the fish consumed worldwide as food is supplied by the developing countries and over half of the fishery trade in value terms originated from the developing countries. This is a very significant achievement for the developing countries engaged in fishery trade, and developing country aquaculture has played a significant role in the progress.

With the global capture fishery production stagnating around 95 million MT, the importance of aquaculture production in the fishery trade has been on the rise. For 26% of international trade aquaculture is very important and for 21% it is starting to gain importance. Looking from another angle, only for 16% of all commodities, aquaculture does not play a role. These species are mainly small pelagics, where culture would be prohibitively costly (Helga Josupit 2006). During the ten-year period from 1994-2004, the substantial increase in the quantity of shrimp traded coincided with the strong expansion in aquaculture shrimp production. During the year 2004 more than 41 % of total shrimp production (about 2.5 million MT) was of farmed origin.

India's current Seafood exports stands at 512 000 tones valued at USD 1.64 billion (2005-06). During the year 2004, India occupied the 19th position among the top exporting countries of the world (FAO). Shrimp is the main item of export from India, with its share being 58% by value and 28% by quantity.

Figure 12.1. Share of Shrimp in India's Total Exports

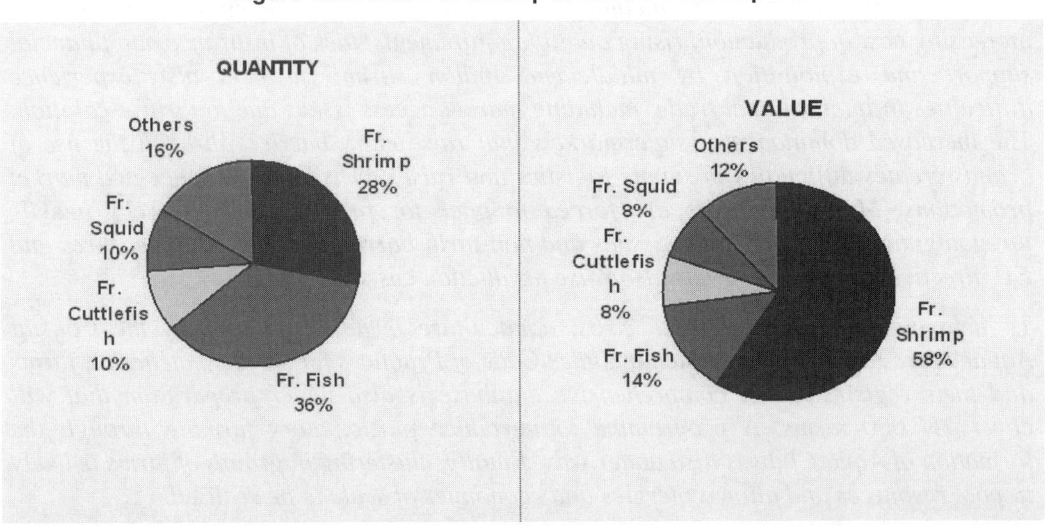

Source: MPEDA

Indian shrimp are exported to European Union (27.93%), USA (31.94%), Japan (20.26%), China (1.62%), Australia (1.93%), Middle East (4.12%) and South East Asia (4.13 %). In 2005-06 The OECD countries accounted for 86% of the shrimp exported from India.

OECD countries accounted for 86% of the shrimp exported from India.

Figure 12.2.Market-wise export of frozen shrimp from India

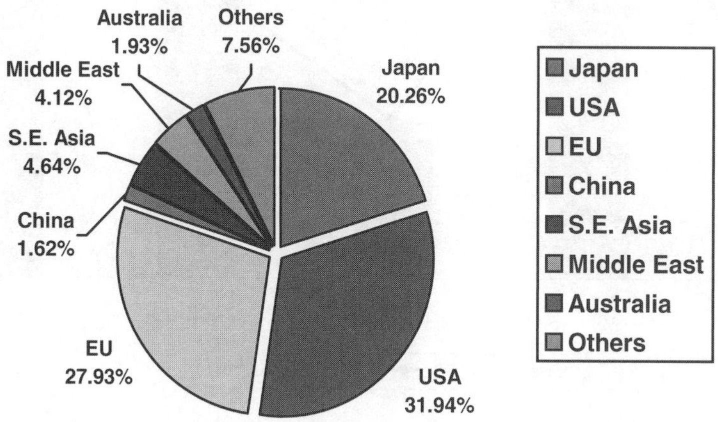

Source: MPEDA

Among aquaculture nations, India occupies the second position in the world with an annual production 2.47 million MT. Since fresh water fish which constitutes 93.1% of the production is mostly for the domestic market and does not figure significantly in international trade, this paper discusses mainly those species, which are exported from the country. Aquaculture is carried on in the country mainly in fresh water and brackish water environments. The shrimp cultured in the brackish water area in the coastal States and fresh water prawn cultured in fresh water are the major crustaceans. The shrimp cultured is predominantly the Black tiger shrimp, *Penaeus monodon* which is native to India. Earlier the white shrimp, *P. indicus* was also being cultured but of late its share is negligible. The main fresh water species cultured is the giant fresh water prawn *Macrobrachium rosenbergii*. Ninety five per cent of the production of shrimp and prawn cultured in the country is exported. India virtually has no aquaculture in the marine waters.

India exports shrimp mainly in the form of headless shell-on, PUD, PD, Peeled tail-on etc. Block frozen/IQF raw materials meant for further processing dominates India's exports. The current level of value addition of shrimp is only 4.63%.

Figure 12.3. Export of Frozen Shrimp to OECD Countries

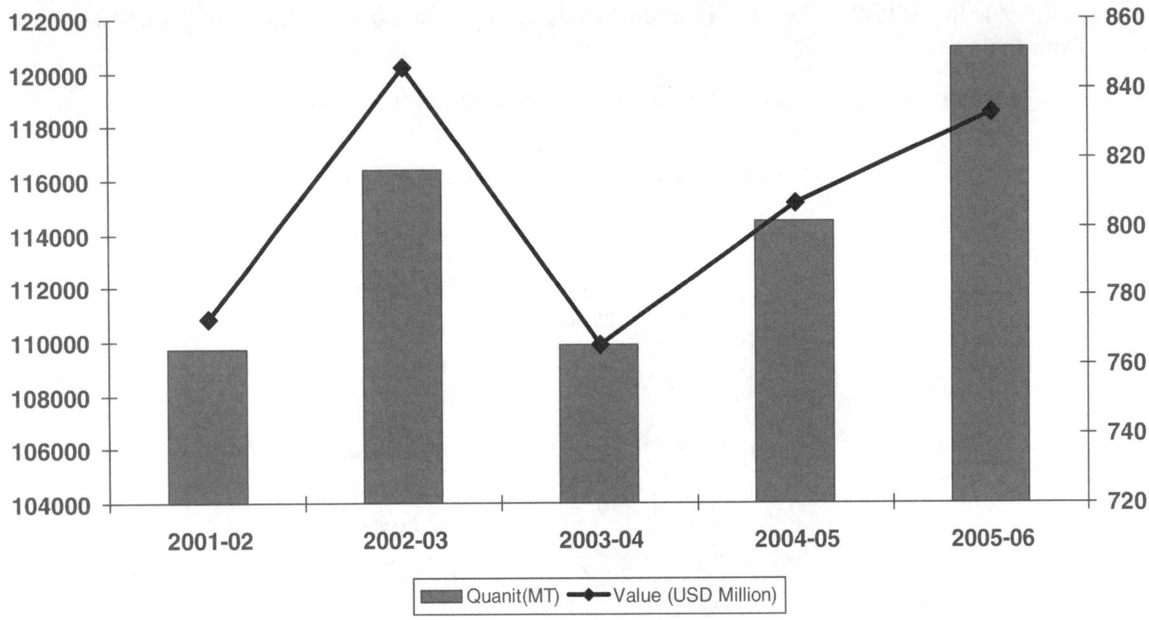

Source: MPEDA

Aquaculture of Shrimp in India

During the year 2005-06, India produced 143 000 tonnes of shrimp and 43 000 tonnes of fresh water prawn in a total farmed area of 185 000 ha. The culture operation took place mainly in the 9 coastal States in the country; *viz.* Andhra Pradesh, Kerala, Maharashtra, Karnataka, Gujarat, Goa, Orissa, Tamil Nadu and West Bengal. Export of shrimp earned the country USD 970 billion during the year 2005-06.

Estimates show that India has over 1.2 million ha of brackish water area suitable for shrimp culture. Out of this, only 190 000 ha (15.8 %) are currently developed for culture. Small farms dominate the culture operation. It is estimated that there are about 150 000 brackish water shrimp farmers in the country. About 90% of this is held by farms below 2 ha in size, 6% by farms of 2.5 ha in size and 2% of 5 to 10 ha and another 2% above 10 ha (Figure 12.4). Therefore brackish water shrimp aquaculture in India is an enterprise of small and marginal farmers who do not have access to capital resources. Almost all the coastal shrimp farms grow only the Black tiger shrimp (*P .monodon*).

Figure 12.4. Composition of Indian Shrimp Farmers Based on Farm Size Holdings

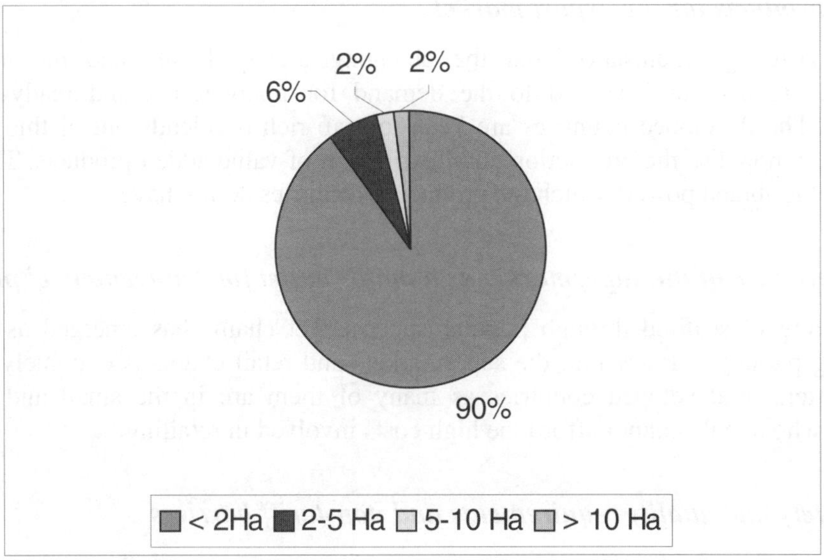

Source: MPEDA

Shrimp farming which gathered considerable momentum in the late 1980s and 1990s played a major role in taking India's seafood exports past the USD 1 billion mark in the year 1993-94. The area under shrimp aquaculture increased rapidly from 65 000 ha in 1991 to 100 700 ha in 1994-95. Production also increased from 35 000 Mt to 82 850 Mt during the same period. However, repeated outbreaks of the white spot disease in the middle of the 1990s dealt a severe blow to the industry. With no immediate remedy available for the deadly disease the industry came to be looked upon as highly risky. As a consequence, financing institutions and insurance companies left the field, leaving the farmers to fend for themselves. During the last decade, culture techniques have undergone many changes and innovations in an effort to contain disease. Today, the farming sector is propped up by feed manufacturers and distributors who supply feed, the major input, on credit, and buy back the shrimp. The shrimp industry is estimated to employ over 2 million people and has been instrumental in ushering considerable prosperity in the coastal rural areas.

Some Characteristics of Globalisation

Before we discuss the challenges faced by the Indian aquaculture, in the context of the globalised environment, let us examine the major characteristics of globalisation in the industry.

The increase in popularity of seafood and its perception as a healthy food

The perception of seafood as a wholesome food has led to increased demand for seafood in the developed countries .The reduction in tariffs brought about by the WTO has facilitated trade on a global scale with many developing countries emerging as major suppliers. Since sea catches are stagnant this increased global demand has to be met from aquaculture. As a result aquaculture is expanding on an unprecedented scale both in extent and intensity.

Greater value-addition and development of a variety of ready-to-eat and ready-to-cook products for the export market

The growing urbanisation and the changing dietary habits and the increasing propensity to eat out have led to the demand for ready–to-eat and ready-to- cook products. The developed countries are keen to reap rich dividends out of this trend by trying to monopolise the production and distribution of value added products. They have considerable 'brand power' which the producing countries do not have.

The emergence of the supermarket as a major outlet for 'convenience' products

Retailing of seafood through popular supermarket chains has emerged as the new marketing paradigm. Entry into the supermarkets and retail chains is extremely difficult for exporters of developed countries as many of them are in the small and medium category who simply cannot afford the high costs involved in retailing.

Food safety and quality requirements and non-tariff barriers

The growing food safety awareness among consumers and the governments of importing countries has led to the stringency in Sanitary and Phytosanitary (SPS) measures adopted by them. Food standards have become non-tariff barriers and increase the cost of compliance heavily.

Anti-dumping measures

Anti dumping measure sanctioned by the WTO rules are often used as protectionist measures affecting the livelihood of those involved in the export of seafood.

Choice of species for farmers is becoming geared to the demand for specific products in international markets

Supply of products which are in great demand creates enormous pressure on exporters and producers to switch over to new products without regard for the exotic nature of such products Such Commercially driven private sector initiatives of unregulated introduction and transfers are evident in many parts of the world.

Growing concern for environmental sustainability

Sustainability of farming operations is increasingly coming under scrutiny of buyers and consumers .Issues like the use of banned antibiotics and other chemicals in aquaculture and the need to get products certified by accredited agencies have led to the emergence of eco- labeling as a tool for differentiating products.

Coping with the Challenges of Globalisation

Increasing consumer preference for seafood in the major developed countries like USA and countries in the European Union despite the declining demand in Japan has been fuelling the demand for shrimp. Imports of shrimp to the USA crossed the 500 000 Mt mark in the year 2004. This unprecedented demand for shrimp has spurred the expansion and diversification of aquaculture. However, the disease problem of Black tiger has become a stumbling block to expansion of production. During this period, the SPF brood stock of *L. vannamei* became available for commercial use. Driven purely by market demand and commercial interest, *L.vannamei,* the Pacific white shrimp, which was originally cultured in South America, became adopted on a wide scale in Asian countries, like China and Thailand and to some extent in Indonesia and Vietnam. Because of its SPF nature the farmers in these countries could undertake intensive production of the species. Table 12.1 shows the production statistics of these countries during the years from 2004 to 2006.

Table 12.1. Cultured Shrimp Production of Selected Asian Countries (2004 to 2006)

Country	Year		
	2004	2005	2006
China	375 000	408 000	405 000
Thailand	325 000	374 159	386 742
Vietnam	290 000	310 420	346 528
Indonesia	190 000	197 000	206 000
India*	126 000	143 000	152 000
Others	90 838	101 204	111 000
Total	1 396 838	1 533 783	1 607 270

Source: GSOL 2006

* Mainly Black Tiger shrimp.

During the last few years, production of cultured shrimp has expanded on an unprecedented scale and with this expansion, international prices have fallen. Figure 12.5 represents the price pattern as experienced by the Indian exporters during the period from 2001 to 2005. From an average price of USD 7.19 in 2001, price fell to USD 6.76 in 2005.

Figure 12.5. Average Unit Value Realisation for Indian Frozen Shrimp

(USD per kg)

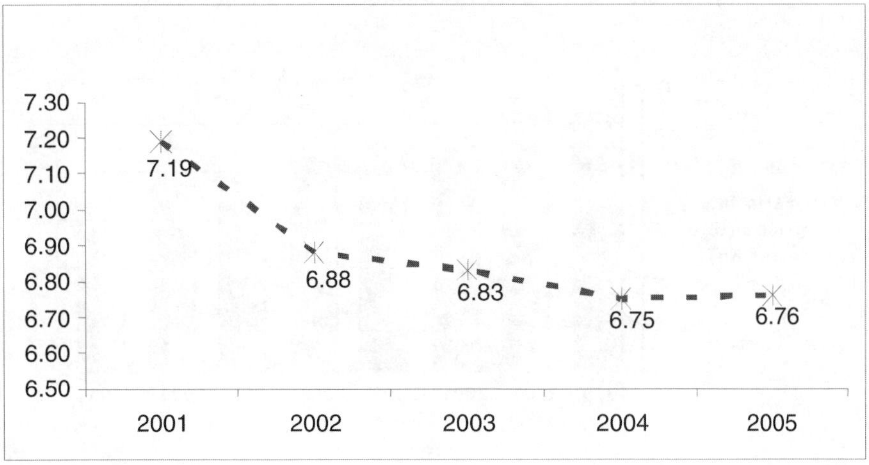

Source: MPEDA.

A country like India is engaged at moderate production levels and markets mainly products of less value addition. However, the global trend towards more efficient of production and virtual commoditization of shrimp has resulted in large scale production of L. Vannamei. The consequent fall in prices has been brought in its wake two adverse consequences *viz.* anti dumping action by USA and reduced margins to farmers and exporters of India.

The unprecedented level of imports into the USA and the stiff competition faced by the US shrimping industry led to the initiation of anti-dumping cases in the year 2003 against India and 5 major shrimp producing countries namely; China, Thailand, Vietnam, Ecuador and Brazil, on a petition filed by the Southern Shrimp Alliance of USA. India consistently took the stand that there was no dumping. Anti-dumping to India was a patently protectionist move, which became necessary when US shrimpers could no longer compete with the exporting countries, which could produce their shrimp at much higher level of efficiency than ever before. The anti-dumping action resulted in imposition of duties ranging from 4% to 10.17% on the Indian exports. This was followed by the imposition of bonding requirements by the US Customs and Border Protection. Bonds were required to be furnished by every importer in order to guarantee the payment of anti-dumping duties, which are to be determined by the US Government on a retrospective basis, one year after the expiry period to which the duty relates. The bond is required to be furnished on the basis of anti-dumping duty, payable on an entered value of shipments made over one year.

Owing to the nature of the anti dumping investigations and the retrospective system followed by US, an exporter in effect, is required to furnish 3 consecutive bonds within in a span of 3 years. This requirement has imposed unbearable burden on Indian exporters and many small and medium exporters, have stopped exporting to the US. The number of exporters to USA declined from 169 in 2002-03 to 107 in 2005-06, and to 77 in 2006-07 (up to December 2006). Owing to the difficulties faced by the exporters, the prices offered to the farmers by the exporters were also reduced (Figure 12.6) and this resulted in closure of many farms, causing major social and livelihood problems to farmers and farm workers in India; particularly in the State of Andhra Pradesh which produces 50% of the aquacultured shrimp in the country.

Figure 12.6. Farm Gate Prices Offered to Shrimp Farmers

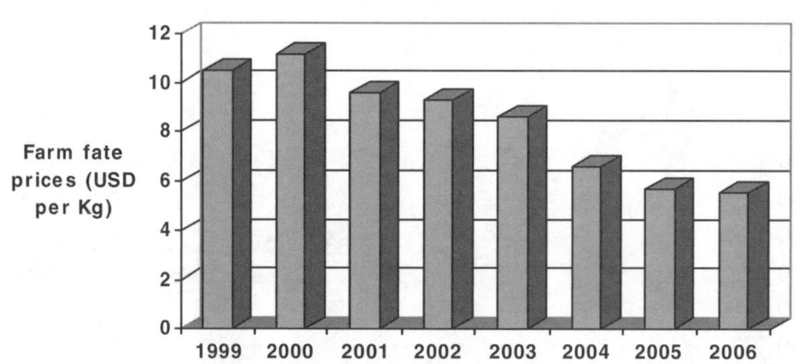

Source: MPEDA

India, having failed to get positive response from the US Customs and Border Protection despite diplomatic efforts, held consultation with the USA at the WTO, the consultations having failed India filed a dispute before the Dispute Settlement Body of the WTO. The bonding requirements have been challenged by the government of India which considers that the enhanced bond requirement under the Amended Bond Directive as applied to importers of the subject merchandise is inconsistent with Articles I, II, VI:2, XI and XIII of the GATT and Articles 1, 7.1, 7.2, 7.4, 7.5, 9.2, 9.3, 9.3.1 and 18.1 of the Anti-Dumping Agreement.

Thus the anti dumping action, which has affected all the shrimp producing countries including India, also brought in its wake a new non-tariff barrier in the name of continuous bonds.

Faced with the hostile tariff and non-tariff in the USA, India's exports to the USA have fallen steeply (Figure 12.7) and a significant share of India's exports has now been diverted to European Union.

Figure 12.7. Declining Exports of Frozen Shrimp to US Market

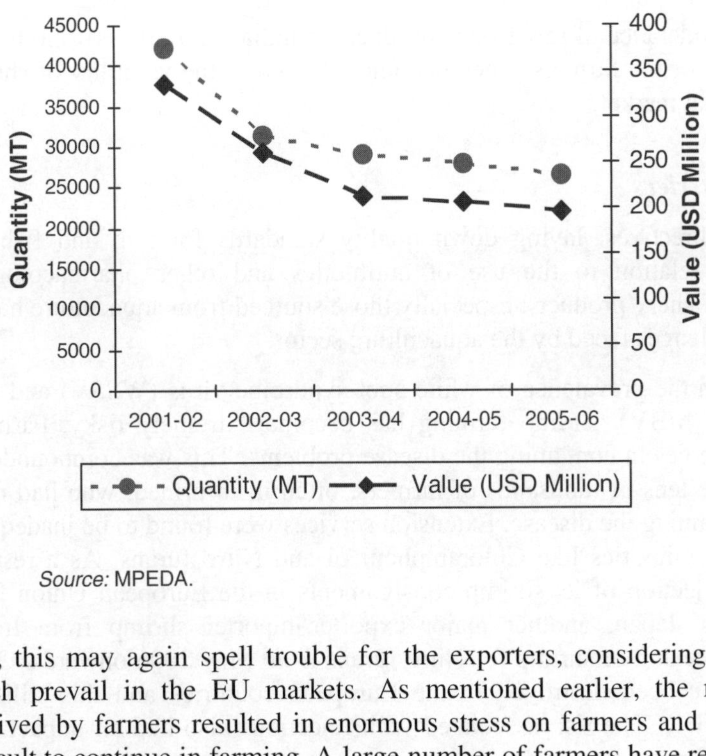

Source: MPEDA.

However, this may again spell trouble for the exporters, considering the non-tariff barrier, which prevail in the EU markets. As mentioned earlier, the reduced profit margins received by farmers resulted in enormous stress on farmers and many of them found it difficult to continue in farming. A large number of farmers have reported to have shifted to agriculture from aquaculture.

Tariff escalations

Since the developed countries are major markets for developing countries, their tariff profiles have a significant impact on economic opportunities for developing country producers and exporters. 68% of the OECD fish imports are subject to tariffs ranging from 0 to 5%, while the EU markets have higher rates of tariffs. The key fish importers have maintained higher tariff rates for most value added processed fishery products from developing countries. For developing countries such escalation as well as maintenance of separate quotas for selected items amounts to restriction of free access to trade. Developing countries have consistently argued that the structure of international trade is biased against them and called for further liberalisation by the developed countries. Combined with domestic factors, developed country tariff structures compel many of the developing countries to limit their exports to low value-added, raw fishery products. The tariff escalation and related import duties restrain some developing countries from diversifying their exports towards value added and more profitable products. Only further liberalisation facilitating reduction in tariff peaks and tariff escalation can provide increased market access to processed and value added products from the developing countries.

The continuation of tariff escalation and low technological progress keep processors tide down to raw product exports. It is sometimes maintained that tariff escalation hinders industrial development and diversification, so that poor countries remain producers of raw materials and goods with lower value added. In fact, the poor countries have a comparative advantage of cheap labour and tariff escalation is an expression of protectionism, in order to defend labour intensive industries in rich countries. (Fisheries Circular No.1016, FAO)

The preponderance of raw frozen products in India's export portfolio has unfavorable consequences for the farmers. They become exposed to the volatility of shrimp prices in the international market.

Non-tariff barriers

The EU directives, laying down quality standards for fish and fishery products; especially in relation to the use of antibiotics and other pharmacologically active substances in fishery products; especially those sourced from aquaculture has been one of the major challenges faced by the aquaculture sector.

Faced with the prevalence of white spot syndrome virus (WSSV) and the Monodon Baculo Virus (MBV), shrimp farming has become extremely risky. Farmers resort to desperate remedies in containing the disease problems. This was compounded by the fact that there were tens of thousands of farmers, often ill informed, who had no idea of the means of containing the disease. Extension services were found to be inadequate. This led to the use of antibiotics like Chloramphenicol and Nitro furans. As a result, India has been facing rejection of its shrimp consignments in the European Union from the year 2002 onwards. Japan, another major exporter/importer shrimp from India has also introduced screening of shrimp for Nitro furans from May 2006 onwards. Often, rejected containers have been destroyed in the European countries and this still continues in countries like UK. The risk associated with rejections also had an impact on the prices offered to farmers.

It is in the area of non-tariff barriers, especially those relating to residue level of chemicals and antibiotics that India has had to face the biggest challenge to the sustainability of its aquaculture operations. To cope with these problems the country has adopted both short term and long-term measures. Firstly, India has introduced legal measures to ban the use of antibiotics in farms, hatcheries and processing plants through notifications issued under the Export Inspection Council Act 1963. Under these measures, maximum residue limits have been fixed. Certain antibiotics and pharmacologically active substances have been banned.

Response of the Government and the Farming Sector

1. *Selection of good quality seeds:* Disease diagnostic laboratories established by the Govt. guided the hatchery operators and the farmers for the selection of pathogen free brooders and seeds respectively. This included the dot-blot test to start with and the advanced PCR technology to follow on. Hatcheries were also had to identify good quality brood stock and the concept of single spawner in single tank etc. to ensure quality seeds.
2. *Farm management measures:* Restriction was imposed by the Govt. on stocking densities in the shrimp farms, so that higher stocking densities do not cause stress to the stock which eventually could invite the disease causing pathogens. Water quality management, feed management, and health management of the stock were also given due consideration.
3. *Modern concepts in aquaculture:* From higher productivity and profitability a paradigm shift was encouraged towards sustainability of aquaculture operations. In order to achieve this, usage of antibiotics in hatchery production cycles as well as farming operations are totally discouraged. Instead pro-biotic mode of operation was given to thrust and promoted to ensure good health and quality. Further, use of reservoirs to pump in the water and treatment of effluents through effluent treatment units were also encouraged. Closed system of farming operations is also being promoted to sustain the farming activity. Farmers were also encouraged to plant mangroves in and around the farms to maintain environment sustainability.
4. *Impact assessments*: While continuing the promotional impetus, organisation such as MPEDA have been also involved in environmental impact assessment to review the situations. For example, MPEDA had entrusted a study with National Environmental Engineering Research Institute (NEERI) to investigate the environment impact associated with the shrimp farming activities in the coastal region. More recently, MPEDA took up a study with NACA, Bangkok for shrimp disease control and coastal management in India. The Indian Council of Agriculture Research Institutes such as Central Institute of Brackishwater Aquaculture.(CIBA) and Central Marine Fisheries Research Institute (CMFRI) also undertaking environmental studies to protect the cause of coastal aquaculture.

Apart from the above initiations from the Government level, the farmers themselves have adopted several measures to counter the disease recurrence and to ensure sustainability in their farms. The field level trial attempts by the farmers for adopting various pond management measures are also reported to be highly successful.

The Marine Products Export Development Authority (MPEDA), has organized village level campaigns against use of antibiotics. So far, more than 2000 such campaigns have been organized in the country. This has created tremendous awareness among farming communities. *Residue Control***:** To comply with the EU requirements, India has put in place a National Residue Control Programme (NRCP), in which the MPEDA, through its field offices, collects shrimp samples from farms, hatcheries and processing plants and get them analyzed for the presence of the notified residues. To make this possible, the MPEDA has set up three state-of-the-art laboratories, having the latest equipments like HPLC MS MS, HPLC GCMS, ICP, etc. These labs have received National accreditation. MPEDA has positioned 11 HPLC MS MS laboratories in various coastal States and spent about USD 5 million for establishing these labs. Further, exclusive recruitment and training of technical staff have been made for running of these states of art laboratories. The MPEDA has been carrying out the NRCP satisfactorily during the last three years. This also forms as a surveillance programme for training the presence of antibiotic residues in cultured products and helps in taking corrective steps in farming operation.

Owing to the measures adopted by the country there has been significant reduction in the number of rejections at the EU borders (Figure 12.8).

Figure 12.8. Consignment Rejections at the EU Borders since 2002

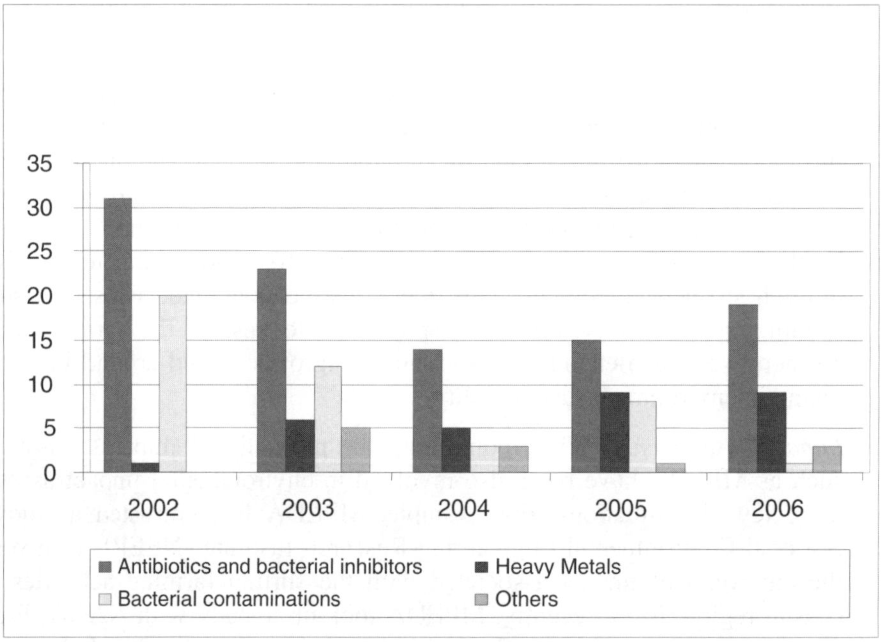

Source: MPEDA

Medium and long term measures

Recognising the need to attack the problems at its roots, the MPEDA feels that the use of antibiotics has to be dispensed with right from the hatchery level. The Authority has, therefore, launched a programme to convert two of its own hatcheries for pro-biotic mode of operation; without the use of banned antibiotics. This technology is expected to be transferred to all the shrimp hatcheries in the country in the near future. The MPEDA has also been empowered by the Coastal Aquaculture Authority of India to register hatcheries. The registration system is expected to discipline the hatcheries. The government of India has also made registration of the farms by the Coastal Aquaculture Authority mandatory. The Coastal Aquaculture Authority has the power to close down the farms, which violate the norms and guidelines laid down for farming.

The MPEDA has also addressed the root cause of the anti-biotic problem, *viz.* disease brood stock of Black tiger shrimp. It is noticed that almost 50% of the brood stock of *P. monodon* found in the coastal areas have been infected with the WSSV virus. Diseased brood stock transmits the virus vertically to the progeny and, if such diseased progeny is used for farming, the farmer runs the serious risk of loss of crop. Therefore, the solution lay in producing specific pathogens free brood stock of *P. monodon*. The MPEDA, through its research arm, the Rajiv Gandhi Centre for Aquaculture (RGCA), has taken up a comprehensive project in the bio-secure Andaman & Nicobar Islands to produce domesticated SPF brood stock. This project is expected to cost about 5.5 million USD.

It is interesting to note here that in keeping with the trend of globalisation, it is a few US Companies, which have monopolised the SPF brood stock of both *L. vannamei* and *P. monodon*, which are not cultured on any significant scale in the USA. It served the commercial interests US Companies a great deal when transcontinental movement of *L. vannamei* was effected.

In India free import of seeds or Broodstock is forbidden. The National Committee on Introduction of Exotic Aquatic Species under the Ministry of Agriculture has to screen any such proposal thoroughly before any permission is granted.

Traceability

The requirement of traceability as an integral part of HACCP and the food safety system of the buyers has also put strain on the export of cultured products. While India has welcomed the introduction of traceability and has taken steps to introduce the system in its processing plants, the enormous problems faced by the traceability system on account of the structure of India's aquaculture industry has often not been given the attention it deserves. Traceability, which can be readily implemented in integrated production and processing units faces serious challenges when applied to thousands of small and marginal farms, because they often sell their products to intermediaries and rural markets.

The primary requirement to achieve traceability at the farm level is a comprehensive database using information technology. A Geographical Information System (GIS) with the use of Satellite Imagery is being attempted by MPEDA in association with the Andhra Pradesh State Remote Sensing Agency with the technical support of the National Remote Sensing Agency (NRSA). This project has already been initiated to survey and map the vast areas under coastal aquaculture in the State.

The twin challenges of environmental sustainability economic viability

Globalisation has brought in its wake environmental concerns. Consumers and buyers have started asking questions about the sustainability of the process by which seafood is produced in aquaculture systems. The impact of trade on environment has always been a matter of concern. Under its influence, retailers and supermarkets have sought to differentiate the cultured products by giving eco-labels to these. Sustainability is sought to be attained through the market mechanism.

The challenge of environmental sustainability, the problem of antibiotic residue and the need to reduce cost of production for sheer survival has led to some innovative thinking in India, which has culminated in the introduction of the system of Participatory Farming by the formation of Aqua Clubs. In the year 2002, the MPEDA and the Network for Aquaculture Centres in Asia Pacific (NACA) joined hands to popularize sustainable farming practices through the introduction of Better Management Practices (BMP). The experiment began with the epidemiological survey conducted as a part of the MPEDA-NACA programme, based on the FAO Code of Conduct for Sustainable Aquaculture

Practices. The study emphasized the need for maintaining pond bottom and water quality, screening of seeds, and optimization of feeding, low stocking density and minimization of water exchange. The experiment, which started in the village of Mogalthur in Andhra Pradesh, organized the shrimp farmers in to groups based on the proximity of their farms with a view to co-ordinating the farm activities in a participatory manner. These groups, which are self-help groups or Aqua Clubs, were found to be extremely powerful in educating the farmers on collective adoption of BMPs. This was also a process of empowerment and inculcation of self-reliance..Instead of the Government officials playing the role of an extension agent, the farmers themselves built up the system of knowledge transfer and collective adoption of practices.

The aforesaid village demonstration programmes expanded in the subsequent years to 28 groups covering an area of 813 ha involving 730 farmers. The programme was extended to States like Orissa, Gujarat, Karnataka and Tamil Nadu. The results of this experiment demonstrated the following:

1. *Successful crops resulting in improved profits*: The project was undertaken in remote village in Andhra Pradesh where mainly small and marginal farmers were operating the farms and losing the crops periodically owing to disease outbreaks. However, owing to the adoption of Better Management Practices (BMPs) as per the directions of MPEDA – NACA project, these farmers could raise successful crops resulting in profits.
2. *Reduced disease prevalence*: One of the most significant outcomes of this project was a reduction in disease prevalence in the members belonging to. aqua clubs. (demo ponds). The decrease in disease prevalence and increase in planned harvest lead to betterment of overall crop outcomes.
3. *Aquaculture operations, without using antibiotics*: No antibiotic was used in the aqua club shrimp farm operations. The use of other chemicals was also discouraged. Farmers are encouraged to prevent the disease problem rather than treating it. Random sampling in farms of aqua clubs did not reveal any trace of the banned antibiotics.
4. *Traceability through Aqua clubs*: Cluster farming through aqua club/ aquaculture societies provides a novel approach of producing high quality traceable BMP shrimp. Co-operation and co-ordination among various stake holders such as hatchery, Nursery, farm and processors is necessary to sustain an efficient traceability system.
5. *Increased cooperation among farmers*: The aqua club activities have increased the interaction among farmers and provided opportunities for mutual help and assistance. Regular exchange of information and sharing of knowledge was made possible through the inter action within their groups, causing increased awareness. This resulted in stronger bargaining power for the club members to purchase the inputs as well as to sell the products. The quality of inputs are also given due importance by the members.
6. *More interaction between farmers and stakeholders*: The MPEDA – NACA project has successfully introduced the concept of "contract hatchery seed production system" to meet the quality requirement of seeds.while procurement. This enabled the adoption of BMPs in the hatcheries for production of good quality seeds and its marketing at affordable rates.

7. *Increased awareness on environment*: The educated members of the aqua clubs are fully conscious of the need for protecting the environment for sustainable aquaculture operations. The motivated farmers have resorted to planting of mangrove plants around the farms. This could act as barrier during cyclones and protect the coast line as well as the farms.

8. *Better price for BMP shrimps*: The BMP shrimp produced by the aqua club farmers can gain a better price in the market, as there is an assurance on quality.

Overall, the MPEDA-NACA Project could demonstrate reduced level of disease incidences while increasing the efficiency of resources such as feed, seed, fuel and finance. Further, the environment risks were also reduced owing to the sustainable management measures adopted.

In order to induce all the shrimp farmers to adopt BMPs developed through MPEDA-NACA programme, the MPEDA has taken up a scheme to institutionalize the concept. Farmers are encouraged to form societies for adoption of code of practices for sustainable farming. The BMPs and code of Practices are based on the FAO and the GAA Guidelines for responsible aquaculture. Based on their commitments to Code of Practices, the MPEDA registers them and extends financial assistance for sustainable practices. Such societies are expected to be established countrywide and will be part of MPEDA's 'Bottom up' approach to promote environmental sustainability and economic viability and full traceability. These societies also have the potential to emerge as producer organization/companies, with better bargaining strength in the market place, under fair trade practice.

In order to strengthen the society-based farming system and to build capacity among the farmer groups for adoption of the latest innovations in farming, the MPEDA has set up an apex outreach organisation called the *National Centre for Sustainable Aquaculture* (NACSA). The NACSA will adopt a highly participatory system of training group coordinators, who will transmit knowledge to the farmers.

Organic farming

The adoption of BMPs provides tremendous opportunities for farmers' societies to graduate to organic farming. The MPEDA has entered into a technical collaboration programme with the Swiss Import Promotion Programme (SIPPO), to built capacity among the farmers for production and marketing of organic shrimp and scampi.

The challenge of alien species displacing native species

As mentioned earlier, the unprecedented increase in production has put an enormous pressure on the price of shrimp. The distinction between species is slowly vanishing and shrimp is emerging as a commodity. The *L.vannamei* shrimp with its SPF character and lower cost of production has emerged as a cheaper option to those who need to produce counts of 31 and above. The corresponding sizes of Black tiger shrimps are not so cost-effective. The consequent spurt in demand for *L.vannamei* has prompted private industries to clamour for the introduction of the species in India without adequate regard for the bio security risks associated with the transcontinental introduction of the species. The Government of India has given permission for two private companies to import SPF stock of *L.vannamei* on a pilot scale. This is a new challenge that countries like India have to cope up with. Should India introduce an alien species to satisfy certain segments of the trade, disregarding the threat of introduction of the Taura syndrome virus? So far India has stuck to the stand that a thorough risk assessment has to be done, before the species is allowed to be imported freely.

Legal framework for regulation of aquaculture

Responding to the challenge of establishing sustainability in coastal aquaculture operation, the government of India has enacted the Coastal Aquaculture Authority (CAA) Act and made it effective in December 2005. Under this Act, all farms situated in coastal areas are brought within the regulatory power of CAA. Coastal area include areas of land within a distance of 2kms from the high tide line of seas rivers creeks and backwaters Farms are required to be compulsorily registered under the CAA Act, which does not permit any new farms to be established within a distance of 200 meters from the high tide line. The CAA also has the power to lay down detailed guidelines for carrying out coastal aquaculture under section 3 of the CAA Act. These guidelines are intended to assist and guide the farmers in adopting Good Management Practices Presently traditional/improved traditional and scientific extensive shrimp farming are permitted in the coastal areas as they involve low stocking densities and limited application of supplementary feed or fertilizers. The advanced technology involving semi intensive and intensive culture operations are not recommended as they involve higher stocking density as well as larger quantities of feed and fertilizers. One of the major stipulations is that the stocking density of post larvae cannot exceed 10 per square metre. This is, by and large followed in the country. The approval of farms having an area of 5 hectares or above is a three- tier process. First a district level committee comprising representatives of technical departments including the department of environment inspects the farms and sends its report to a state level committee (SLC). The SLC finally recommends the farm to the Coastal Aquaculture Authority.

The mandatory guidelines for selecting a new site for coastal aquaculture activity involve the following

- Mangroves, agricultural lands, salt pans, ecologically sensitive areas such as sanctuaries, marine parks etc should not be used for coastal aquaculture.
- The aqua farm should be located at least 100 metres away from any human settlements.
- To maintain a distance of 100 metres from the nearest drinking water source.
- The farm should not be located across natural drainage canals and flood drains.
- To maintain a distance of 50 – 100 metres from the nearest agriculture land.
- Spacing to be provided between adjacent shrimp farms.
- The water area of the farm shall not exceed 60% of the total land.
- New farms to be permitted in already farming areas only after studying the carrying capacity.

Some of the ancillary activities like shrimp hatcheries and shrimp feed mills are also regulated. The authority is also vested with penal powers and can punish violators with fines or imprisonment which may extend to three years.

The challenge of diversification

With the changing product profile in markets, and the multitude of ready-to-eat / ready-to-cook fish based products which crowd the retail outlet and supermarkets, the challenge facing India is to diversify. The success stories of countries like Vietnam (*Pangasius* spp.) And China in diversifying to new species has attracted the world-wide attention. As a consequence, there is tremendous pressure on India to diversify into finfish species. However, diversification in the short term is difficult because of technological reasons. The MPEDA, in order to develop new technologies has embarked upon several R&D projects through its research arm, the Rajiv Gandhi Centre for Aquaculture, targeting species like Indian Sea bass, the Mud crab, Groupers, Cobia, etc. Development of new technologies will throw up new opportunities like cage farming, for which India has enormous potential..With coastal land becoming more and more expensive, India has to diversify into offshore cage farming of marine finfish. India has the potential to emerge as a major cage culture country in the near future.

India exports significant quantities of Mud Crab (*Scylla Serratta*) in live form to the South East Asian countries .Crablets are collected from mangroves and estuaries, fattened in tanks and exported in live form. This provides livelihood a large number of people including rural women. The MPEDA embarked upon a programme to produce crab seeds on a large scale in its hatchery and has been highly successful. This is a break-through considering its usefulness in relieving the pressure on the wild stock of crabs .This is indeed one of the success stories in India's march towards sustainable aquaculture.

Value-addition

With the changing complexion of the markets and the increasing importance of the supermarkets and the retail chains, the Indian aquaculture has to gear itself up for the production of ready-to-eat and ready-to-cook products. This calls for considerable investments in the processing sector, which needs technological up gradation for production of value-added products. The flow of FDI into the fishery-processing sector in India has not been very encouraging. However, with the prospect of aquaculture production expanding and diversifying, FDI can become an engine of future growth for the processing industry. However, market access to the retail chains is still a formidable problem as the expenses involved in display warehousing and promotion are beyond the reach of the Indian processors, who are mainly in the small and medium sector. India's products need to gain better Brand Equity in the international markets.

The future promotion efforts of the MPEDA will therefore, have to be in the following areas:

1. Promotion of brand equity of Indian products
2. Promotion of India has a seafood-processing hub.
3. Negotiation with buying countries for lower tariffs for value-added products.

Thus, the foregoing account illustrates how a fast developing country like India has had to face enormous challenges in the area of aquaculture and how the institutions in the country have responded to these challenges.

Opportunities

While the challenges of globalisation may have spelt trouble for the stakeholders of the aquaculture industry, they have nevertheless presented opportunities of different kinds. The key opportunities are: *(1)* the increased demand for various fishery products and the opening up of markets; *(2)* the opportunity to diversify aquaculture; *(3)* the opportunity to benchmark practices with the most advanced in the world; *(4)* capacity building; and *(5)* the acceptance of sustainability as an integral part of aquaculture practices.

The increased demand for various fishery products offers considerable trading opportunities to the exporter and production opportunities for the aquaculturists. It is projected that by 2020, 41% of the finfish consumed in the world will come from aquaculture (*Fish to 2020,* International Food Policy Research Institute 2003). Countries like India, blessed as they are with abundant natural resources like land and sea for aquaculture see in this a great opportunity to expand activities. Besides expanding land-based aquaculture, the country also has enormous opportunity for promoting cage culture in offshore areas, estuarine waters, reservoirs, rivers etc. Along with this expansion, comes the opportunity for diversification of species. As mentioned earlier, India has tremendous potential for developing the Indian Seabass (*Lates calcarifer*), the Groupers, (*Epinephelus spp*), Cobia and a variety of other marine fin fishes. Research and development are already underway to adapt technologies suitable for the farmers. Efforts by private entrepreneurs to bring in technology for species like tilapia etc. have begun. In fact, this diversification exercises is part of India's vision for producing at least one million tons of brackish water species by the year 2012 in a sustainable way.

The changing profile of products and the emerging power of the retail segment are opening up opportunities. Entrepreneurs can access these markets by diversifying into ready-to-eat and ready-to-cook products, provided they are able to overcome the entry barriers like the high investment required, gaining certification from reputed international agencies like the BRC, etc.

It is true that SPS requirements as well as technical barriers have placed considerable strain on the exporters and producers. However, these barriers have also afforded the opportunity to upgrade the systems against the benchmark of international standards. India's current experiment in upgrading its laboratories to international standards is a clear example of this development. The MPEDA has itself has upgraded its 3 laboratories to ISO 17025:2005 standards and become state-of-the-art. The PCR testing system for disease diagnosis in the country is currently undergoing a process of up-gradation through a technical collaboration arrangement between the MPEDA, NACA, the Australian Centre for International Agricultural Research (ACIAR), and the Central Institute of Brackishwater Aquaculture, India. There are a number of other examples of this kind. Up-gradation to international standards has also been accompanied by the high degree of capacity building among technical staff of hatcheries, farms and laboratories and the process is continuing. The urge to learn the latest and the most sophisticated systems is intense and to facilitate this, frequent training programmes abroad in the best laboratories in the world has become a matter of course.

With the emergence of sustainability as a major issue in aquaculture, India has been in the forefront in the promotion of sustainable aquaculture. It was one of the first countries, which experimented with the BMPs and the 'aquaclub' concept through the MPEDA-NACA project, which started in the year 2002 a time when aquaculture certification was still in its infancy. With sustainability gaining strength, Indian aquaculture is set to undergo the substantial change in the direction of cluster-based farming adopting good aquaculture practices. The issue of organic certification has also gained popularity. Effluent control in bigger size farms has become a norm. India is currently attempting the ultimate in sustainable farming namely producing the SPF brood stock and domesticating the same for long term sustainability of its brood stock population and conservation of marine resources. It is also expected that with the pressure on fish meal resources of the world, production of feed with less and less of fish meal would also capture the attention of Indian research institutions.

Chapter 13

Aquaculture in a Global Age – The European Perspective

Richard Bates, DG Fish, European Commission

Global aquaculture production experienced an annual growth rate of 9% between 1995 and 2004. However, over this same period, European aquaculture production only increased by 3% annually and has stagnated since 2000. Decreasing prices as a result of global competition, plus limited availability of space in coastal areas and complex environmental legislation, have all contributed to this situation.

Although fisheries are an exclusive Community competence, competence for aquaculture is shared with the 27 Member states. Richard Bates highlighted that aquaculture in the European Union is governed by the following policy areas:

- *Environment: water quality, biodiversity.*
- *Health and welfare: hygiene regulations, Animal Health Directive.*
- *Food Quality: organic farming regulations will be applied in 2009.*
- *Structural Policy: to support economically restricted areas and improve diversification.*
- *Trade policy: liberalisation of trade in fish products, tariffs.*
- *Common organisation of the market.*
- *Research policy: technology in aquaculture, consensus guidelines for sustainable production, food safety.*
- *Maritime policy: an area that is under development.*

In 2007, a review will get underway to establish whether the EU can be more proactive in providing a better framework for European aquaculture development as a follow-up to a Commission Communication in 2002 on a strategy for the Sustainable Development of European Aquaculture. In this respect there is a need to address the image problem of aquaculture production that in many places is seen as having negative environmental effects (e.g. use of medicine, escapees, fish welfare).

This presentation will introduce aquaculture governance and policy in the EU by considering relevant case studies on salmon and sea-bass. The presentation will end with a view of the value chain and the way forward.

The EU is celebrating its 50th anniversary. The 1957 treaty of Rome was the starting point for the European Economic Community, now the European Union. The original treaty is now called the EC Treaty. A second treaty on the European Union considers other issues such as security policy, justice and home affairs *i.e.* non-economic issues. A proposal has recently been submitted to merge these two treaties into one single piece of legislation, although this proposed Treaty establishing a Constitution for Europe has been opposed by the populations of France and the Netherlands. The EC has an overall pro-globalisation policy, although some Member States see this as advancing too fast. Nevertheless, the peace and prosperity that the EU has brought is still highly appreciated.

The EU began with 6 Member States, which became 27 in 2004. Ten member states are former Warsaw Pact countries. The EU is characterised by four freedoms; the freedom of the movement of people, goods, services and capital within the borders of the EU. This initially raised immigration concerns which have reduced. Presently, the EU is the world's leading trade power with a single market of almost 500 million people. In addition to the 27 EU Member States, three other states are members of the single market. These European Economic Area (EAA) countries are Norway, Iceland and Liechtenstein, who by virtue of the EEA treaty of 1995 are closely associated with the EU but are not members.

The EU works for fair trade and seeks to harness globalisation, particularly through the World Trade Organisation. The EU aims to ensure that its trading partners, particularly in the developing world, can join the trade system, thereby ensuring a better world. The EU's globalisation policy has been summarised in a document published last year, titled, "Global Europe Competing in the World" (available at the homepage of the Director General for Trade at (www.europa.eu.int). This document shows how the external agenda for creating opportunities in a globalised economy ties in with the internal agenda, Lisbon Strategy for Growth and Competitiveness, with aims to further economic development in the EU.

Although aquaculture continues to grow worldwide, capture fisheries have levelled off, declining by 10%, at EU level, in recent years. Meanwhile, global consumption of fish doubled in the period from 1973 to 2003. The FAO predicts that aquaculture will provide the majority of seafood to meet growing per capita consumption in the years ahead. The big question is how aquaculture can benefit from this perspective while growing sustainably. Recent statistics show that aquaculture constitutes 17% of the volume of EU production and 27% of value. The main producing countries are France, Spain, Italy, the UK and Greece. The three branches of aquaculture in Europe include marine fin fish farming, fresh water fin fish farming and a traditional shell fish sector where oysters and mussels in particular are grown. This sector has not grown by much in recent years. The challenges are decreasing prices due to high global competition, the limited availability of space in coastal areas and complex environmental requirements.

In the Treaties, fishing is an Exclusive Community Competence. The Commission is the executive body that proposes legislation. However, fisheries rules are generally decided by the Council of Ministers and in other areas of legislation by the Council with the European Parliament. Aquaculture, however, is a shared competence between the EU and Member States but an EU policy statement does exist: Communication from the Commission of 2002, "A strategy for sustainable development of European aquaculture".

A review is currently underway to see if the EU can be more proactive in providing a better framework for aquaculture development. Although there has been an annual growth rate of 9% globally over the last 10 years in aquaculture, this has not been reflected in the EU where, from 1999, aquaculture only had a 3% growth rate and has been stagnating since 2002. Marine fin fish farming has grown slightly, to equal tonnage from fresh water farming, aided by a small decline in freshwater fish farming. Shellfish farming has been more or less static.

A number of different policy areas impact on EU aquaculture. EU environmental policy has a strong focus on sustainable development, preservation of water quality, and shortly, the use of alien species. The latter is to install a permit system for new species brought into Europe to protect biodiversity and to ensure that in the rush to expand, develop and diversify, irreparable damage is not caused to biodiversity. Health and welfare policies also affect aquaculture. Directives were originally introduced in 1991. Regulations are different from Directives in that they are applicable in the actual original wording in all of the member states. As the EU has 23 official languages, identical legislation will exist in all these languages. Instead, Directives are a means to achieve a goal and subtle differences may exist in legislation between different Member states. Recently, a new set of legislation has been agreed that updates and simplifies the original legislation, for example, in the areas of food hygiene (three new Regulations) and of animal health (a new Directive). The new legislation for hygiene applied from 2006 and for animal health will apply from 2008. Current health legislation has been closely aligned to the CODEX Alimentarius and animal health legislation with the OIE (International Organisation on Animal Health). On welfare issues, the Commission has been working in tandem with the Council of Europe - a non EU body comprised of 40 Member states – that has a convention on farming and the welfare of animals used in farming. However, the Council of Europe has slowed its work in this area. The Commission has requested scientific advice from the European Food Safety Authority on different aspects of animal welfare for aquaculture. More general food quality issues are generally determined by industry. The Commission does have new legislation on the way for organic farming and this will deal with aquaculture, including seaweed. The legislation, due to come into force in 2009, will be applicable to all production, both from the EU and any third country wishing to sell products as organic in the EU.

Structural policy supports the development of aquaculture in the Community. A particular focus for the programming period 2007-2013 is on diversification and also on aquaculture environmental measures. Support is primarily for small scale industry.

In trade policy and the related common organisation of the market, there is a general trend towards liberalisation of trade policy. The existing scheme will be modified in line with the World Trade Organization, where negotiations have currently stalled. The common organisation of the market was originally based on rules for the vegetable market, introduced in the early 1970s. Current rules date from 2000 but have the complex job of balancing the needs of primary producers with the needs of other actors in the food chain.

Research policy is very important and the EU currently spends about EUR 20 million per annum on aquaculture research. The Seventh Framework Program on Research was launched in March, 2007. The EU is a leader in aquaculture technology and has a long history of public and private research in this area. The Commission's recent Green Paper on Maritime Policy, points towards better linkage between the various EU policies relevant the sea to ensure that their impact on each other is taken into account.

To illustrate the opportunities and challenges of globalisation in aquaculture two case studies are informative. The first case study is that of salmon - the most globalised sector of aquaculture. Norway was the originator of salmon farming and is still the dominant player. The industry has been marked by supply outstripping demand at certain periods. An ongoing dispute with Norway regarding levels of production has not yet been resolved and is currently at the WTO dispute resolution board. In the meantime, anti-dumping safeguard rules have been applied on and off for the more than a decade. Within the EU, the UK is the largest producer and Ireland has a limited amount of production. The industry is very globalised with Chile, Canada, Australia and the USA dominant players. As an example of its globalised nature, the top five companies in Norway produce 38% of total salmon production, while the top four companies in the UK (two of which feature in the Norwegian list) are responsible for 66% of UK production. Several European companies are also active in America, both North and South. An important issue for salmon is that the cost of regulation is often higher in the EU.

Sea bass and sea bream are an example of globalisation at the diagonally opposite end of the EU. Greece, Turkey (a candidate for EU membership) and Cyprus (now an EU member) are dominant producers. After a market crisis, the industry consolidated and the top 20 firms in Greece produce 80% of supply. Greek companies are diversifying, both outside and within Europe. For example, a Greek company is spending EUR 11 million on building a re-circulating sea bass and sea bream plant in Wales.

Finally, value chain issues: 68% of global wild catches are for human consumption. Significantly, almost one third is for fish feed manufacture. Aquaculture is the fastest growing public food sector in the world but has encountered an image problem. This is very important for the future of aquaculture as it globalises. Concerns include fish feed supply, use of medicines, contaminants, fish welfare, escapees, impacts on ecosystems and water pollution. This has evoked questions surrounding the need for a sustainability label linked to sustainable growth. This is something the EU is currently considering, although the presence of other labels is also something to think about.

To conclude, closer cooperation is required between all global actors within the aquaculture sector. Lessons can also be learnt from other sectors of the food industry, particularly the experiences of poultry and beef where a history of cooperation between the different sectors and actors along the value chain exists. Secondly, there is a need for a global effort to tackle the image problem, including better communication and the use of technological innovation to surmount problems. A better understanding of consumer demand is necessary. Consumers want seafood that is safe and wholesome and food safety cannot be compromised. And finally, there is a need to better distribute the benefits at every step of the food chain from farmer to retailer. Climate change is also something to consider; the issue is not simply one of food miles or how climate change can be averted, but what the impact of climate change is on aquaculture. For example, rising carbon dioxide levels in water prevents oysters and mussels from forming shells and a species may grow in a different geographic area in the future.

Chapter 14

Organic Aquaculture – Opportunities for Emerging Markets of the Environmental Challenge for Exporting to Europe

Markus Stern, Swiss Import Promotion Programme (SIPPO), Switzerland

Markus Stern stated that consumer preference is the driving factor behind the growth in organic aquaculture. In 2006 the worldwide market for organic food represented a USD 30 billion industry, and is booming. Market research has shown that the market is sustained by the perceived health benefits of organic aquaculture and changing consumer demand towards more healthy diets.

Organic aquaculture requires a system of inspection, standards and technical guidelines, intensive monitoring of impacts, integration of communities, processing according to organic principles, natural breeding procedures, prohibition of GMO material, limitations in stocking density, the use of organic feed and fertiliser, prohibitions in the use of pesticides and fertilisers, restrictions on energy consumption and the use of natural medicines only. To get fish farmers in developing countries into organic farming is a challenge. However Markus Stern thought that this could be achieved through collaborative approaches, through transfer of technology and with the support of aid agencies.

Shrimp is the most important product traded in the international seafood business and has been a success story for the past 25 years. Shrimp farming is a multi-million dollar business that supports a large number of associated industries and most of production comes from emerging markets.

Alongside increased volumes of production, trade and consumption, demand is also increasing for improved sustainability, social acceptability and human health and safety. This not only affects the international trading environment and puts pressure on producers to focus on their production methods in order to address these concerns, but also requires producing countries to develop and implement adequate and appropriate policies that provide an environment conducive for responsible production and trade.

Until recently, farmed shrimp had a bad image in Europe. In September 2001, the EU found chloramphenicol residues in Chinese, Vietnamese and Indonesian shrimp. Both the EU and US became concerned about antibiotic residues in shrimp products and began strict inspections for chemicals in all shrimp imported from Asian countries, while shrimp from China and Vietnam were banned.

More and more consumers demand information on the products they are consuming, including information such as health and the effect of production on the environment. For these reasons, demand for organic products is growing worldwide and is a market with tremendous prospects for growth. In particular, it offers creative producers and exporters from developing countries some excellent opportunities for success. The greatest opportunities lie with premium shrimp. Premiums exist for:

- Prohibiting the use of antibiotics,
- Cooling, storing and delivering the best shrimp to market,
- Informing the buyer of farm standards.

Image, particularly product image, is crucial to the success of a company. Reputations can be destroyed within seconds, but take years to build again. It costs a premium to build an image for a shrimp producing country and to inform the consumer accurately. In other words, there is a premium for ensuring healthy products. This premium is paid by satisfied consumers.

What is the most important thing for a company to succeed? It could be the dynamism of the managing director, the motivation shown by its employees, a new outstanding product or the company's healthy financial resources. While all of these factors are important, the most important thing for the success of a company is to have customers.

Sustainability marketing is today's buzzword to win over customers. Future success in business is determined by three aspects: ecological, social and consumers. All three contribute to management success and a good business result. One without the other will no longer bring success in the future. However, the driving factor is always the consumer. This is reflected by new market trends that are leaning heavily towards sustainability marketing. The European population is becoming increasingly conscious of natural and fair production. Nobody wants products that damage or pollute the environment and that are too far removed from what is natural and original and where workers in the production process have poor working conditions and/or are poorly paid.

Marketing is everything that promotes or hinders sales. It includes negative headlines in the press, disregard for the environment, price wars, poor working conditions in aquaculture farms and factories, disregard for hygiene regulations, poor quality and uncertainty about a product's origin. The markets, and therefore requirements, are changing. Providers must adapt to market changes if they want to succeed. If all these activities fail to harmonize, the result will be poorly functioning gear wheels; in some instances, they will be going in opposite directions. Such an export machine will never work.

The rules for export markets and particularly European countries are a labyrinth for exporters, especially emerging markets and markets in transition, as they are fraught with an ever-increasing number of pitfalls and new requirements. These include European norms and standards, quality requirements, certificates, traceability and design and market trends. There are also import regulations specific to each country such as charges, quotas, social accountability, environmental protection, health and safety regulations, total quality management, etc. In terms of import conditions and import requirements in particular, the following points are of utmost importance nowadays:

Health and safety

(To ensure zero risk for the safety and health of consumers): HACCP, Good Agricultural Practice (GAP), Good Manufacturing Practice (GMP), EurepGap, IFS, BRC, organic certificates, phyto-sanitary certificates.

Patents

Social accountability

SA 8000, Fair trade (IFAT, IFOAM, Max Havelaar)

Environment / waste management

Green dot (some countries), ISO 14001, Eco-labelling, FSC, MSC

Quality management

ISO 9000 series, ISO 22000

Import duties

Levies (anti-dumping), taxes, VAT, import licensing, quota, bans.

Standardisation

Technical specifications, performance of goods and services

Organic production is one of the cornerstones of the future and the international market for organic food is booming, with a value of approximately USD 30 billion in 2006. The two largest markets for organic food are the United States and Europe, with sales of around USD 14 billion in each, followed by Asia and Oceania. The organic market is no longer a niche market; organic products are now commonly available in nearly every big supermarket in Europe. Neglecting the organic market sector would be a strategic mistake for two reasons:

- there is a tremendous growth potential in the global organic market; and
- change is occurring in consumer behaviour globally.

The organic market is open to any kind of fish and seafood from aquaculture farms, such as Salmon, Shrimp, Tilapia, Cod, Trout, Carp, Mussels, Seabrass/bream, and Pangasius/Catfish.

Why do consumers prefer organic products?

- ¾ of consumers prefer organic products because they are health-conscious;
- 60% of consumers prefer organic products for environmental reasons;
- ¼ of consumers prefer organic products because they taste better.

In this context, it is very important to know that the percentage of consumers who buy organic products with the aim of making a personal contribution to environmental protection has increased in recent years. For successful marketing, it is important to know the market. How many consumers buy organic products?

- 20% of consumers regularly buy organic products;
- almost 50% of consumers occasionally buy organic products.

This adds up to 70%, or more than 2/3 of the population - around 340 millions of consumers in Europe. However, the European consumer is also very price-sensitive:

- Between 50 and 60% accept a surcharge of more than 15% for organic products;
- 1 in 3 accepts a surcharge of up to 15% for organic products;
- Only 10% do not accept a surcharge for organic products at all.

Organic aquaculture is not just a small change of production. It is more than that – it is a process and a philosophy requiring standards and technical guidelines. Organic inspection and certification includes the whole chain, from the prime material to the retailer, including hatchery, storage, processing and transportation, in several steps:

1. Exchange of information.
2. Pre-evaluation visit.
3. Inspection.
4. Contracts.
5. Certification.

An internal control system must also be introduced. The relevant certification committee decides on the admission and certification of new farms. It also decides on the annual renewal of certification. Following EU regulation 2092/91, organic products from third countries must be produced, processed and certified in accordance with equivalent standards. In the future, independent bodies for: *training, inspection* and *certification* may be necessary.

Recommendations include:

- Regional Knowledge/Training Centers.
- Regional Certification Bodies.

SIPPO is the Swiss Import Promotion Programme under the patronage of *SECO*, the State Secretariat for the Economy of the Swiss government. SIPPO supports private businesses in emerging markets and markets in transition that are endeavouring to access the Swiss market as well as markets in the European Union. The aim of such economic and trade promotion is to support emerging markets and markets in transition in their attempts to integrate into the global economy. SIPPO services include business branch-related market information, advisory services for products and marketing, promotion in Switzerland and assistance at selected European trade fairs. The SIPPO programme is part of Switzerland's development and foreign trade policies. It complements other technical cooperation programmes that have elements of trade.

SIPPO has been involved in several Organic aquaculture projects since 1999:

Table 14.1. Organic Aquaculture Projects

Vietnam	Black Tiger shrimp	1 198 farmers (1 farm)	Certified
Ecuador &	White shrimp	4--5 Farms	Certified
Peru	Trout	1 Farm	Certified
	Scallops		In process
Bangladesh	Black Tiger shrimp	1 000 farmers	In process
			300-500 certified 2007
India	Black Tiger shrimp & FW Prawns	1 500 – 3 000 farmers	Just started
Indonesia	Not yet defined		Planned
Mozambique	Not yet defined		Planned

Vietnam: Organic certification (BIO SUISSE and Naturland) since 2004. Buyers are mainly from Switzerland and from Sweden, but interest also comes from other countries such as Germany, France and the USA. Organic Pangasius from Vietnam are also available in COOP Switzerland (PPP – GTZ/Binca).

Ecuador: White Shrimp (P. Vannamei). Market access support through SIPPO. Buyers are mainly in Switzerland, UK, Germany and Canada/USA.

Peru: White Shrimp (P. Vannamei). Market access support through SIPPO with Ecuador (COOP and MIGROS Switzerland). Trout: one farm in the lake of Arapa region ("Andes" Highlands) started the process with support from PROMPEX after workshops with SIPPO in Peru in 2002. It has been organically certified since 2005. Scallops: during the process, organic guidelines were agreed by SIPPO in accordance with Naturland guidelines.

India: Project just begun. Areas include the state of Kerala and the state of Andrah Pradesh. Enlargement to West Bengal and other states by MPEDA is foreseen.

Indonesia: Awareness creation carried out through workshops/meetings, projects in planning.

Mozambique: Awareness creation carried out through workshops/meetings, projects in planning.

SIPPO's main activities are in export promotion and match-making between exporter and importer. However, an exportable product with certificates is necessary. Therefore, SIPPO also has influence on the production side. SIPPO deploys other partners or partner organizations specialised in aquaculture or related technology. For sustainable organic production, local partners have to been trained and to be involved in the whole programme:

In *Vietnam*: the National Fish & Seafood Association (VASEP), the regional association CASEP and the Ministry of Fishery.

In *Peru*: direct partners are PROMPEX (Peruvian Export Promotion Organization) and farmers.

In *Bangladesh*: Close cooperation with the national NGO Shushilan, Bangladesh Frozen Food Exporters' Association and the Ministry of Fishery.

In *India*, the direct partner is MPEDA, the Marine Products Export Development Authority.

In 2000, SIPPO launched an organic shrimp project in the Ca Mau Province of Vietnam in order to favour sustainable and responsible shrimp farming with the German certifying body Naturland. Aquaculture is quickly expanding in Vietnam and the intensification of production of Black Tiger Shrimp in the South could lead to ecological disaster and mangrove destruction. 1 198 farms were converted to organic production and certified, enabling them to obtain a premium on top of the world market price for this certificate. ¾ of the premium is shared among the farmers, middle men and the factory. ¼ remains in an Eco-fund. This money is used to pay for annual inspections and certification, marketing activities, further training, re-forestation and other activities. In 2006, COOP Switzerland imported USD 3 million of organic black tiger shrimp from Camau. A 20% premium on USD 3 million is USD 600 000. ¼ of this *i.e.* USD 150 000 is for the Eco-fund.

Presently, the supply of breeders and post-larvae shrimp is from uncontrolled catches in the wild. So far, nature has been generous and catches of wild parents were sufficient to supply shrimp farms with post-larvae. Therefore, no considerable effort was made to master reproduction in captivity of *Penaeus monodon* in Vietnam and in South-East Asia. In Latin America, especially in Ecuador, hatcheries have already been built and organically certified. Due to the fast development of this industry, however, the availability of wild breeders has decreased significantly. The price for a wild-caught Black Tiger female shrimp is up to USD 400. This makes it very difficult for small hatcheries to purchase breeders. The sustainable development of this sector cannot be built on such unpredictable supply. It is for this reason that standards for organic shrimp production by Naturland require farms to be fully independent from wild caught post-larvae. The purpose of this project, therefore, is to develop shrimp larvae production based on breeders raised in captivity. The specific objectives of the project are:

- to master the artificial reproduction of Black Tiger shrimp within a controlled environment;
- to set up a system of breeder production on an experimental scale and to implement a technology transfer to allow Vietnamese producers to ensure their future;
- to introduce organic certification of pilot hatcheries in Ca Mau province and thus to supply organic shrimp farms with appropriate larvae.

Infofish

Infofish is an intergovernmental organization that markets information and provides technical advisory services for fishery products in the Asia Pacific region. Its head office is in Kuala Lumpur and it has recently started an organic aquaculture programme in

Myanmar, Thailand and Malaysia. SIPPO supports this programme with technical assistance and with import promotion in the EU and Switzerland.

There are four broad objectives for the present project:

1. To contribute to the sustainable development of the aquaculture sector by ensuring the safety of organic/ aquaculture products from the Asia-Pacific region.
2. To increase the knowledge about certification and marketing of organic aquaculture products.
3. To facilitate transfer of technology to small and medium scale sectors of production and marketing of organic / aquaculture products.
4. To encourage investment in sustainable, eco-friendly aquaculture and domestic marketing and export-processing of produce.

The end result of the projects would include:

- Consolidation and possible growth of the market share for organic/ aquaculture products in the major markets.
- Consolidation and growth of domestic markets and regional markets for organic/ aquaculture products from target countries.
- Gradual reduction in rejections and detentions of aquaculture exports from the three target countries and Asia-Pacific in general.
- Gradual increase in market prices for organic/aquaculture exports from the target countries; improved awareness of product safety/environmental aspects among all sectors of the aquaculture industry and necessary skill levels to address related issues.

As Table 14.2 shows, 12 aquafarms exist in three countries, with a total volume of more than 100 metric tons. In addition to Black Tiger shrimp farms, freshwater prawns and Tilapia are part of the organic aquaculture programme. This programme, managed by INFOFISH, should be a guide for all emerging markets.

There is a tremendous sales market in Europe, but not enough products. When entering the European market, the following aspects have to be taken in consideration in addition to the trend towards ecological products (Marketing factors):

Aging: The European population is getting older, which influences what products consumers buy.

Health: Europeans are very health conscious. All consumers are interested in seafood origin in addition to health benefits.

Wealth: European consumers can afford to buy food of good quality at acceptable prices.

Economic situation of the country: When the economic situation in a country improves, as in Europe, consumers are willing to spend more money, on more luxury products and lifestyle, e.g. in going out to eat. This also influences the shrimp market in restaurants.

Exotic products: To eat is not only a necessity, but also a pleasure. In Europe, people travel more and more. Therefore the demand for exotic food is increasing.

Table 14. 2. Details of Participating Companies and their Planned Organic Production

		Current Production Volume (mt/yr)	Potential organic production (mt/yr)		
			Yr 1	Yr 2	Yr 3
Myanmar					
Toe Myint Aung Co Ltd	- Tiger shrimp	5	3	5	8
Pale Nadi Co Ltd	- Tiger shrimp	30	3	10	20
Pyae Phyo Kyaw Company Ltd	- Tiger shrimp	14	3	5	10
Arsha Thai International	Freshwater Prawn	24	3	10	20
Thailand					
Samutsongkram R&D Center	- Tiger shrimp	1	1	1	1
Sureerath Farm	- Tiger shrimp	300	10	15	20
Bunjonk Farm	- Tiger shrimp	4	1.5	2	3
Surakij Farm	- Tiger shrimp	2	2	2	2
Malaysia					
Shrimp Farm, Sebatu, Melaka	- Tiger shrimp	40 (pot)	1.5	3	4
Brackishwater Aquaculture Research Centre, Gelang Patah, Johor	- Tiger shrimp	10	2	3	4
MoA Freshwater Prawn Farrm,, Sik,	Freshwater Prawn	10	2	4	6
Tilapia Cage Culltturre Farrm,, Terengganu	Tilapia	-	5	10	15

Finding the right segment: In response to consumer demand, import and trading companies make their purchase of shrimps more and more dependent on environmentally and socially compatible origins and provide this information to consumers. This is called product responsibility and product transparency. Consumers refrain from buying shrimp until they are given a cast-iron guarantee that the above-mentioned requirements are met. Potential producers and exporters of organic aquaculture products must focus on market research, target markets, species, quality, consumer groups and the demand of consumer groups.

Organic aquaculture products are an excellent market opener as demand is higher than supply. In addition, products are competitive and there is a price premium for the aquafarmer. As it is more difficult for an importer to switch to another supplier, firm contracts and long-term cooperation are the result. Natural production sites help to protect the environment and to improve the image for the producing country:

- Organic aquaculture production follows market trends in Europe.
- There are excellent opportunities for smaller producers in rural areas.
- Organic aquaculture production favours "natural" production methods.
- Organic aquaculture production can often be applied in developing countries through current production, but needs certification.

Threats:

- In the start-up phase there are some additional costs, for instance for inspection and certification.
- Production requires more awareness to obtain high quality products.
- Transportation costs might be higher due to smaller volumes.

COOP is the second largest retailer in Switzerland. COOP buys the entire harvest of the SIPPO organic shrimp programme in Vietnam. Nevertheless, demand continues to exceed supply and COOP cannot provide all of its retail stores with organic shrimp from Ca Mau. Other distributors and retailers in Europe are keen to buy organic shrimp.

Most producers in emerging markets, when asked why a Swiss importer would buy the product in question, would answer due to price. However, a Swiss buyer would respond differently:

- First, there must be confidence in the partner.
- Second, trust in the commercial and political structure of the country must exist, as well as the company or organisation and other organisations such as logistics companies, banks, etc.
- The buyer or importer wants to be informed about the place of production, about any changes in the company, in the quality, in delivery, etc.
- The fourth point is reliability, delivery time and delivery conditions, quality control, etc.
- Finally, the average buyer is interested in a long-term business relationship and is interested in future strategy, vision, business plan or marketing plans.

A combination of confidence, trust, communication, reliability and vision is what makes up the price. An organic aquafarm normally fulfills all these requirements, not just because organic seafood is a product, but because it is also a philosophy.

If companies or aquafarms wish to sell in Europe, they have to act like Europeans. In this sense, they must take the market seriously and give the market what it wants. If the product/company is aligned with the market, if ecological and social responsibility is a focus, if sustainability marketing becomes the vision and mission of the company and the central theme is one of communication, then lasting success follows.

Europe has almost 500 million potential fish and seafood consumers and consequently, almost 500 million opinion makers. Switzerland is located at the heart of Europe but is not a member of the European Union. However, Switzerland is heavily linked to the EU by various bilateral agreements. This means that Switzerland is taking on almost all of the EU's laws and decrees such as import regulations and food requirements. 70% of the European population regularly or occasionally consumes organic products – this is approximately 340 million consumers. If every third month they consume 200 grams, the volume per year will be 816 000 metric tons.

Building an image for a shrimp-producing country and providing consumers with accurate information has its price. Caring about the health of those who enjoy eating shrimp has a price. Creating loyalty among buyers and consumers for organic shrimp has a price. The price is the organic production, but also a premium from the buyer in Switzerland or in the EU. Organic shrimp farming is a sustainable form of aquaculture. Add social responsibility for ecological and economic reasons and it becomes one of the successful ways to go into the future. Furthermore, it is a tremendous opportunity for emerging markets.

Chapter 15

Meeting the Demands and Challenges of Globalisation of Trade in Aquaculture: The Role of a Regional Inter-Governmental Body

Sena De Silva, Network of Aquaculture Centres in Asia-Pacific (NACA)

Asia is the largest global aquaculture producer and exporter of aquaculture produce. Major exporters are Thailand, India, the People's Republic of China, and Vietnam. However, Mr. De Silva outlined that the bulk of the fish farms are small scale individual family farms, often clustered together, which also tend generate a lot of synergies.

The challenge for these small-scale aquaculture holdings is to ensure that they produce environmentally-friendly, socially-responsible aquaculture that meets food safety requirements and are profitable. The Network of Aquaculture Centres in Asia-Pacific (NACA) operates in the Asia-Pacific region to enhance capacity building through the sharing of expertise and communication through relevant programs conducted in conjunction with national organizations and farmer groups, as well as dissemination of information through electronic and other media. NACA helps to provide access to technical services and respond to stringent market requirements by developing Best Management Practices along the supply chain, piloting traceability systems, developing guidelines for certification, and connecting responsible small scale producers to markets.

Summary

The Network of Aquaculture Centers in Asia-Pacific (NACA), an inter-governmental organisation of 17 member nations and dedicated to aquaculture development in the region, has been in existence for over 18 years. It is the largest such body dedicated to aquaculture development globally, and represents nations that contribute well in excess of 80% to global cultured seafood supplies. Its primary mandate is to ensure sustainability of the sector with special emphasis on improving the livelihoods of small scale farmers, and contributing to poverty alleviation and food security in the region.

The Asian aquaculture sector, as in almost all the other primary production sectors in the region, essentially consists of small scale rural aqua farmers, family owned and operated, often congregated in clusters in given areas, and generating synergies and acting harmoniously. The aquaculture sector of many Asian nations currently contribute more to the GDP than the corresponding capture fisheries sector, and accordingly farmed seafood products are gradually bypassing the latter in global trade. However, with increasing demands on food quality, food safety, traceability and associated certification procedures the Asian aquaculture sector is confronting new issues which it has to comply to.

It is in the above regard, and in the context of the aqua-farming systems in Asia, that an organisation such as NACA, with its wide network of stakeholders, has a pivotal role to play in facilitating the global market needs. In this paper the above points are discussed in detail and examples are provided how NACA, as an inter-governmental organization, and in conjunction with other national and international organizations, has initiated, very successfully, programs of activities that would lead to small scale seafood producers in Asia to comply to global trade requirements and standards.

Introduction

Aquaculture is recognised as the fastest growing food production sector globally, and Asia continues to dominate this sector over the years. Currently the sector accounts for over 40% of all food fish (seafood) consumed, and also contributes significantly to global fish trade (FAO 2007a). It is also generally accepted that in view of the unlikely increase in capture fisheries, the short fall in food fish supplies to meet the demands of a growing population as well as increasing levels of per caput consumption in most countries (Delgado *et al*., 2003), will have to be met with through aquaculture.

Trade in fish and fishery products has increased globally. This change is also associated with a shift in production and consumption patterns, the major one in both instances being that the role of developing countries has become more prominent (Kurien 2005). Over a period of twenty years (1983-2003) in a number of developing countries, fish exports have become predominant over other commodities such as coffee, tea, meats, amongst others. (Kurien 2006). The growing demand for aquaculture products, particularly in respect of export markets, have increased the need to cater to increasing consumer demands on food safety, quality control standards, traceability and associated certification and ethical attributes (Singer and Mason 2006).

The great bulk of Asian aquaculture is extremely diverse and small-scale, often consisting of family owned and managed units, clustered in suitable areas, operating in harmony. In this regard aquaculture in Asia is no different to most of the other agricultural practices in the region; large, mechanised farms are not the norm. In such a scenario the problems and challenges encountered in meeting the modern day food quality and certification requirements are rather different to systems encountered in large scale, vertically integrated, farming enterprises, as often found with, for example, *salmonid* farming in Norway.

In this synthesis an attempt is made to evaluate the role of a regional, inter-governmental agency, dedicated to aquaculture development in the Asia-Pacific region, in facilitating and complying with the ever increasing demands of the markets, perhaps being intensified through globalisation of trade. However, all such compliance also needs to have minimal negative economic impacts on small scale farmers, the backbone of the aquaculture industry in Asia.

What is the Network of Aquaculture Centres in Asia-Pacific (NACA)?

NACA is an inter-governmental agency consisting of 17 member governments that has been existence for over 18 years. Its primary mission is to facilitate aquaculture development in the region, in particular the well being of small scale, rural aqua farmers, and ensuring sustainability, improvements to livelihoods and environmental integrity, and contribute to poverty alleviation. NACA apart from its member governments, works in close cooperation on aquaculture related research and development activities with regional and international agencies such as, for example, Food and Agriculture Organization (FAO), Asian Development Bank (ADB), World Bank (WB), Association of South East Asian Nations (ASEAN), Australian Center for International Agricultural Research (ACIAR), International Development Research Center, Canada (IDRC), Asia-Pacific Economic Cooperation (APEC), and so forth. It is pleasing that NACA has recently opened up a dialogue with the Organization for Economic Cooperation and Development (OECD) in regard to the promotion of south-south cooperation between Asia and Africa, and the establishment of a "NACA Like Network" in Africa, an element considered essential for effective facilitation of aquaculture development in Africa by most (FAO 2007b).

Aquaculture in Fish Trade

In order to consider the trade of aquaculture produce it is important and relevant to deal briefly with the nature of Asian aquaculture, in particular to get an understanding of the challenges the sector would have to encounter with fisheries trade globalisation. Aquaculture currently accounts for nearly 43% of all sea food consumed (FAO 2007). The great bulk of the production feeding this demand is in Asia (Figure 15.1), led by PR China, India, Vietnam and Thailand, all of which are also major exporters of aquaculture commodities. Indeed, eight Asian nations are amongst the top ten aquaculture producers in the world (Table 15.1). The data in Table 15.1 also indicates the overall predominance of PR China in global aquaculture production.

Figure 15.1. Global Aquaculture Production

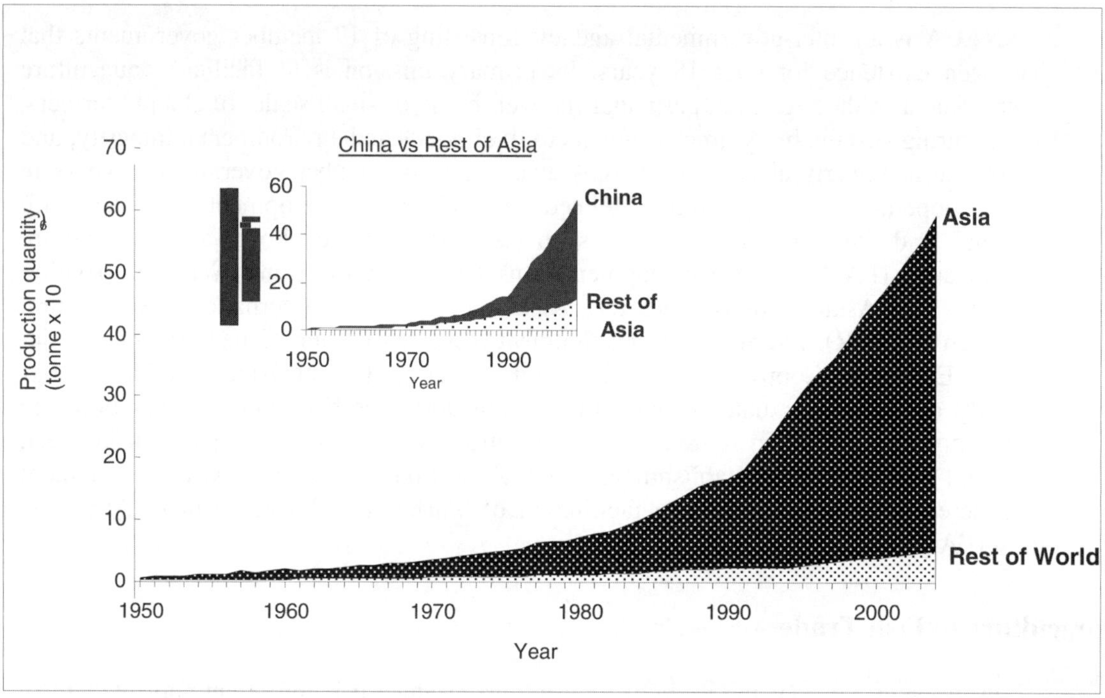

Table 15.1. The Top Ten Nations Leading Globally in Aquaculture Production (x103 t)

Nation	Production
China	32 414
India	2 838
Vietnam	1 437
Indonesia	1 197
Thailand	1 144
Bangladesh	882
Japan	746
Chile	698
Norway	657
Philippines	557

Source: Based on 2005 data, FAO Statlstics, 2007.

The fish production in Asia, over the last few decades has witnessed a gradual shift of increasing dominance of aquaculture (Figure 15.2), and in nations such as the People's Republic of China, the biggest supplier of food fish to the globe, the contribution of aquaculture to total fish production has reached nearly 70% (Figure 15.3).

Figure 15.2. Changes in Capture Fisheries and Aquaculture Production over the Years, and the Percent Contribution of Aquaculture to Total Asian Production

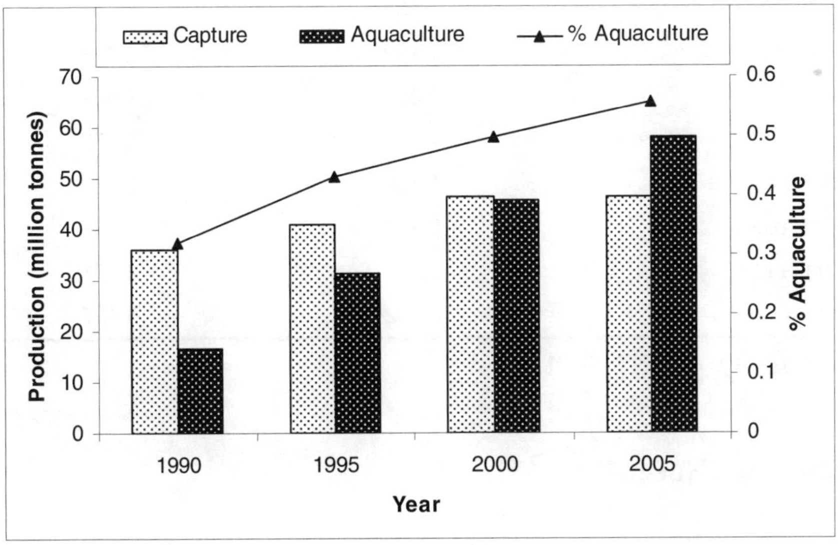

Figure 15.3. Changes in Capture Fisheries and Aquaculture Production (1990-2005) in PR China, and the Percent Contribution of Aquaculture to the Total

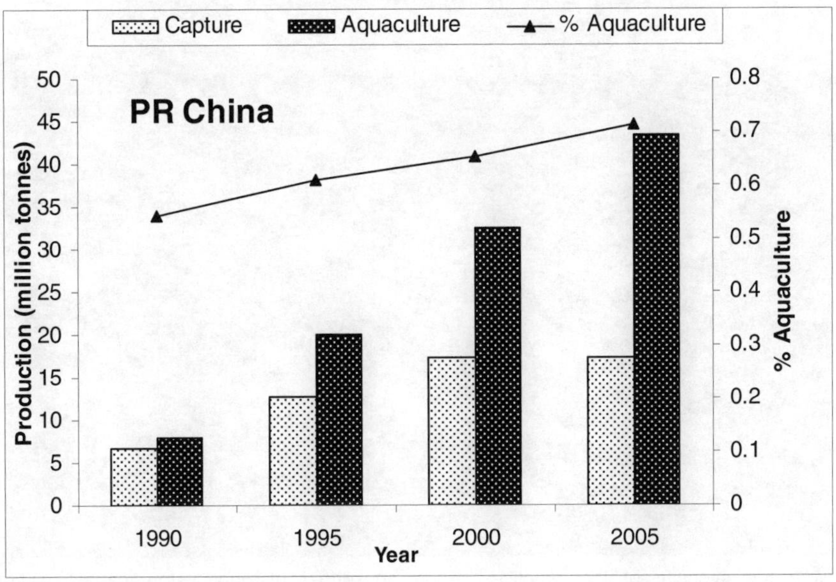

Obviously, the above changes will impact on the fish trade in Asia, when cultured commodities will become more and more important, and will continue to contribute increasingly to the GDP of such nations, by passing that from capture fisheries.

Table 15.2.Production Value as Percent of GDP

Nation	Capture Fisheries	Aquaculture
Bangladesh	1.884	2.688
PR China	1.132	2.618
Indonesia	2.350	1.662
Lao PDR	1.432	5.775
Malaysia	1.128	0.366
Philippines	2.184	2.633
Thailand	2.044	2.071
Vietnam	3.702	3.497

Source: Sugiyama et al., 2004

Asian Aquaculture in a Nutshell

Figure 15.4. Cage Culture Practices in XinCun Bay, PR China

Note the intensity of the practices, each household utilizing a few cages; a feature of Asian aquaculture.

Asian aquaculture is very different in its produce and *modus operandi* to those of the developed world. Asian aquaculture practices tend be rural, small scale, often family owned and operated practices, but clustered in suitable areas thereby acting in harmony and generating synergies. Asian aquaculture is rarely vertically integrated. Asian aquaculture is also characterized by the wide variety of species cultured and exported. For example, in the Asia-Pacific region, 204 species belonging to 86 families are cultured, as opposed to 336 and 245, respectively, globally (Subasinghe 2006).

The small scale aquaculture practices impose a very different scenario in regard to adoption of certification procedures, maintaining food quality standards, traceability and the like, all criteria that are increasingly being imposed by the global market.

The Challenges Resulting from Globalisation Facing the Sector

The globalisation of the fisheries sector has yielded new markets and stimulated aquaculture production of some commodities – such as shrimp, salmon and catfish - on an unprecedented scale, within a decade and half. At the same time, major challenges are now required to remain competitive in the aquaculture sector in Asia due to increasingly strict food safety standards, trace-ability, certification, and other requirements related to global marketing and trade. The increasing number of certification schemes, often demanded by larger importers and retailers, and the uncertainty that has resulted as to the certification authorities/ bodies is another recent concern. Equally important are unilateral decisions taken by certain importing nations which are seen by many as an indirect means of protectionism. As aquaculture has become increasingly prominent in world trade, it is inevitable that the number of trade issues concerning the sector have also increased. Antidumping, chemical residues and shrimp disease problems are recent examples.

In such instances, the Asian aquaculture sector has a forum to appeal to, *i.e.* the World Trade Organization (WTO), although this has rarely been used for aquaculture cases by developing nations. Thailand has recently filed its first submission against the United States with the WTO dispute-settlement body, accusing Washington of unfair trade practices by slapping an anti-dumping duty on Thai shrimp. An earlier case, with catfish exports to the USA from Vietnam (Vietnam could not go to the WTO because it was not a member at the time) created significant damage to livelihoods of many farmers in Vietnam when first applied, led to rapid diversification, and ultimately a more competitive industry. Today, Vietnam catfish is a fast growing entity that yields nearly 500 000 tonnes per year, and predicted by some to yield one million tonnes by year 2010, and currently providing employment, direct and indirect, to 17 000 persons in the Mekong Delta (Hung and Merican 2006). Even at present the catfish farming has already bypassed rice cultivation, in economic terms, as the main primary industry in the Delta.

The Role of an Inter Government Organisation in Complying with Requirements Emanating from Globalisation

Misconceptions

Often the role and the capabilities of small scale fish farmer is misconstrued and misunderstood, particularly in the industrialized countries. It is often suggested that the small scale, rural fish farmer is incapable of complying with global market requirements. It is far from the truth. It has been clearly shown that the farm output is not related to "literacy level" *per se* but to years of farming experience (Veerina *et al.* 1999). What is often forgotten is that small scale farming is a generational practice and adoption of the best farming practices, which enables access to global markets, is to their advantage.

Adoption of certification processes and assurance of product quality are in the best interest of small scale farmers; it ensures their livelihoods in the long term, and impacts on their living standards. That said, the process of development of standards and certification schemes should include a "voice" from small-scale stakeholders, ensuring the necessary change builds on the traditions and knowledge of the sector, rather than risking conflict and social impacts arising from application of systems which are inappropriate or poorly adapted to Asian farming conditions.

Role of NACA

NACA as a regional inter-governmental organization is committed to ensuring that the small scale fish farmers are able to benefit from fisheries globalisation, and not disadvantaged or excluded from markets by the process. Accordingly, NACA has been in the forefront of a number of initiatives that would lead to the assurance that the produce of small scale farmers are marketable globally, and in producing such a product the farmers will benefit and not be negatively impacted economically.

One of the main initiatives that were undertaken was the disseminating of better practice models in shrimp culture for example. This initiative in association with the Marine Products Export Development Authority, India (MPEDA) has set a very good example in the way of mobilizing small scale farmers through aqua-clubs in the adoption of the better practices, which have resulted in the production of certifiable products, increased yields and profitability.

The organisation of farmers in India has provided a basis for certification and improved market access. The progress made over the years is summarized schematically in Figure 15.5, and the details of implementation are given by Umesh (2007). This program has led to major policy changes in Indian aquaculture, with the setting up of a relevant institution, The National Center for Sustainable Aquaculture (NaCSA) in March 2007.

In a similar vein NACA in conjunction with the Commonwealth Scientific and Industrial Research Organization, Australia (CSIRO), and the Australian Center for International Agricultural Research (ACIAR), MPEDA, India and many laboratories in India, both private and government, initiated an inter-calibration exercise for evaluating the quality of shrimp post-larvae (PL), and the competence of individual laboratories. This exercise has had major impacts leading to the improvements from individual laboratories and most of all have stimulated interest in almost all shrimp producing countries seeking an extension of the program. It is conceivable that all shrimp producing countries become a part of this program thereby ensuring that good quality PL, free of viral diseases, become available to farmers; in essence a certification procedure is being put in place in all shrimp producing countries, which has potential to contribute to ensuring the production of disease free shrimp to the global market.

Figure 15.5. Schematic Representation of the Progress in the Adoption of Best Practice in Shrimp Farming in India

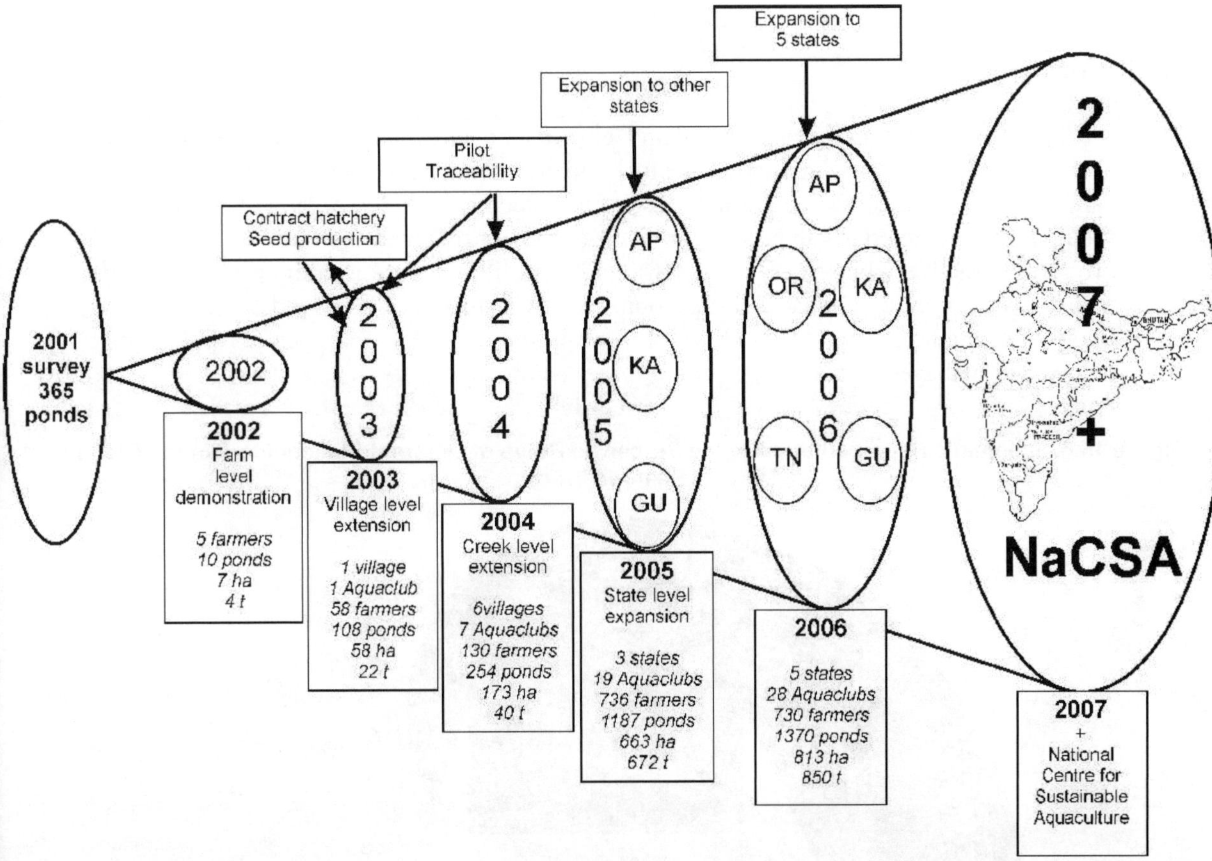

Note: The Inner circles (AP, KA, GU etc.) refer to different States.

NACA, in conjunction with the Royal Government of Thailand and FAO, took a recent initiative with a view to development of guidelines on aquaculture certification, to establish acceptable standards and *the modus operandi* for adoption of "certification procedures" for aquaculture products. In essence, there is a certain degree of confusion and uncertainty in applying certification to aquaculture products; the primary one being how one organises small scale farmers- the back bone of aqua-farming in Asia- to comply with such procedures.

The above issue confronts all governments in the region, especially in the context that some of the major seafood importers and retailers are deciding, unilaterally, to adopt certification as a requirement for aquaculture products, without sufficient consideration to the ongoing activities in this regard in the region. NACA, in cooperation with the Thai Government and FAO, is creating a platform for stakeholders to work together to establish credible aquaculture certification standards and procedures suitable for the small-scale sector.

The above examples are ones directly relevant to the current issues. Amongst others are NACA's involvement in the coordination and facilitation of the aquatic health management program, in conjunction with member governments, FAO and IOE. This program has been ongoing for over ten years, and NACA has become now the central point for all aquatic disease reporting from Asian nations.

Conclusions

The modern day consumer needs and globalisation in trade have resulted in major changes in the primary production sector, and the aquaculture sector, which currently supplies over 40% of the seafood needs, is no exception in this regard. It is however, thought that the emphasis that is being given to the consumer in the trade of aquaculture produce is relatively biased (Figure 15.6), and pays little attention to the producer needs and status even though aquaculture accounts for over 40% of all seafood consumed, and likely to exceed 50% by 2010, of which 80% is produced by small scale, rural, family owned and managed farms.

Figure 15.6. Schematic Representation of the Perceived Roles of Different Stakeholders in the Globalised Seafood Trade

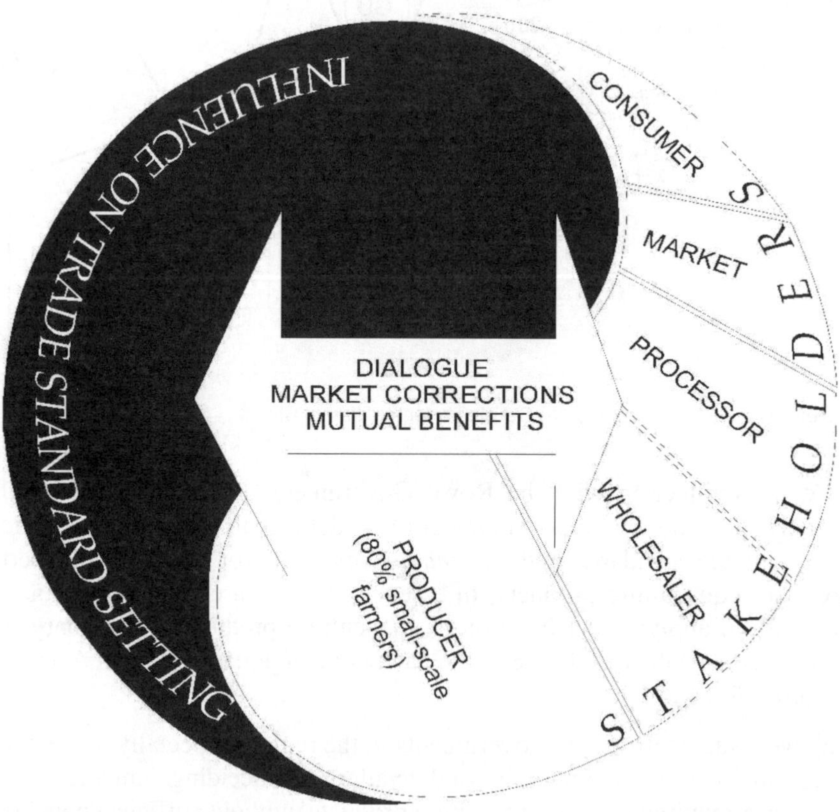

In the above context, organizations such NACA have a pivotal role to play in bringing all stakeholders together and to set a "level playing field" to bring about equity; the task being made that much easier as the dealings of such organisations are at the "coal face", and are viewed with little or no apprehension by small scale producers as well as governments. Moreover, such organizations also have the ability to bring a neutral view to the "table" making negotiations much more amicable, fair and equitable.

Bibliography

Delgado, C.L., N. Wada, M.W. Rosegrant, S. Meijer and M. Ahmed, (2003), "Fish to 2020. Supply and Demand in Changing Global Market". International Food Policy Research Institute, Washington, D.C., pp. 226.

FAO, (2007a). *The State of Fisheries and Aquaculture 2006.* FAO Fisheries and Aquaculture Department, Rome, pp. 162.

FAO,(2007b). Committee on Fisheries. Report of the Third Session of the Sub-Committee on Aquaculture. FAO Fisheries Report 816, pp. 85.

Hung, Le Thanh and Z. Merican, (2006). "Freshwater fish culture in Vietnam for the global white fish market". *AquaCulture* Asia-Pacific, 2 (3), 14-15.

Kurien, J., (2005), "Responsible fish trade and food security". FAO Fisheries Technical Pape*r* 456, pp. 102.

Singer, P. and J. Mason, (2006), *The Way We Eat- Why Our Food Choices Matter.* Rodale Holtzbrinck Publishers, NY, pp. 328.

Subasinghe, R., (2006). *State of World Aquaculture: 2006.* FAO Fisheries Technical Paper 500, 134 pp.

Sugiyama, S., D. Staples, and S.Funge-Smith, (2004), *Status and Potential of Fisheries and Aquaculture in Asia and the Pacific.* RAP Publication 2004/25, FAO, Regional Office for Asia and The Pacific, Bangkok, pp. 53.

Umesh, N.R., (2007), *Development and Adoption of BMPs by Self-help Farmer Groups.* Aquaculture Asia, XII (1), January 2007, 8-11.

Veerina, S.S., M.C. Nandeesha, S.S. De Silva and M. Ahmed, (1999), "An analysis of production factors in carp farming" in Andhra Pradesh, *India.* Aquaculture Research 30, 805-814.

PART 6

Linkages Along the Value Chain

Chapter 16

An Investor's Perspective

Kristjan Davidsson, Glitnir Bank, Iceland

In his presentation Kristjan Davidsson stated that the global fisheries sector is of great interest to investors, be it fisheries, aquaculture, processing or distribution. This interest is growing; especially the aquaculture sector has great future potential. The estimated total value of the seafood value chain is approximately USD 400 billion.

He reminded participants of the rewards that can accrue through carefully designed measures to ensure sustainability, stability and continuity in regulations and rules that form the management system a country develops for its seafood industry. Such rules include industry fundamentals like clear and secure ownership rights and investment regulations. In addition to this regulatory regime, which is of crucial importance to an investor investing in industries using natural resources regulated on national level, conditions investors evaluate are the normal conditions set for any business case; structure, cash-flows, management team etc. Growing emphasis is put on sustainable management of the fisheries resource base and it is to be noted that increased knowledge coming from science adds a more secure investment environment. Consequently scientific basis and advice for management regulations and decisions is a necessity, not to be underestimated.

The growing interest from investors in the seafood sector has lead to increasing professionalism and growing skepticism towards looking at the fisheries sector as a part of the political toolset to be used in e.g. rural policies. Increasing demands from the market for sustainability, security and continuity of deliveries has both supported growth in aquaculture and it also explains the increasing mergers and acquisitions, an industry consolidation that has taken place recently in the fishing industry and will continue to grow in the foreseeable future.

Mr. Davidsson presented the Icelandic fisheries sector as a case study. The introduction of the current system, which is based on individual transferable quotas, has meant more professional operations and better economic result, with less need for political intervention in times of crisis, which was a feature of the Olympic race system applied earlier. Through improved operational results, increasing strength and better negotiation position of the participants, the consolidation of fishing quotas and increased harvesting and processing revenues are clearly a benefit from the investor's point of view, as well as being important for the viability of the industry.

Not least is the aquaculture sector an important investment opportunity. However, any investment decisions must of course take into account, in addition to the business environment in the industry, the legal and regulatory environment of the geographical area under consideration for investment. Hence the risk profile for investors in for example salmon aquaculture in developed countries is quite different from the risks investors face in shrimp aquaculture in developing countries. It is essential to understand both the operational and legal environment of operations before making investment decisions. Investors in developing countries are often confronted with a risky investment environment and consequently the demand for return will be different than for an investment in a developed country.

It is thus of great importance for governments, aiming to attract investors to their seafood industries, to take measures for eliminating or reducing as possible any risks, be it in their legal framework or other applicable factors, as reduced risk normally means less demand for return on invested capital, in other words; the industry in question will have better access to capital and also the capital should be cheaper, the more secure the investor is of his capital, be it a bank with a credit or an investor with equity.

Picture money or capital as a vehicle or car; if used correctly it can move from one place to another. In the fisheries context, capital can move a fish farmer from one hectare to five or from one cage to twenty. However, banking, financing and investing is much more like operating a car rental. Investors expect payment of the rental cost and any interest accrued, customers who drive the vehicle carefully, the vehicle returned, and especially at the end of the agreed period of rental. Nevertheless, it is still not enough to have cars and people; roads, traffic rules and other rules are also required. Infrastructure and legal frameworks are, in a similar way, necessary for effective execution of business. Like adequate roads and sensible traffic rules, legal frameworks ease the trip from one place to another. Bad roads and bad rules hinder effective traffic: they lead to waste or even destruction of resources and time - or cars and money in our analogy. In every market, different types of car rental, operators or agencies exist, from a rickshaw to a local car rental agency, multinational chain or the provision of limousines. The type of offer depends on market characteristics, traffic rules, laws and infrastructure. The rental rules may also vary. For example, basic rules include a driver's licence, exclusion of convicted or wanted criminal, a minimum age etc. Essentially, rules are there to help the business.

Prices vary according to market demand and supply, the infrastructure available and traffic rules. Bad roads will damage cars so good roads will lower the price. If no foreign cars are allowed, prices rise. Unfortunately, it is a fact that many car dealers lobby their governments to create obstacles to competition with the intended or unintended consequences of raising prices and bringing about protectionism. This mostly harms customers and, in many cases, also the car rental agencies as less business will be available. In general, the best off are those with the best rules. A final point using this analogy is that one person may exist who wishes to own a million cars, even if he is only able to drive one at a time. The same exists with money: people want to own hundreds of millions of dollars even if they only can spend a limited amount or use a limited amount, although some do give it away.

Likewise, the seafood industry is a big and growing business and for that reason it is attractive to investors. When a business is growing, investors are more likely to make more money. That is why the seafood industry holds great opportunities for investors; demand for protein is growing; a focus on fish as healthy is also increasing. Much of this increase in supply in order to meet demand comes from aquaculture. Obviously, aquaculture is a huge growth story for investors loaning capital or directly investing themselves. However, at the moment the industry remains fragmented although it is moving towards consolidation, which will bring an enormous business opportunity. Furthermore, the growing emphasis on sustainable management means investors will know when they will "get their car back". This provides a business opportunity.

On the issue of consolidation, it is worth comparing the meat industry with the seafood industry. The combined revenue of the top ten companies in the meat industry is about USD 8 billion. In the seafood industry, the revenue of the top ten companies is only around USD 2 billion. In meat companies, operating revenue *i.e.* earnings before interest, taxes, depreciation etc., is 10%; it is half that in the seafood industry. However, of interest to investors is that mergers and takeovers are increasing in the seafood industry, roughly calculated to be worth close to USD 20 billion.

This is the sort of consolidation occurring in the Icelandic seafood industry. The top ten Icelandic fishing companies have found that with the introduction of a quota system where quota can be traded freely as any other commodity, operating results have been steadily increasing. Of course, other factors are relevant but generally, the quota system has improved the situation. This is because elsewhere, politicians require election every set number of years. In Iceland, it is the fishers who have taken responsibility for their decisions, with resulting effects on earnings.

The greatest issue for policy makers worldwide is to enable the seafood industry to operate as a business. Much like the paper producing industry, the food producing industry should be a business and not a political tool in rural policy for example, where other tools such as improved communications and education would be more appropriate. Political issues do remain of course, such as the framework for infrastructure i.e. the 'traffic rules', but fundamental issues include the need for security and safety to attract investment. Likewise, sustainability is necessary for long-term investment. Another responsibility for government is the role of taxation as the legal rules that apply in a market will affect all the players in that market. Consequently, investors consider sets of parameters that include the legal operating environment, the physical environment, human capital *e.g.* the management team, available capital and profitability before making decisions.

Glitnir Bank has been active in the seafood industry for 100 years, providing and selling services from corporate banking to credits, leasings and bonds, as well as acting as an advisor to other sellers and buyers. In other words, they obtain rent from their cars. In all, around USD 2-2.5 billion is currently being invested through Glitnir Bank. However, Glitnir Bank only operates in about 20 countries. Where the industry is allowed to be an industry *i.e.* rent the cars or be the car rental agencies, and where policy makers understand that their role is to provide the infrastructure and 'traffic rules', then industry will flourish.

Chapter 17

Microfinance, Small Scale Fisheries and International Fisheries Trade in Selected APRACA Member Countries

Thirapong Tangthurasunan, President, Bank for Agriculture and Agricultural Cooperatives (BAAC), Indonesia

Thiraphong Tangthirasunan highlighted the importance of microfinance for the development of small scale aquaculture and fisheries in the Asia–Pacific region. In this region, fishing is a priority sector, but the availability of financing for the sector is limited due to the high risks involved. Also, industry structure is geared towards small-scale family enterprises. Most governments do not directly lend to fishing companies; this leaves banks and other financial institutions as potential lenders.

Thiraphong Tangthirasunan has outlined the fisheries policy settings of Cambodia, India, Thailand, Philippines and Indonesia in this chapter. In these countries the fisheries sector is seen as a priority sector for governments. Policy approaches common to the five countries include: capacity building across the value chain to enhance technology and sustainable management, support towards institutional building and a strengthening the sector's legal framework to protect the small-scale operators. In this regard, microfinance is viewed as a practical solution to the growing demand for financial services by the poor. There is now evidence of certain banks' shift in business culture to attune their services toward low-income rural households and small enterprises. Likewise, there is empirical evidence that microfinance, provided by various types of banks, enables the poor to better manage their consumption and their business risk, gradually build their assets, develop their enterprises, enhance their income capacity and generally improve their quality of life.

Introduction

This chapter looks into the situation of small scale fisheries and international fisheries trade in selected APRACA-member countries namely Thailand, the Philippines, Cambodia, Indonesia and India with particular focus on the availability of microfinance services for the development of small scale fisheries and international fisheries trade. Information contained in this report consists of data from country papers as well as from review of related literature.

The role of fisheries: For livelihood and food security

The livelihood of millions of people in Asia depends on aquatic resources. Small-scale fisheries provide employment for millions of fishers directly engaged in fishing activities, including rural aquaculture, and for millions more working in fisheries-related activities such as fish processing and marketing, boat building and net making. Including family members, hundreds of millions of rural people in developing countries depend on fisheries for their livelihood. Due to an increase in overall fishery production and trade over the last few decades and a corresponding increase in employment in the fishery and aquaculture sector, an estimated 38 million people were directly engaged in fishing and fish-farming as a full-time, or more commonly part-time occupation, compared with 28 million a decade earlier (FAO 2002). An estimated 90% of the 38 million people recorded by FAO globally as fishers and fish farmers are classified as small-scale (FAO 2004). Asia, in particular, has a total of some 25 million fishers and fish farmers, which is more than double the number in the 1970s and 80% of the world's total (IFAD 2004). There are millions of other rural people involved in seasonal or occasional fishing and/or aquaculture activities that are not recorded as 'fishers' in official statistics.

Table 17.1. Number of Fishers and Farmers (in 1000s), by Region

Continent	1990	1995	2000	2001	2002
Africa	1 917	2 238	2 585	2 640	2 615
North and Central America	767	770	751	765	762
South America	769	814	784	760	770
Asia	23 654	28 552	30 770	31 493	32 821
Europe	654	864	821	746	746
Oceania	74	76	86	81	81
World	27 835	33 314	35 797	36 534	37 795
Of which fish farmers:					
Africa	...	105	112	115	111
North and Central America	53	74	74	69	65
South America	16	88	92	92	93
Asia	3 698	6 003	8 503	8 720	9 502
Europe	11	36	37	39	39
Oceania	Neg.	1	5	5	5
World	3 778	6 307	8 823	9 040	9 815

Source: FAO, 2002.

Furthermore, the importance of small-scale fisheries as a major source of animal protein cannot be over-emphasised. In many least developed countries of Africa and Asia, fish accounts for more than 50% of the total animal protein intake. In almost all these countries small-scale fisheries provide over three-quarters of the domestic fish supply. As fish is generally more affordable to poorer members of society, a greater amount of this protein source is consumed on a per capita basis than any other type of protein

(Tacon 2001). Thus, in Southeast Asia, possibly a billion people rely predominantly on fish for animal protein.

The demand for fish is therefore increasing, not only due to an increasing population but also to a greater awareness of the importance of fish in the diet (Delgado *et al* 2003). There is a general consensus that traditional sources of fish such as global capture fisheries have peaked (FAO 2002) and the future of wild-caught fishery production appears to be uncertain. Currently, 47% of fish stocks are described as being fully exploited or close to their maximum sustainable limits (Delgado *et al* 2003; FAO 2002). Others are in a state of decline, or are completely exhausted. Recent studies based on trawl surveys in eight Asia-Pacific countries by the WorldFish Centre indicate that the situation may be far more serious than these figures suggest, and that substantial degradation and over-fishing have occurred. According to the surveys, coastal stocks have declined by as much as 40% in five years (Silvestre *et al* 2003). Consequently, it is believed that the amount of fish available for the region's fishers is now only a fraction of what was available before the industrialization of fishing (Sugiyama *et al* 2004).

Coastal populations that were once almost entirely dependent on inland or coastal capture sources of fish have therefore seen their resources decline, and once cheap and plentiful wild fish has become less available and less affordable (Wilfredo *et al* 2006). In some locations around coral reefs, fishers are turning to lucrative yet destructive practices such as the use of explosives (so-called blast fishing) and cyanide to stun and capture fish (Burke *et al* 2001). There are also numerous reports of conflicts over diminished fishery resources and increased illegal fishing activities as fishers from one community, region or country encroach into the territories of their neighbours (Bulcock and Savage 2005). Destruction of habitats, sedimentation, pollution and dive tourism impose major threats to aquatic resources.

Current trends in international fisheries trade

Despite the fisheries crises around the globe, the international trade in aquatic products has grown significantly over the last few decades due largely to increased demand supported by improvements in technology, transport and communications (FAO 2003). The global appetite for fish has doubled in the past thirty years, with total fish consumption rising from 45 million metric tons in 1973 to more than 91 million tons in 1997. Trade tends to flow from the less-developed to the more-developed countries with mainly tuna, small pelagics, shrimps and prawns, rock lobsters and cephalopods as major products. Developed countries imported over 32 million tons of which 68% was fish for human consumption, while developing countries imported 19 million tons of which 47% consisted of fish for food. Fish export values have consequently increased from USD 15 billion in 1980 to USD 56 billion in 2001 (Macfadyen *et al* 2003). A large percentage of fisheries and aquaculture production now enters international marketing channels and chains, with more than 37% exported in 2000 in various forms. Developing countries, predominantly in Asia, play a major role in this trade and fisheries and aquaculture are therefore significant contributors towards national economies across the regions (Sugiyama *et al* 2004). In 2002, Thailand, which had been the main exporter of fish and fish products since 1993 reported export values of USD 3.7 billion, 9% lower than in 2001 and 16% below 2000 values. Thus, also in 2002, China overtook Thailand for the first time to become the world's main exporter of fish and fish products, with exports valued at an estimated USD 4.5 billion.

The discussions that follow will focus on selected APRACA-member countries particularly Thailand, the Philippines, Cambodia, Indonesia and India.

Small Fisheries Development in APRACA Member Countries

Poverty situation

The level of poverty remains high among developing countries in general. An estimated one billion people around the globe live on less than one dollar a day, and 70% of the world's poor are women. It is further estimated that about 1.9% of the world's population derive their livelihoods from fishing and fishing-related activities, in both inland and marine environments and the vast majority can be found in Asia (FAO 2004). The majority of these fishers are small-scale, artisanal, coastal operators and among the poorest in society (FAO 2002). Income generated by the fisheries sector is generally lower than those from other sectors and within the sector itself small-scale fishers earn the lowest incomes (Silvestre *et al* 2003). While economic growth has helped to reduce the number of poor people in the world, the positive impacts of growth on poverty have been less than expected, partially because of inequitable distribution of the benefits, population increases, political instability, and in some parts of the world, the devastating effects of the HIV/AIDS epidemic.

Within Asia, poverty in coastal areas is a defining characteristic of countries such as Bangladesh, India, Indonesia, Myanmar, Pakistan, Philippines and Vietnam (IFAD 2002). However, the extent of poverty in coastal communities is difficult to measure and while there have been many studies on poverty in farming and urban areas there have been few that have focused on the fisheries sector (FAO 2002).

Box 17.1. Poverty Incidence among Fisheries Workers in the Philippines

In 2003, poverty incidence among fisheries workers was 50.8%; compared to total poverty of 33% or poverty among farmers at 46.6%. Moreover, except in Regions II and VIII, poverty among fishers is higher than regional poverty; this includes the poorest regions (ARMM, Caraga and Bicol). Regions with high shares of the total fisher population also have very high rates of fisheries poverty.

Table 17.2. below describes the extent of poverty among some APRACA-member countries:

Table 17.2. Poverty Status of Some APRACA Member Countries

Country	Poverty Status
Cambodia	The poverty rate – that is, the percentage of the population living under the poverty line – is estimated at 35% for 2004. It is highest in rural areas and lowest in Phnom Phen.
India	Vast majority of India's poor people live in rural areas. Rural poverty is estimated at 42.7% with 43.3% of India's rural people belonging to Scheduled Tribes and Castes (Mukherjee 2006)
Indonesia	The number of those below the poverty line has risen by 3.95 million from 35.1 million in February 2005 to become 39.05 million by March 2006, or 17.75% of the population of the country. Over 70% of fishers are poor. It may be over 80% in some areas. Poverty levels in coastal communities are generally considered to be around 80% of the population (Suspita 2006). There are 36 million poor people in Indonesia (Jaya 2006).
Philippines	Poverty remains to be the biggest problem in the Philippines despite decline in poverty incidence of families from 27.5% in 2000 to 24.7% in 2003; and the poverty incidence of the population from 33% in 2000 to 30.4% in 2003. Almost three out of four (73%) of the total number of poor in the country reside in the rural areas (NAPC 2004). The poverty level in rural areas is much higher at 48.8% against 18.6% in urban areas; about 5 out of 10 rural residents are poor compared with almost 2 out of 10 urban residents.

Importance of fisheries

Given the relatively high poverty levels in these countries, the fisheries sector plays a very significant role in their national economies. The contribution of capture fisheries and aquaculture to the Gross Domestic Product (GDP) ranges from 0.540% (India) to 2.633% (Philippines) in the case of aquaculture and 2.044% (Thailand) to 10.030% (Cambodia) in the case of capture fisheries.

Table 17.3. Contribution of Fisheries to Gross Domestic Product (GDP

Production value as % of GDP			
Capture Fisheries		*Aquaculture*	
Cambodia	10.030	Philippines	2.633
Indonesia	2.350	Thailand	2.071
Philippines	2.184	Indonesia	1.662
Thailand	2.044	Cambodia	0.893
		India	0.540

With respect to fish production in these countries, the trend for the past two to three decades has generally been increasing. In the Philippines, output has grown by an average of 3.4% per year. Using FAO categories, the largest share continues to be contributed by marine capture (56% in 2004). The smallest share is provided by inland capture (only 3.7% in 2004). Aquaculture averaged 8.7% over the last 30 years, compared to only 2% growth for marine capture. In India, the total fishery production increased nine times from 0.70 million tons in 1950-51 to a maximum of 3.76 million tons in 2005-2006. The growth of inland fishery has increased at a higher rate, from 0.218 million tons during the 1950-51 period to 2.81 million tons in 2005-2006. The fisheries sector, thus, has been sustaining India's population of over 1 billion representing over 16% of the global population.

Box 17.2. Cambodia's Fisheries

Cambodia's coastal waters are very productive and rich in fishery resources and the diversity of systems and species remains high. Its inland water habitats are also one of the most productive resources on earth primarily because of the Mekong River and the Tonle Sap River with the Great Lake at its upper stretches. The potential fish catch is reported to be five times higher than that of most lakes and rivers in tropical areas. Fish is therefore an important staple along with rice and a main source of protein for the rural Cambodian people. Although there is a lack of reliable data on fish exports, fish is the main export commodity of Cambodia.

In Cambodia, the growth of fisheries production has also been rising in general but there has yet to be a systematic collection of marine fish production data by the country's Department of Fisheries. It should be noted that among the top ten major producers of marine fisheries in the world, Indonesia ranks 6th; India, 8th; and Thailand, 9th (Table 17.4).

Table 17.4. World Marine Capture Fisheries Production by Major Producing Country in 1995, 1999 and 2000

(millions of metric tons)

Country	1995	1999	2000	
	Tons	Tons	Tons	Ranking
China	11.0	15.0	14.8	1
Peru	8.9	8.4	10.6	2
Japan	5.9	5.1	4.9	3
United States	5.2	4.7	4.7	4
Chile	7.4	5.0	4.3	5
Indonesia	2.7	3.7	3.8	6
Russian Federation	4.1	3.8	3.7	7
India	2.7	2.8	2.8	8
Thailand	2.8	2.7	2.7	9
Norway	2.5	2.6	2.7	10
Sub-total	53.2	53.8	55.0	n.a.
Other countries	31.5	30.9	31.0	n.a.
Total	84.7	84.7	86.0	n.a.
Major ten producers as a percentage of total marine capture fisheries production	63.0	64.0	64.0	n.a.

Source: FAO

Fisheries are also a major source of agricultural employment. In 2004, the sector employed 1.4 million persons, equivalent to nearly 12% of agricultural employment or

4.3% of total employment in the Philippines. In Cambodia, the fishery sector employs approximately 5% of the labour force, about 1.3 times more than the country's garment industry. Where the diversity of systems and species remains high, such as in Cambodia, aquatic resources offer considerable opportunities to coastal people to diversify their livelihood in order to meet their changing requirements (IMM *et al* 2005). Coastal fisheries provide employment to two million people in Indonesia (Silvestre *et al* 2003). The types of livelihood are complex and vary tremendously from full-time small-scale operators to those involved in seasonal and migratory positions in the processing and marketing industries (IMM *et al* 2005).

Type of fishing activities

Since yields from capture fisheries are not expected to increase significantly in the next few years, emphasis is being placed on the aquaculture sector's ability to provide increasing quantities of aquatic products. Production from inland aquaculture and marine and brackishwater-based aquaculture are both increasing and now account for 30% of total aquatic production (Delgado *et al* 2003). Aquaculture production in Asia-Pacific Region is diverse, but in terms of volume it is still dominated by freshwater fish production (39%), followed by aquatic plants (29%), crustaceans (13%), marine and diadromous fish (13%) and mollusks (7%). In terms of value, crustaceans such as the tiger prawn dominate, accounting for 49% of production followed by freshwater fish (35%) (Wilfredo *et al* 2006).

The shifts in production from fishing to aquaculture, and the growth in the international trade in aquatic products, are often believed to offer the potential to contribute towards poverty alleviation and food security through the creation of jobs and alternative sources of food. From the country papers and existing literature, aquaculture practices or activities considered potentially pro-poor, meaning, those that the poor are able to undertake, are as follows in Table 17.5.

Table 17.5. Common Aquaculture Activities

Country	Aquaculture Activity
India	Mud Crab Fattening Shellfish Culture Shrimp Processing
Indonesia	Traditional Milkfish Production in *Tambaks* (Ponds) Traditional Prawn Culture Mud Crab Fattening Shellfish Culture Sea Cucumber Seaweed Culture Shrimp and Finfish Hatcheries
Philippines	Shellfish Farming Milkfish Production in Cages and Pens Tilapia Production Backyard Grouper Production in Cages Seaweed Culture
Cambodia	Shrimp Farming Mollusc Culture Catfish, Tilapia and Carp Production in Cages Seaweed Culture Crocodile Farming Fish seed Production

Another important fishing activity that has become increasingly popular for the last decade is processing of fish products "because fish is highly perishable". More than 90% of internationally traded fish and fish products are in processed form a significant proportion of which come from the Asia-Pacific Region. Examples of processed fish products include fish paste, fish sauce, smoked and dried fish or shrimps, frozen fish balls/fillets, chilled shrimp meat/crab meat, dried sea weed, among others. In Cambodia, the value of preserved, processed, and exported commodities of both the inland and marine fisheries is very significant in the fisheries economy, estimated between USD 34 300 000 and USD 40 400 000 (Tana and Todd 2001).

Box 17.3. Socio-Economic Value of Fish Preservation and Processing

Preservation of marine products is of great socio-economic importance to the coastal poor. Preserved fish products ensure adequate protein during low fishing periods. Subsistence fishers use their abundant catch of small fish to make fermented fish paste and smoked fish with the assistance of family members. Large fish are used to make fermented fish or salt dried fish. Other important processing activities include drying of small shrimp, squid, ray and shark and preparation of shrimp paste.

Policies for development of the fisheries sector

The fisheries sector in Asian countries has long been considered as a sub-sector of the agricultural system as a whole. Thus the fisheries sector policies are mostly inclusions of broader agricultural and national economic policies. However, the significant contribution of fisheries to the gross national product (GNP) and to scarce foreign exchange resources as a result of continued increases in production of fish and fish products has been recognised. The fisheries sector is therefore now one of the priority sub-sectors and in many instances policies are being formulated which target this particular sector.

In the Philippines, the legislative framework for fisheries is mostly contained in two Republic Acts, the Local Government Code of 1991, and the Fisheries Code of 1998. The Local Government Code places resource management, including coastal and inland fisheries under the jurisdiction of local governments. It also devolves to local governments the provision of aquaculture support services and the operation of fish ports. On the other hand, the Fisheries Code reserves aquatic resources in Philippine waters for the sole use of Filipino citizens. Municipal waters are inland waters and marine waters up to 15 km. from the shore. Waters up to 10 km from the shore is reserved for municipal fishers; the municipal or city government may however permit small to medium scale commercial fishing in the 10-15km region. Other laws pertinent to fisheries are the Agriculture and Fisheries Modernization Act (AFMA) of 1997, the National Integrated Protected Areas System Act of 1992 and the Wildlife Resources Conservation Protection Act (RA 9147).

In Indonesia, the Presidential and Ministerial Decrees regulate policies and provide guidelines for new policies. Past policies of the country reveal that it promoted fisheries development to support national economic growth by encouraging domestic consumption and exports of fisheries products, and by promoting private and foreign investment in the fisheries sector. Presidential Decree no. 23 of 1982 provides high priority to small-scale farmers and cooperatives to develop mariculture, and allows both foreign and domestic private investment to encourage modern technology adoption, although in some instances *e.g.* shrimp hatcheries, foreign investment is restricted. The rescue program

"PROTEKAN 1999-2003" following the economic turmoil of 1997-98 identified the fisheries sector as one of the areas where stocks are still under-exploited, and identified a potential for aquaculture development. As part of the rescue program, capture fisheries are being developed and supported to improve product quality, diversify products, and develop infrastructure. It is expected that the development of the capture fishery will contribute one-fourth of foreign exchange earnings.

As a leading producer of fish and aquatic resources in Asia and in the world, Thailand carries special importance in reviewing its fisheries policy. The Department of Fisheries is the principle agency in formulating fisheries policies and implementation of these. However, the sector is not only affected by its own policies, it is also guided by complementary national economic and agricultural policies. Current national fishery development policy has five principle components that form the core for the formulation of strategies and action plans: *a)* development of fisheries and related organisations; *b)* fishery resources and environmental management; *c)* aquaculture development; *d)* policy beyond Thai waters; and *e)* fishery industry and business development.

Consistent with the policy, the government has put much emphasis on gear exchange to encourage the fishers to change their illegal destructive fishing habits. In general, the private sector is the principal stakeholder in investment in the fisheries industry. The government acts as facilitator, facilitating raw material acquisition, product certification, standardisation of products to attain international standards, sale promotion, etc.

Thailand's Board of Investment (BOI) promotes investment in agriculture and agricultural products. The BOI listed aquaculture (except shrimp culture), deep sea fishing, fish feed manufacturing, trading centres for fisheries products, agro-industry processing zones, and aquariums and ocean marine services as priority activities for investment promotion.

In Cambodia, many changes have taken place in the economic policy of the country which has affected both freshwater and marine fisheries. The current fisheries reforms impacting the Cambodian economy are based on 'resource benefit re-distribution' as enunciated by Prime Minister Hun Sen in October 2000. The principle of resource benefit redistribution is to redesign the 'jurisdictionals' for different groups of stakeholders through which fishing corporations and local communities share the benefits extracted from the fisheries resources. Thus, in 2000, the government released 56% of leased lots to the fishing community, aemed at reducing conflicts and sharing more resources with the poor. Some fishing communities were established by the government and some of them were created by non-governmental organisations.

In India, the National Fisheries Development Board has been set up in order to develop the fishery sector through modern tools of research and development including biotechnology. It has been registered on July 2006 under the Andhra Pradesh Society Registration Act 2001 and has become operational. The Coastal Aquaculture Authority Act which provides the rules to protect as well as develop the fishery sector was established in 2005.

Box 17.4. The King's New Theory

Part of Thailand's development policy is to make small-scale farmers/fishers sufficient, self-reliant and competitive. Following the economic crisis of 1997, "The King's New Theory" was implemented in order to encourage small-scale farmers who own about 2.5 hectares of land to engage in integrated farming to cover rice and other crops, livestock, poultry and fish feeding. With appropriate technology and local experience/wisdom, the small-scale farmers will gradually improve and will be ready to become commercial farmers.

Experience in international fisheries trade

Thailand used to be the leading exporter of fish in the world in terms of value but was overtaken by China in 2002. Other countries like Taiwan, Chile, Indonesia and Republic of Korea are among the top ten fish exporters of the world. Although many of the developed countries contribute to the world export of fish, their net export is negative, and has been declining since the early 1980s (WorldFish Center 2000). Other countries like India, Philippines, Vietnam, and Bangladesh also share a significant amount of fish and fish products in world exports.

Thailand remains the number one shrimp exporter. About 306 000 tons of shrimp products have been exported to other countries in 2006. In order to ensure good quality of Thai seafood and aquaculture products, the Department of Fisheries conducts regular inspections. Today, the Department is recognized by the food and safety control authorities in various countries including the EU, Argentina, Japan and China. Since 1998, the Department of Fisheries introduced two very important programs for quality shrimp production particularly the Good Aquaculture Practice or GAP and the Code of Conduct for Responsible Fisheries on Aquaculture or CoC, to raise the standards of the Thai shrimp aquaculture industry to high international levels.

In the Philippines, there is a good number of high value fish species preferred by consumers in developed countries. These species – grouper, shrimp, lobster and tuna – are the icons of "luxury consumption". Exporting these products generates foreign exchange which can be used to purchase inexpensive food fish from the international market to meet domestic food requirements. In 2000, the Philippines exported a total of 199 719 tons of fish and fishery products valued at USD 506.6 million.

Exports of fish products from India have been increasing steadily for the past three decades. The share of India in the global trade has increased from 6.1% in 1992 to 6.5% in 2003. It is second to China with respect to inland fish production.

Cambodia's experience in international fisheries trade took off in 1990. Its major export products are frozen shrimp and freshwater fish. KAMFIMEX, the state owned enterprise is the sole licensed exporter of fish products. The fish is transported to Cambodia's sole seaport in Sihanoukville, from where it is sent to other countries. According to KAMFIMEX, the most lucrative markets for export are United States and Australia, followed by China, Hong Kong, Singapore and Malaysia; Thailand and Vietnam.

In Indonesia, exports of fish products in 2001 amounted to 0.48 million tons, bringing in a foreign exchange earning of USD 1.63 million. The export value of fish products in

2002 declined to USD 1.57 million with export volume reaching 0.51 million tons; but in 2003 export value swelled to USD 2 billion with export volume of 696 thousand tons.

Rural Finance/Microfinance for Small Scale Fisheries

For discussion purposes, it is important to define and distinguish rural finance from agricultural finance and microfinance. First, as defined by the World Bank, rural finance is the provision of a range of financial services such as savings, credit, payments and insurance to rural individuals, households, and enterprises, both farm and non-farm, on a sustainable basis. It includes financing for agriculture and agro-processing. On the other hand, agricultural finance is defined as a subset of rural finance dedicated to financing agricultural related activities such as input supply, production, distribution, wholesale, processing and marketing. Finally, microfinance is the provision of financial services for poor and low income people and covers the lower ends of both rural and agriculture finance as seen in the diagram below:

Figure 17.1. The Relationship between Rural, Agricultural and Micro Finance

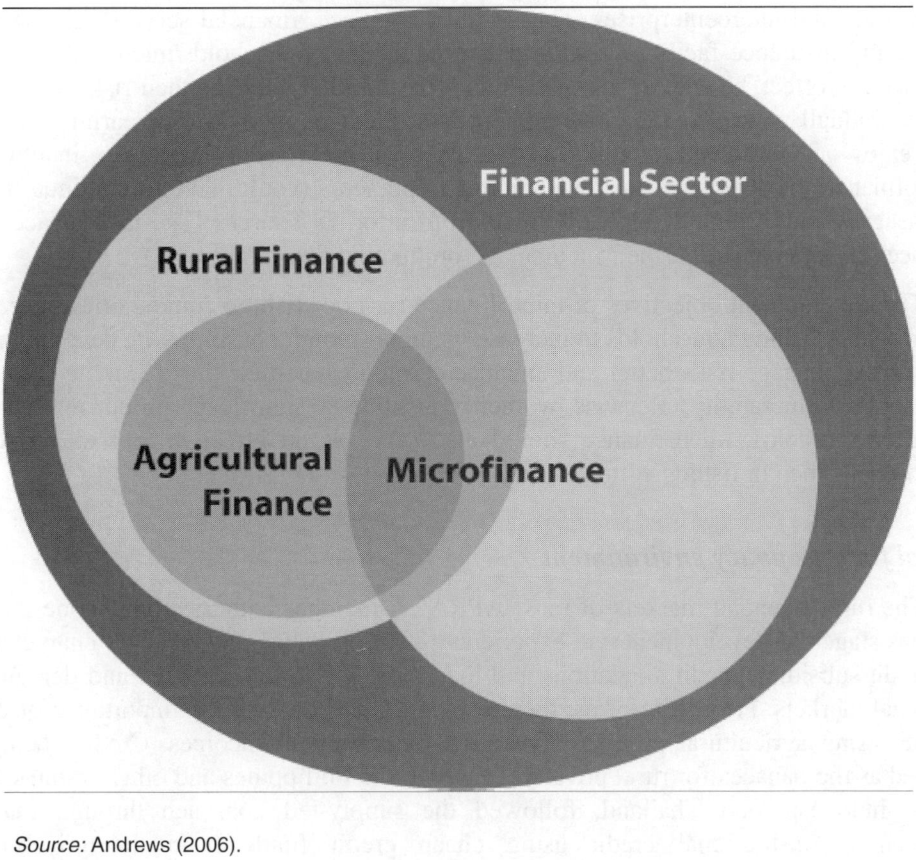

Source: Andrews (2006).

Importance of rural finance/microfinance for small fisheries development

Among APRACA member countries, credit is deemed necessary to complete the package of support services for the sustained development of the small fisheries sector. There is empirical evidence that microfinance can help the poor especially in terms of smoothing and increasing income (Robinson 2001). Remenyi and Quinones (2000) noted, for instance, that in Indonesia, an average annual increase of 12.9% in the income of borrowers was observed compared to the 3% income increase among non-borrowers. Moreover, among microfinance borrowers in Bangladesh, a 29.3% annual average rise in income was recorded while a 22% annual average increase in income was estimated among non-borrowers. Sri-Lanka also showed a 15.6% rise in income among borrowers and a 9% rise among non-borrowers. In the case of India, a 46% annual average rise in income was reported among borrowers while non-borrowers only showed a 24% annual average increase in their incomes. Morduch and Haley (2002), thus, indicate that under the right conditions, microfinance is an instrument that can address the needs of a broad range of the population including vulnerable and disadvantaged groups, such as small-scale fishing and fish farming households. Development practitioners and policymakers, therefore, view microfinance as a practical solution to the growing demand for financial services by the poor and to the reality that most formal financial institutions do not serve the poor because of perceived high risks, high costs involved in small transactions, perceived low profitability and inability of the poor to provide the required physical collateral (Asian Development Bank (ADB) 2000). Most if not all of these financial institutions have a business culture that is not geared to serve the poor, low-income households and microenterprises. Through microfinance, financial services like savings, credit, and insurance facilities can be delivered to poor households/micro-entrepreneurs who will, in effect, be able to smoothen their consumption, manage their risks, build their assets gradually, develop their microenterprises, enhance their income earning capacity and enjoy an improved quality of life. Without permanent access to institutional microfinance, most poor households/microentrepreneurs would have to continue to rely on meagre funds from savings and credit from informal sources. The lack of access to finance services constrains their income and production capacities.

The development objectives of microfinance for poor fishing communities, therefore, are to enable fishing households to increase income, smooth consumption, develop micro-enterprises, manage risks better and enhance earning capacities, thus reducing economic and social vulnerability. Because women constitute a significant proportion of poor fishing households, microfinance should also serve as an effective tool to assist and empower women in fishing communities.

Rural finance policy environment

The rural financial markets of most APRACA-member countries have gone through various stages of development and experience - from a policy environment characterized by credit subsidies, credit allocations and loan targeting to a liberalized and deregulated financial markets. Prior to reforms, there was much uproar over the importance of credit in increasing agricultural productivity and rural household incomes. Credit, then, was viewed as the panacea for rural poverty. In effect, the Philippines and other countries like India, Indonesia and Thailand, followed the supply-led approach through massive infusion of institutional credit using cheap credit funds from the government. (Llanto 2005). In the Philippines, for instance, various commodity-specific agricultural credit programs such as the *Masagana 99* for rice, *Masaganang Maisan* for corn, *Gulayang Pangkalusugan* for vegetables, and *Bakahang Barangay* for livestock, among others were implemented through rural banks as conduits. Rural banks were 'gifted' with 'cheap' funds (*i.e.*, at very low interest rates) via the Central Bank's rediscounting window as an incentive for them to lend to small farmers. Without the 'incentive', these banks would normally not lend to these farmers because lending to the latter is generally perceived as risky and quite costly to administer. Moreover, some government non-financial agencies did the lending themselves, providing loans directly to program beneficiaries but without putting in place an appropriate loan collection mechanism.

Similarly in India, government has intervened heavily through the well-known 'melas' in the 1980s, in which large volumes of funds were imprudently issued as subsidized loans to the supposedly weaker segment of society and loan waivers offered until 1991. Subsidised credit programs were likewise implemented in Indonesia through its two nationwide programs specifically created to intensify agriculture, stimulate rural non-farm enterprises and to increase rural employment: *(1)* Bimas Rice Intensification Scheme, and *(2)* the Small Investment and Permanent Working Capital Schemes. The Bank Rakyat Indonesia (BRI) unit *desas* were selected to channel subsidised credit to rice farmers.

For most of these countries, however, the positive effects of a supply-led strategy were short-lived since farmers started to default on their loan repayment, government funds dwindled and a great number of rural banks closed shop and went bankrupt. While some farmers blamed the weather, others simply refused to pay because of the notion that government funds are a dole-out and need not be returned. In addition, the government agencies that implemented credit programs did not exercise due diligence in loan collection. And because of too much dependence on cheap funds from the government, most rural banks neglected deposit mobilisation and demonstrated leniency in loan screening and collection which led to poor repayment rates.

Given this experience, governments started to liberalize and deregulate the financial markets. Having an enabling rural financial policy environment was the goal of every government. Greater reliance on market principles especially in pricing as well as allocating credit funds was therefore the basic framework for reforms. In effect, interest rates were deregulated and subsidies gradually removed. Majority of the APRACA-member countries thus, liberalised and deregulated their rural financial markets although in varying degrees. Countries like the Philippines, Cambodia and India, for example, follow a full market-determined interest rate policy regime wherein all financial institutions are given the freedom to set their own interest rates. The importance of such a policy is to allow banks and other lending institutions to cover their costs in lending so that their operations can be sustained and credit to the rural areas can continue to flow. However, while the Philippines has totally removed its anti-usury law, India has maintained the same law and thus, still keeps a close watch over the level of interest rates in order to ensure that the rates charged by the institutions do not become usurious according to the provisions of the law. In Cambodia, the government has also stopped putting ceilings on interest rates but it (the government) continues to negotiate with microfinance institutions (MFIs) for them to bring down their interest rates.

In Thailand, banks are said to be allowed to charge their own rates but interest rate ceilings are imposed for Thrift and Credit Cooperatives at 19%; agricultural cooperatives and informal organizations at 15%; BAAC and GSB at 19%. While Indonesia has adopted market-oriented interest rates as a policy in credit delivery among microfinance institutions, it continues to practice subsidized credit and loan targeting under special programs of the government for specific sectors.

Most of the countries have stopped direct lending by government non-financial agencies and therefore, provides financing services through banks and other financial institutions.

Available rural finance/microfinance services for the farming/fisheries sector

Since these countries continue to recognize the importance of rural finance/microfinance in poverty alleviation especially in the small-fisheries sector, they have worked towards the strengthening of their financial institutions in order to improve the delivery of credit (as well as savings) services to the poor. For instance, in Thailand, the Bank for Agriculture and Agricultural Cooperatives (BAAC) is focused on providing financial services to farmers and fishermen as well as to agricultural cooperatives and farmer/fisher associations. In addition to credit, BAAC supports projects that link production, processing and marketing systems in order to increase production efficiency, increase value added and income of farmers/fishers. BAAC also includes training or institution building as a critical component of the credit package.

Box 17.5. BAAC's Full Package of Assistance

BAAC has been coming up with measures to reduce risk in lending to the fisheries sector. Since 2003, BAAC has been promoting among these farmers, the guidelines issued by the Department of Fisheries including the Good Aquaculture Practice (GAP), Code of Conduct (CoC) for responsible aquaculture including Organic Aquaculture Approach. BAAC aims to educate and enrich skills of client-farmers/fishers. Moreover, starting in 2005, BAAC has participated in "Contract Farming System" arrangements. In this case, marketing of products is assured even before production. While other agencies provide technical assistance, BAAC provides the loans according to their regulations.

Like BAAC of Thailand, the Land Bank of the Philippines is the institution mandated to provide small farmers and fishers in the Philippines access to credit and other financial services. Through its Agrarian and Domestic Banking Services Sector, LandBank designated and deployed Small Fisherfolk Specialists in its Regional Development Assistance Centres to focus on serving the credit needs of small fish farmers. Land Bank also launched the LandBank Support Program for Fisheries which aims to provide loans to small fishers. Under this program, about P873 million and P882 million were allocated in 2004 and 2005, respectively. In terms of retail credit channels, LandBank utilised accredited cooperatives, countryside financial institutions such as rural banks and the Quedan and Rural Credit Corporation (QUEDANCOR).

In Cambodia, the financial sector is still in its infancy. Banks have to this point played only a small role in savings mobilization and financial intermediation and their operations are generally confined to Phnom Penh (Fukui and Llanto 2003). Nearly 40% of the people have no access to formal bank branches and only 6% of total banking sector advances are for agriculture or related activities. Cambodia does not have any specialised public bank, or other financial vehicle, that extends retail services to farmers. Such an institutional *vacuum* in formal financial supply seems to characterize the basic environment of Cambodian rural finance; there is high anticipation for micro-finance development to fill the huge gap in demand and supply of rural financial services. NGOs which became active in the late 1970s with the humanitarian approach have gradually moved out of relief and rehabilitation works and redefined their roles in interventions in the development process. Among those interventions, micro-credit undertakings have thrived since the early 1990s, and most of the country's 24 provinces now have micro-credit services provided by NGOs (reportedly numbering 72 in total). Cognizant of serious lack in rural financial services, the government has adopted a series of policy measures, supported by international donors[1] which have been in effect since the late 1990s. While institutions have yet to be substantially developed, currently, specialized banks engaged in micro-finance and NGO MFIs are integrated into the regulations of the NBC. With technical and financial supports by donors, the NBC developed banking legislation that includes registration and licensing requirements and some prudential standards for larger MFIs. As of the end of 2001, there were 32 registered NGOs and MFIs, the top five of which dominated more than 80% of the aggregated loan portfolio, including ACLEDA Bank (specialised bank), the EMT and Hatthakaksekar (licensed NGOs), and PRASAC and Seilanithih (registered NGOs). In total the rural financial institutions reach around 420 000 borrowers, which represent about 23% of the rural households. There is no available information on specific programs for the fisheries sector.

Box 17.6. Growth of ACLEDA Bank

The rapid growth of ACLEDA Bank is phenomenal: after only about three years in operation as an NGO micro-finance project, ACLEDA transformed into a micro-finance specialized bank in 1995, and it, as of the end of 2002, had 75 branches in 14 provinces and a workforce of 863. Its loan outstanding had grown to USD 27 million, serving more than 80 000 borrowing customers.6 Fully supported by international donors since its establishment, ACLEDA Bank has been expanding its range of products in retail banking to include loans to small and medium enterprises, cash management and money transfer services, and deposit services to the public. Savings mobilized by ACLEDA's deposits are yet small (USD 5.7 million in 19 070 accounts as of the end of 2002), but have been growing rapidly in recent years.As regards the RDB, a governmental wholesale conduit to MFIs, the loan outstanding (to five financial institutions) was USD 1.9 million as of January 2002, with an additional USD 20 million available in credit from the Asian Development Bank (ADB) to be on-lent to MFIs.

Source: Fukui and Llanto (2003)

In Indonesia, it is worthy to note that following the collapse of the Bimas, the Bank Rakyat Indonesia (BRI) unit system became more responsive to the needs of farmers. The unit Desas was transformed into full service rural banks; each unit was treated as a discrete profit or loss center within BRI; and the units were evaluated on profitability rather than on hectares covered or money lent. One of the unique features of the unit Desas is that they make individual loans based on collateral, usually in the form of land, and loans are made for one to three years. Local village officials are involved in the screening by acting as character references for the borrowers. The transformation of the BRI unit Desa system produced spectacular results in outreach and financial performance. Among the many types of formal microfinance in Indonesia, therefore, BRI Unit and Rural Banks (BPR) have been recognised as the microfinance leaders that serve micro, small and medium enterprises (MSMEs).

1. (i) establishment of the Credit Committee for Rural Development (CCRD) in 1995;

(ii) introduction of a framework in the Banking Law to enable eligible NGOs and other rural finance providers to become regulated micro-finance institutions;

(iii) creation of a unit in the National Bank of Cambodia (NBC) to supervise and monitor MFIs; and

(iv) establishment of an apex institution to provide financing for MFIs, namely the Rural Development Bank (RDB).

Box 17.7.Bank Rakyat of Indonesia

BRI is a state-owned bank which ran a programme of directed subsidized credit for rice farmers until 1983. The Unit Desa (BRI-UD) or Village Bank, was established in 1984 as a separate profit center within BRI, under a General Manager who reports directly to the BRI Board of Directors.

BRI-UD consists of a nationwide network of small village banks. The founding objectives of BRI-UD were to replace directed agricultural credit with broad-based credit to the rural population involved in any type of economic in any type of economic activity; to replace subsidized credit with positive on-lending rates with spreads sufficient to cover all financial and operational intermediate costs; and to provide a full range of financial services (savings as well as credit) to clients. All these objectives were achieved only a few years after the programme's inception, and BRI-UD's phenomenal success in savings mobilisation became its distinguishing achievement. Various other financial institutions including the Badan Kredit Kecamatan (BKK) which target the extremely poor rural population are operating along similar guidelines in the Indonesian rural lower-income market.

In India, one of the most successful microfinance programs in the country is the Self Help Group-Bank Linkage Program. Under this scheme, a number of NGOs organised the poor into informal self-help groups (SHG). An SHG is a small, economically homogeneous group of rural poor, generally not exceeding 20 members, voluntarily coming together for mutual help and benefit. There are three main linkage models involving the banks, NGOs and SHGs that promote microfinance in India. Through the National Bank for Agriculture and Rural Development (NABARD), these linkage models were implemented and could perhaps be the largest microfinance program in the world in terms of outreach.

More recently, microfinance interventions among fishers have been concentrated mainly in the States of Kerala, Tamil Nadu and Andhra Pradesh in South India and a few initiatives in Orissa and West Bengal in East India. In particular, the South Indian Federation of Fishermen Societies (SIFFS) is a NGO working in the marine fisheries sector in the States of Tamil Nadu and Kerala. The SIFFS provides loans for fishing equipment, facilitates marketing of fish caught by the members and promotes savings.

Box 17.8. Self-Help Group-Bank Linkage Program in India

This aims to make improvements in the existing relationships between the poor and the banks, with the intermediation of NGO's, which either plays the role of promoters of SHGs or of financial intermediaries. Model I – the bank itself is the self-help group promoting institution (SHGPI) that trains and provides credit directly to SHGs; Model II – the NGO acts as the SHGPI that trains and helps the SHG link with the banks; Model III – the NGO acts as the financial intermediary between the bank and the SHG. As of March 2002, Model II was the most prevalent linkage mechanism, involving around 75% of the more than 460,000 SHGs organised in India, followed by Model I (16%) and Model III (9%)

Extent of outreach: Agriculture and fisheries credit performance

Despite continued efforts by governments to continue to find ways of making credit for small farmers/fishers adequate, timely and sustainable, marginal farmers and fishers continue to have limited access to credit. Based on studies done by the Agricultural Credit

Policy Council (an APRACA-member institution), the following findings indicate that the provision of loans to the agriculture and fisheries sector needs to be improved further:

1. Formal banks' lending to agri-aqua-agra production declined in real terms, from less than P 100 billion in 1980 to P 44 billion in 1986 and P 86 billion in 2004 (see Figure 17.1).
2. The ratio of agricultural production loans to total formal loans granted also declined from 7% in 1986 to 4.0% in 1997 and dropped further to 0.9% and improved slightly to 1.1% in 2006 (Table 17.7).
3. Informal sources of finance continue to dominate and have not changed much from 1981-1982 to 2001-2002 at a 60:40 ratio in favour of informal credit. In fact, the ratio has become worse in 1986, 1988-1989, 1990-1991, and 1997-1998. The usual indictment is that formal finance is persistently inaccessible while informal finance is still limited and/or very costly to compensate for its high risk and high transaction costs (Table 17.8). The government has meddled too much and has pursued more wrong, than right, interventions (such as directed credit to non-financial government institutions, subsidized credit, credit rationing), thereby resulting in many pitfalls and expensive blunders.
4. Credit from non-bank financial institutions (*e.g.*, lending investors, credit unions, and cooperatives, self-help groups, NGOs, microfinance institutions) has been generally increasing in terms of loan volume and outreach, but they provide loans mainly for non-agricultural purposes and therefore, has not reached out to small fishers.

The same findings seem to hold true across other countries in Asia including Indonesia where lending to the small fisheries sector is still considered a highly risky business by the banking industry; and in India where credit to fisheries from the banking sector is but 1% of the total credit that goes to agriculture.

Major Issues, Constraints, and Opportunities in Aquaculture Credit

Based on country papers and review of related literature, the major issues in aquaculture credit are as follows:

- Very limited access to both formal and informal credit among the enterprising, labouring, and low-income fish farmers (earning less than P 100 per day).
- Lack of viable and profitable aquaculture projects and the attendant risks (technology packaging; weather aberrations such as typhoons, floods and drought; business and financial risks, peace and order, market risks).
- Limited the provision of credit to fish farmers by the rural financial institutions. In fact, LandBank, despite its Small Fisherfolk Financing Program, which has at least P800 million for credit allocation to small fish farmers, could only reach less than 4 000 fish farmers in 2004 and 2005, representing a meagre 0.24% penetration ratio vis-à-vis the more than 1.6 million small fishers in the country.
- On the demand side, fish farmers are not considered creditworthy mainly because *(i)* they lack physical and livelihood assets, including land (capital) that

can be used as collateral; *(ii)* they do not possess technical or technology expertise (capacity); *(iii)* they tend to be more individualistic, are not functionally organised, are geographically dispersed, and have no critical mass to warrant economies of scale in production, marketing, and post-harvest handling operations (condition); and *(iv)* they do not have good credit history, and the banks do not have a reliable database on their financial, livelihood, and character standing (capacity to pay).

- Further aggravating the lack of credit access is the very high cost of credit.

The Biggest Challenge: How to Make Microfinance Work in Agriculture and Fisheries

Therefore, the biggest challenge faced by policymakers and practitioners today is how to make microfinance work in agriculture and fisheries. A major difficulty in making microfinance work for agriculture and fisheries projects is the fact that microfinance loans are paid in small amounts and collected regularly in small amounts while agriculture and fishery projects need lump sum amounts and have long gestation periods which make it difficult for borrowers to pay small amounts on a regular basis. Making microfinance work in agriculture can be achieved by first, identifying the challenges faced by banks and other financial institutions providing agricultural credit in order to know what to address and how to address these challenges. Andrews (2006) came up with a list of factors unique to rural and agricultural markets that constrain both the supply and demand for credit, as follows:

- High transaction costs for both borrowers and lenders.
- Generally lower population density and dispersed demand.
- Often limited economic opportunities available to local populations.
- High risks faced by potential borrowers and depositors due to the variability of incomes, exogenous economic shocks and limited tools to manage risk.
- Seasonality – potentially affecting both the client and the institution.
- Heavy concentration on agriculture and agriculture related activities exposes clients and institutions to multiple risks, both idiosyncratic (one household) and covariant (entire region or country).
- Lack of reliable information about borrowers.
- Lack of market information and/or market access.
- Weak institutional capacity – including poor governance and operating systems, low staff and management skills.
- "Crowding out" effect due to subsidies and directed credit.
- Lack of adequate or usable collateral (lack of assets, unclear property rights).
- Risk of political intervention, which can undermine payment morale through debt forgiveness and interest rate caps.
- Inhospitable policy, legal and regulatory frameworks.
- Undeveloped legal systems, inadequate contract enforcement mechanisms.

- Undeveloped or inadequate infrastructure.
- Land held may be too small to be sustainable or located too remotely to be reached efficiently.
- Individuals may be dependent upon only one crop with no other external sources of income.

There could be no blueprint for a program or scheme that could address these challenges because the design, methodology, products and delivery of services need to be modified to fit the specific country, region, cultural context, operating environment and even the challenges affecting an institution. Institutions must be able to identify and consider the peculiarities of their clients as well as the different factors affecting their respective areas of operation and understand that what works on Southeast Asian farms may not be applicable in Sub-Saharan Africa on small holder plots of land and *what is successful in highly populated rural areas of Bangladesh is not easily replicable in the more remote rural environments of Nicaragua, Peru, or even Afghanistan* (Andrews 2006).

Nonetheless, some lessons learned from the implementation of past credit programs in the Philippines (Kraft 2001) which may be considered when designing a microfinance program for agriculture and fisheries sector are as follows:

- In the design of a credit program, it is imperative *"to establish linkages and networking."* This requires well-planned social marketing of the project, effective institutional development and training program in community organizing, technology packaging and transfer, and strengthening of fishers, cooperatives and associations, particularly their savings and lending capability and creditworthiness as well as support to project development and packaging, overall project execution and management, and market linkage. There is also *an urgent need to implement a comprehensive aquaculture-based financing strategy and practical demand-led schemes and innovations.*

- As an integral component of well-phased and appropriate policy reforms and conducive climate for investment in aquaculture and potentially viable alternative livelihood projects, *the following support interventions will help improve the investment climate and viability of aquaculture finance: (i)* provision of public or social goods (support infrastructure facilities (*e.g.*, farm-to-market roads, rural bridges, irrigation and drainage facilities, electrification, transport and logistics systems, post-harvest facilities, potable water supply) which reduce production and marketing costs; *(ii)* technical assistance via technology transfer and commercialization in terms of the research, development, and extension (RD&E) support continuum; and *(iii)* social, educational, and health services that enhance the productive capacity of fisherfolk.

- Institutional arrangements that will encourage the greater participation of private banking institutions in aquaculture credit, through the institution of positive changes and reforms at the demand and supply side of credit, will minimise lending costs and risks. Moreover, appropriate institutional arrangements are required in mobilising credit supply and availability through a national network of rural viable credit cooperatives.

- Improving credit delivery and repayment performance of fisheries project involve a total household cash flow approach to developing technically and economically viable land-based income-generating projects that will augment income from aquaculture on a sustainable basis. Such improvement in credit delivery will enhance creditworthiness and credit repayment performance.

Bibliography

Bulcock, P. and W. Savage (2005) STREAM Media Monitoring Reports (January to December 2006) Bangkok: NACA-STREAM.

Burke, L., E. Selig and M. Spalding (2001). "Reefs at Risk in Southeast Asia". World Resources Institute, Washington D.C., USA.

Delgado, C. et al. (2003). *Outlook for Fish to 2020*, International Food Policy Research Institute (IFPRI),Washington, DC, USA.

IMM, CFDO and CBNRM LI (2005). *Understanding the Factors that Support or Inhibit Livelihood Diversification in Coastal Cambodia.* An output from DFID-funded research in Cambodia. Exeter,UK, IMM Ltd.

Llanto, G. M. (2005). *Overcoming Infrastructure Constraints to Economic Growth in the Philippines,* Philippine Institute for Development Studies**,** Makati City, Philippines.

Macfadyen, G., R. Banks, M. Phillips, G. Haylor, L. Mazaudier and P. Salz (2003). "Background paper on the International Seafood Trade and Poverty". Prepared under DFID (U.K.)-funded by EC-PREP project (EP/R03/014) *International Seafood Trade: Supporting Sustainable Livelihoods Among Poor Aquatic Resource Users in Asia.* Poseidon AquaticResource Management Ltd (UK), Network of Aquaculture Centres in Asia-Pacific and STREAM Initiative.

National Anti-Poverty Commission (NAPC) (2004), *Thrusts of NAPC for 2004-2010*, Manila, Philippines.

Remenyi, J. and Quinones, B., (2000), *Microfinance and Poverty Alleviation.* Pinter, London, U.K.

Silvestre, G.T., Garces, L.R., Stobutzki, I., Ahmed, M., Santos, R.A.V., Luna, C.Z., Zhou, W., (2003). "South and South-East Asian coastal fisheries: their status and directions for improved management: conference synopsis and recommendations". In: Silvestre, G., Garces, L., Stobutzki, I., Ahmed, M., Valmonte-Santos, R.A., Luna, C., Lachica-Aliño, L., Munro, P., Christensen, V., Pauly, D. (Eds.), *Assessment, Management and Future Directions for Coastal Fisheries in Asian Countries.* WorldFish Center Conference Proceedings 67, WorldFish Center, Penang, Malaysia, pp. 1–40.

Sugiyama S., Staples D., and Funge-Smith, S.J. (2004). "Status and potential of fisheries and aquaculture in Asia and the Pacific" RAP Publication 2004/25.. Bangkok: FAO Regional Office for Asia and thePacific.

Suspita, A. (2006). "The current status of Mariculture in Indonesia and Livelihood Opportunities for Coastal Communities. The future of mariculture: A regional approach for responsible development of marine farming in the Asia-Pacific Region". Bangkok, Thailand STREAM.

Tana, T. S. and Todd, B.H. (2001). *The Inland and Marine Fisheries Trade of Cambodia.* Phnom Penh, Cambodia: Oxfam America,

Tacon, A. G. J. (2001). *Fish Meal and Fish Oil: Global Supply, Demand and Outlook.* Report to the International Food Policy Research Institute, Washington,D.C., USA.

Wilfredo G., Funge-Smith S., Rimmer, M., Phillips, M.J., Sim S.J., Kongkeo H. and Bueno, P.B. (2006) *An Overview of Aquaculture in Asia-Pacific and the Outlook for Mariculture in Southeast Asia.* Bangkok:NACA

Chapter 18

Globalised Out?
A Case for Fish Trade in Developing Countries

Stephen Mwikya, Consultant, Kenya

Stephen Mwikya presented an overview of the challenges that developing countries today face in the era of globalised fisheries markets. 60 out of 77 ACP (African-Caribbean-Pacific) countries have significant fish stocks, as well as significant socio-economic and socio-cultural linkages that relate to these fish stocks. However, these linkages are threatened through risks to local fishing communities, by difficulties in entering the global value chain and through access agreements. Access agreements can also undermine stock sustainability in developing countries. The presence of subsidies in some developed countries' fleets aggravates the problem.

The desire of many African countries is to benefit from their natural resources and subsequent trade in fish stocks. Instead, many find that access negotiations are skewed, that MCS (monitoring, control and surveillance) networks are expensive and that rules may be undermined as a result. In addition, many face challenges in retaining value-addition within their own countries, as well as problems gaining market access due to tariff escalation.

For Stephen Mwikya, the policy challenge for developing and developed countries with respect to globalisation in fisheries is to construct an international policy framework that enhances development, taking into account socio-cultural linkages in developing countries' fisheries. This is a shared responsibility but one that would ensure that developing countries are able to obtain a fair value for access to their resources, capture more value along the value chain, and adopts technologies that enhance value addition.

Introduction

In line with Millennium Development Goal number one, those of eradicating hunger and extreme poverty, developing countries are increasingly seeking to develop their natural and human resource potentials. Of the 77 African, Caribbean and Pacific

Countries (ACP), 60 have significant natural capture fisheries resource potential that is either subject to international trade, or can be commercialised easily. Fisheries are the only major natural resource in some of the ACP countries, especially Small Island States. In some cases such as Mauritania, Kiribati and Tuvalu, revenue from this resource accounts for more than 40% of Gross Domestic Product. This level of dependency on fisheries requires both national and international policy directed at ensuring that developing countries capture maximum value from trade in their fisheries resources.

Fish in developing countries, perhaps more than any other natural resource, is highly globalised. Fishing in Exclusive Economic Zones (EEZ) of most developed countries is almost entirely conducted by Distant Waters Fishing fleets from developed countries, with host countries being reduced to side-spectators awaiting some financial compensation from these foreign fishers. This is made possible by decades of subsidies by Distant Waters Fishing nations to their DWF fleets, during which they build their competitive edges beyond the reach of host developing nations. In the processing sector, imposition of stringent sanitary and phytosanitary standards and high bank interest rates have made it almost impossible for local investors to become competitive exporters. Many exporters in developing countries are therefore from developed countries, where they are able to access affordable credit. Developing countries have been gradually getting globalised out of their fisheries.

The expectation that developing countries will use trade in their fish as a tool for economic development presupposes that fisheries trade applies perfectly the laws of supply and demand, where there are many buyers and sellers, with no integrations amongst capture, processing and retail levels of the chain significant to cause monopolistic tendencies. Well, as discussed throughout this paper, this is not the case in developing countries fisheries. On the outset, 'globalised out?' seems an illogical question to pose, since capture fish belongs to whoever has legal right to its ownership. It first belongs to the State, since no one directly nurtures its growth, then the fisherman, the trader, processor, supermarket and eventually the consumer. This may be considered the value chain of fish, where every 'owner' has a right to bargaining for a price commensurate to the value they confer, and also an obligation to ensure every other level of the chain gains value from the commodity.

Fisheries in developing countries, especially coastal and inland capture fisheries, mainly consist of artisanal or small scale fishers. Many of these countries practise 'open access' policy, where as many fishers as may wish gain access to fishing on the basis of being riparian communities. This poses a particular challenge to fisheries management in ensuring that exploitation is kept at sustainable levels and in some cases has led to over exploitation especially in shallow or small inland water bodies. Their efforts are however aided by the fact that majority of these fishers use low technology crafts that are unable to exploit deeper waters which are almost exclusively left to industrial fishers. About 30% of fish harvested by small scale fishers is lost through spoilage due to lack of preservation facilities, and efficient infrastructure to move the produce to marketing centres. This presents the first policy challenge to developing countries: should it be normal practise to permit fishing access to persons who have no capacity to preserve and present their catch to the market in a wholesome and timely manner, or should a state encourage globalisation of its fisheries in order to drive standards up, minimise post harvest losses and maximise rent collection? The question is made more pertinent considering that the fishers do not pay for the access, and therefore this represents a resource rent loss to the state, even though the fishers benefits. To what extent should the Government assist (*e.g.* in preservation of fish) those who are already privileged to have free open access?

Generally, those who have the privilege of being granted access to a fishery on behalf of the rest of the public have an obligation to ensure that it contributes to economic development.

Further, fish processing such as drying and smoking in developing countries represents a price value loss, not addition. Dried (and often smoked) fish in many parts of Africa usually retails at about the same or even less price that fresh fish. This is despite the fact that drying fish results in about 60-70% loss in weight (moisture content of fish is about 70%), and therefore, to break even on raw material cost alone, dried fish should be at least twice the price of fresh fish, on weight basis. These technologies are mainly practised to save fish from spoilage by attaining shelf-stable products that may be easily distributed widely. This is not unique to processing in developing countries, because fish canning may also be regarded as a price value loss (the world market price for canned tuna is about USD 1/kg, and that for fish for canning is also about USD 0.9-1.2/kg). Considering the urgent need to use fisheries as a tool for economic development, any transformation that results in loss of value needs to be discouraged.

With the exception of Lake Victoria in East Africa, the bulk of international trade in African, Caribbean and Pacific Countries is based on capture fisheries in their Exclusive Economic Zones (EEZ). Almost all fishing in these developing countries waters is carried out by Distant Waters Fishing (DWF) Fleets from nations such as Spain and France (the European Union), Japan, USA, Chinese Taipei, South Korea and China. These foreign fleets operate under fishing access arrangements with the host countries which involve payment of a sum of money for (sometimes a defined amount of) fish. On average, financial compensations to ACP countries for fishing access is between 3-6% of the value of catch. This is a small compensation level, if resource rent principles as applied in other natural resources such as minerals, forestry and crude oil are considered, and also bearing in mind that investment in fishing is much smaller when compared to others such as mining. Even though ACP countries are aware of these unfair compensation levels, they have so far been unable to bargain for better terms from the DWFN (Distant Waters Fishing Nations). This represents the first major distortion in international trade value chain in ACP fish.

Tropical fish fillets, whether tuna, tilapia, Nile perch or others almost invariably are bought from developing countries exporters at about USD 4/Kg fob, and sold at about USD 14/Kg (supermarket price). This implies that about USD 10 value is captured at the developed countries level, even though there is hardly any value processing in these destination markets. The net effect of this is that net margins (including labour and energy costs) in developing countries are rarely more than USD 0.10-0.15/kg. This in turn leads to low landing prices for fishers, which in turn encourages them to fish more to make ends meet, and this threatens stocks sustainability, and enhances poverty. Developing countries traders are unable to 'follow their fish' to market in developed countries and capture these lucrative margins because of restrictions in trade in services imposed by these markets, such as movement of persons from developing countries to developed ones for genuine business.

Fisheries trade also suffers from restrictions targeted at value addition. Several markets impose tariffs targeted at punishing value addition and rewarding export of raw fish. This policy is aimed at using raw material fish to support employment in developed countries markets. This may be a legitimate right of buyers, but becomes morally unjustifiable when very little value is captured at developing countries level where fish happens to be the only economic mainstay. Inability by many developing countries to meet stringent sanitary and phytosanitary standards (which become more important during value addition) has often been cited as the main reason why developed countries are justified in importing raw material fish for further processing in their home factories. There has also been an argument that several of the final products from fish demanded by consumers in developed countries are so specialised that the technology and skills required are unavailable in many developing countries. The issue of quality control standards is disputable because several establishments in developing countries have now been able to attain standards comparable to those in the most demanding markets. Some establishments in ACP countries have been able to, in collaboration with firms in export destinations, make specialty value added products that are cost effective, and delivered to the markets in a timely manner. The Maldives are now making 'katsuobushi', a traditional Japanese smoked and dried tuna flakes, and fresh Lake Victoria nile perch fillets arrive in European supermarkets in less than 48 hours from capture. To a great extent, the private sector in developed and developing countries have been trying, in the absence of policy support from either side, to bridge this trade divide by direct buyer-seller collaboration and capacity transfer efforts to facilitate flow of fish for trade.

Fisheries in developing countries will continue to make only marginal contributions to economic development unless developing countries are able to bargain for fair value for its access, capture more value along the value chain, and adapt technologies that enhance value addition. This paper asks the difficult question: To whom does fish belong in developing countries' waters? If it's obvious that it belongs to these countries, do they capture sufficient value to prove it? The paper discusses factors that limit developing countries (mainly in African Caribbean and Pacific Countries) from maximising their benefits along the fish value chain and value addition. In other words, how can developing countries benefit from the process of globalisation and how these countries can be linked into this process and benefit from it? In order to present a holistic view, the paper discusses value chain issues both at developing countries level and in export trade. It also offers suggestions on aspects for consideration in drafting policies aimed at maximising economic benefits to the ACP from their fisheries.

Value Chain in Domestic and Regional Trade

There is insufficient data on volumes of fish harvested artisanally or by small scale fishers and traded domestically or regionally in developing countries. In reality however, almost all lakes, rivers, coastal and reef fishing in most of these countries is harvested by this segment of fishers. Some of this fish such as tilapia, Nile perch and marine fishes such as kingfish and sea-bream is exported to developed countries destinations by factories that bulk and process them into high quality products. A greater volume of fish harvested artisanally by small scale fishers is however traded in markets around fishing areas, and also in major towns in regional countries, often after being processed in shelf-stable forms such as dried and smoked fish. In terms of value chain, this trade can be categorised into artisanal and small-scale commercial trade. This trade greatly helps in poverty reduction in developing countries because it often involves many persons in its various stages, when compared to industrial fishing and developed countries export fish trade.

Artisanal fish trade

Under artisanal fish trade, the fish is usually harvested for domestic consumption, often with limited trading among immediate riparian communities along the water-body. The fishers use non motorised crafts, and volumes harvested are limited to average sales volumes in these surrounding markets. The fishers are quite selective on species harvested, since consumption of some species is considered a taboo. In Ethiopia for example, artisanal fishers around Lake Tana (the country's largest lake) deliberately avoid harvesting catfish (which occurs in abundance there) because its consumption is disallowed by communities around the lake, and instead go for tilapia and nile perch. A typical fish marketing chain for artisanal fisheries is illustrated in Figure 18.1 below:

Figure 18.1. Artisanal Fish Value Chain

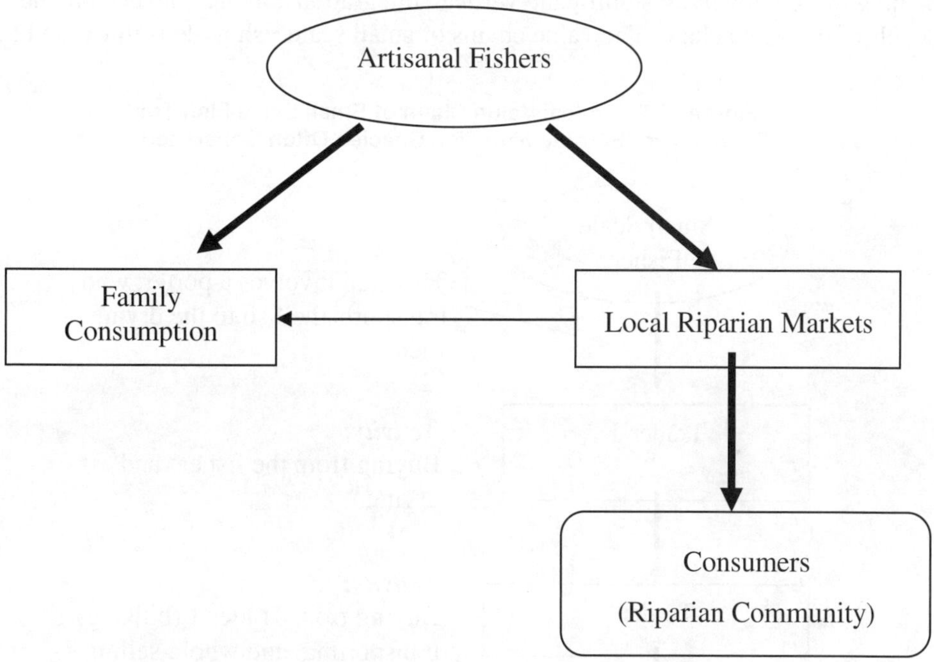

There is hardly any fish preservation in artisanal fish trade, and the catch is usually sold on the same day. Some limited volumes may be processed into dried and smoked forms for marketing in distant (beyond 10 km from water bodies), even though this level of trade is minimal in these fishing communities.

Small scale fish trade

There is still no exact definition of artisanal and small scale trade, but it is now generally agreed that the two differ mainly in extent of commercialization, and types of crafts used. A small scale fisher is a professional trader, whose main thrust is to market his catch, and not feed family as is the case in artisanal fishers. He or she will often spend most of the day fishing, and will seek to sell catch as widely as possible, hence may harvest species his/her immediate community may not normally consume. Such fishing includes small scale octopus and squid fishers in North African Countries of Morocco and

Mauritania and lobster fishers in Lamu (Kenya) and Somalia. Crafts used in small scale fishing vary in size and sophistication, but are usually limited to mainly manual or single outboard engine vessels often not exceeding 15 metres in length.

Small scale fishers often use various means of transportation to transport their catch as fast as possible to markets. These include transport boats, bicycles, passenger transport vehicles and even small trucks. Unfortunately due to unhygienic handling facilities in the boats and landing sites, and lack of ice and other preservation facilities, about 30% is usually spoilt. The level of spoilage increases the further the fishers have to travel to reach market centres. This in turn affects price, as fresh fish attracts a much higher price when compared to semi-spoilt fish. In Bujumbura market (Burundi), fish sold at 4pm (from Lake Tanganyika) retails at about 50% of its value at 6-9 am on the same day.

In terms of value chain, small scale fishing may be distinguished into two types; that which is segmented, with different persons involved at each level of the chain, and fishing which involves significant vertical integration among players on the different levels of the value chain. The value chains of small scale fish trade is illustrated below:

Figure 18.2. Typical Value Chain of Small Scale Fish Trade in Small Sized (Sardine-like) Fish Species Often Consumed Dried

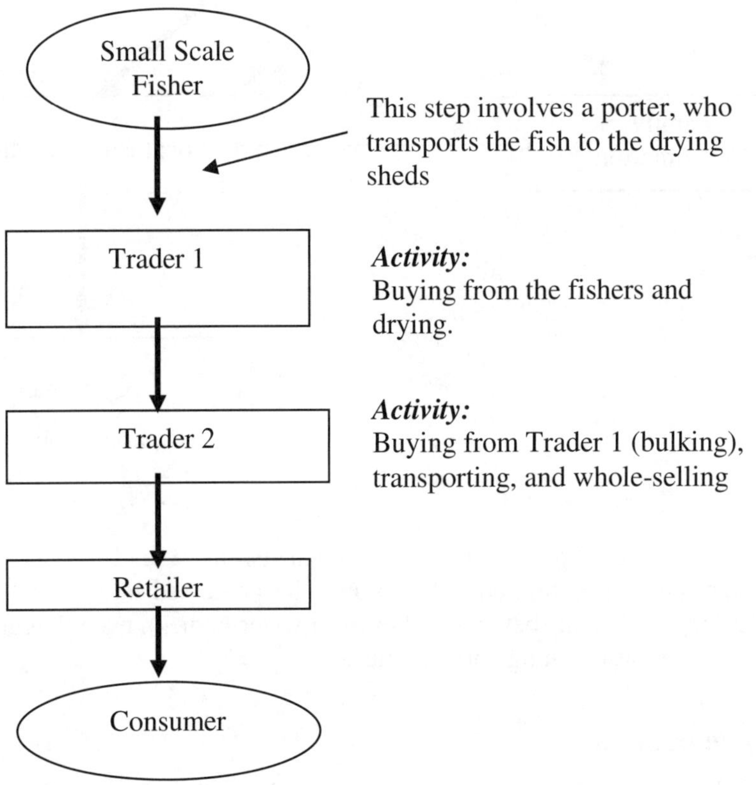

Figure 18.3. Value Chain for Small Scale Fishers of High Value Export Fish, or Fresh-Consumed Domestic Species (e.g. Tilapia, Nile Perch, Marine Fin Fish, Crustaceans, Lobsters and Cephalopods)

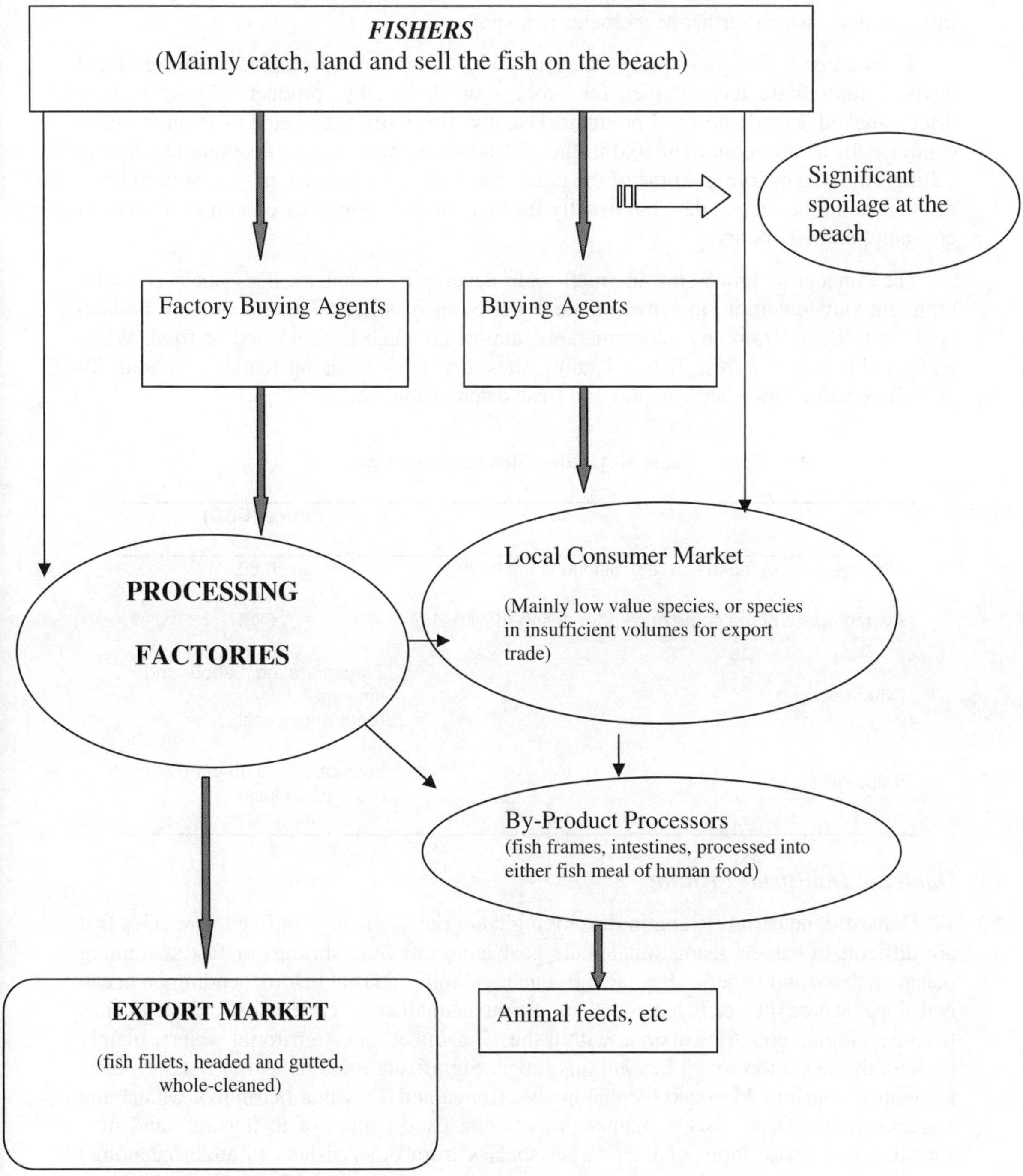

Semi industrial fish processing in developing countries

Consumers in developing countries prefer fresh fish to processed forms. Fresh fish is preferred whole (gutted) which is either steamed or deep fried. In tropical conditions of an average ambient temperature of 30°C, fish spoils rapidly (the average shelf-life of tropical fish kept at room temperature is 6-8 hours after catch (depending on hygiene of handling facilities). If fish was caught by a set-net instead of hook and line, it may have been dead much longer before removal from water, therefore its spoilage would have started in the water. There is hardly any ice or other preservation facilities at landing sites, and marketing centres are usually far removed (at least 10 km) over a poor road infrastructure, which further exacerbates fish spoilage.

To counter the high spoilage rates of fish, traders in developing countries have devised appropriate technologies for processing shelf-stable products. These include dried, smoked, salted and fried products. Usually, fish is first marketed as fresh at major centres until about noon, then sold in the afternoon to small scale processors for drying, salting, smoking or frying. Some of the more stable (to spoilage) fish may be sold to these processors on the second day, but usually fin-fish which is unprocessed will spoil beyond consumption in 24 hours.

The concept of processing in small-scale developing countries does not necessarily represent value-addition. In some remote fisheries such as Lake Turkana in Kenya which is at least 400km from any major markets, almost all catch is sold dried or fried. When compared to price of fresh fish at landing sites, drying or smoking represents about 50-70% price value loss when compared to fresh catch (Table 18.1).

Table 18.1. Dried Fish Value Analysis

	Price (USD)
Average price of 0.5kg fish (e.g. tilapia) when fresh	0.60
Average price of the same price after drying (now 0.2kg)	0.15
Other costs:	Transportation, Labour and time spend drying (opportunity cost)
Value gain/ Loss	Loss of USD 0.45 per 0.5kg (75% value loss)

Domestic Industrial Fishing

Domestic industrial fishing in developing countries is mainly confined to species that are difficult to harvest using small scale gear and craft (*e.g.* shrimps or fast swimming pelagic fishes), and fisheries beyond a 0-3 nautical miles coastal belt (depending on ocean bed slope) where the sea is too rough for artisanal/small scale craft. Most industrial craft from developing countries operate within the 12 nautical miles territorial waters, mainly because the sea is too rough beyond this range. Significant local industrial fisheries exist in countries such as Morocco (cephalopods), Kenya and Tanzania (shrimps), Ghana and Senegal (tuna). These fishers mainly process and export most of their catch, and may sometimes purchase more of their target species from other fishers to attain economic volumes. These fishers also are the main suppliers of marine fish (freshwater fish is mainly supplied by small scale fishers) to markets in developing countries.

Access Fishing

African, Caribbean and Pacific countries do not fish significantly in their Exclusive Economic Zones (EEZs). Fishing in these waters is mainly carried out by Distant Waters Fishing (DWF) fleets from developed countries such as Japan, Spain and France (EU), South Korea, China, USA and Chinese Taipei, which are commonly referred to as Distant Waters Fishing Nations (DWFN).

These countries gain access to fishery stocks through access arrangements which may be summarised into the following types:

1. Government -to- Government access agreements, as is the case between the EU and several (16) developing countries (bilateral agreements), and also the USA-Pacific Island Countries multilateral agreement (between the USA and 17 Pacific Island Countries).
2. Government- to-Private Sector agreements: as is the case between Pacific Island Countries and Japan Tuna Association
3. Rights based licensing of firms that have local base in the host country: as is the case in Namibia
4. Licensing (based on period, not catch levels) of foreign vessels without any specific policy on access fishing policy (*e.g.* in Tanzania, Kenya)

There are several pros and cons on the nature of these agreements (ICTSD 2006), but this paper confines itself to their value aspects. The value of these access arrangements for fish in developing countries EEZ is based on a payment of defined sum of money for either amount of catch, or period of access. The EU agreements compensate developing countries with EUR 100 per metric tonne of tuna caught. The USA agreement with the Pacific Island Countries (17 counties in the Western and Central Pacific Ocean) is based on a lump-sum payment per year (USD 21 million in 2005) regardless of amount of fish (tuna) caught. In general, according to an EU study (IFREMER 1999), financial payments for fisheries access agreements to developing countries accounts for about 2-17% of the catch value (Table 18.2), with tuna agreements financial compensations averaging about 2.6% when downstream value added activities are considered.

This is a low level of compensation for a natural resource by any measure, especially considering that the level of investment required for EEZ fishing is much lower than that for mining (crude oil or minerals), and which attract much higher resource rent levels (about 30%). How are these figures arrived at, and why do the developing countries agree to such terms if they consider them unfair?. In order to answer this question, one needs to consider the process of negotiating access agreements between developed and developing countries. It is not true that developing countries with EEZ fish stocks float out an open 'tender' for whichever DWFN to come forward and bid for a fishing license. This would be the logical thing to do in an open market system, but in reality it does not happen. Access agreements are arrived at through a negotiation process between the host country and DWFN such as the EU or USA, or with a private sector association (such as the Japan Tuna Association). In general, it is the country seeking access that initiates the negotiations, rather than the host country seeking a buyer for its fish, often because the developing country rarely knows the amount of fish available for access. These negotiations are almost invariably private, but the terms of the agreements may be published (as is the case with EU and USA agreements) or kept secret (as is the case with agreements with Japanese private sector). This lack of openness (both during negotiations and also about the final outcome) renders the process vulnerable to manipulations and possible corruption.

Table 18.2. DWFNs' Tuna Off-takes from WCPO and Payments (2003)

	US	Japan	China	Korea	Chinese Taipei	EU
Off-take (2003) MT	94 003	366 783	35 985	208 592	235 188	n.a.
Fleet number	16PS	157LL 35PS 35PL	106LL 8PS	150LL 27PS	153LL, 34PS	5LL 3PS
Financial Compensation/ Economic Benefits	USD 21 million to 17 countries	5% catch value	5% catch value	6% catch value	6% catch value	EUR 100/ton (about 12% catch value)

Source: Off-take and fleet number data, IOTC (2003) and FIAS (2000), adapted from ICTSD 2006. (PS: purse seine, LL: longline, PL: pole and line).

Many Distant Waters Fishing Nations do not regard payments made for access fishing as a trade. The EU for example refers to such payments as 'financial compensation'. Most agreements include components of development aid to the host country's fisheries sector. Also, even though it is not explicitly stated as a consideration in the agreements, the DWFN seeking access may already be providing substantial development aid to key sectors such as education and health in the host country. Further, it is unrealistic to expect a small developing country, which may not even have resources to carry out stock assessment of its EEZ fisheries to effectively negotiate a favourable bilateral agreement with the well endowed negotiations machinery of the EU, USA or Japan. There is clearly a need to implement international policies aimed at ensuring that access fishing is a normal trade, based on market competition, and contributing to economic development of developing coastal and island countries.

Value added associated with access fishing

There are substantial direct and indirect value added activities associated with fisheries access agreements in both host countries and Distant Waters Fishing Nations (DWFN). These include employment, boat and net repair and maintenance, services such as refuelling, re-supply with freshwater, and supply of essential commodities such as food and personal items for crew. On average, host countries capture about 10% of the overall value added value from access fishing (FIAS 2000; FREMER 1999).

Employment perhaps stands out as one of the most important direct value added effects of access agreements. Access agreements in the Pacific Island Countries are said to account for about 10 000 jobs to the islanders (FIAS 2000). In the African continent, EU fisheries access agreements accounted for about 2 400 jobs in 1998 (IFEMER 2000). To the developing islands and coastal countries, jobs on board DWFN vessels operating in their waters represent the only connection between coastal populations and these foreign vessels. Since these fleets mostly do not land their catch in the countries they operate in, they risk looking exploitative and unjustifiable to local populations unless the coastal communities have some of their own people employed on board.

Many fishing access agreements have clauses requiring that locals be employed on board the foreign vessels, but unfortunately an enforcement mechanism is usually not included. In the case of EU-ACP fishing access agreements, the Rules of Origin with regard to crew state that a vessel is qualifying if it has 50% or more crew from EU or ACP employed in substantive positions (Coutonou Agreement 2004). This implies that a vessel operating in ACP countries can have 100% EU crew according to this multilateral agreement, and this denies ACP countries sufficient grounds to enforce more employment of their nationals in these vessels. Since many agreements lack an enforcement mechanism on local crew, this loophole is usually exploited by the vessel captains who are often reluctant to take local crew on board, citing reasons such as language barriers and lack of local crew with the necessary skills in EEZ fishery operations. These are genuine concerns on the part of the ship captains, but they could be resolved by training. A captain would however like to know that if he invests in training local crew (which obviously would take years), he has sufficiently long term possibility to operate in the waters to recover costs incurred. The main problem here is that there may not be policies in place in the two partner countries guaranteeing long term sustainability of these fishing arrangements. In the Pacific Islands however, there are organized training programs for both local fishing crew and observers which are funded mainly by the proceeds from the fishing access agreements. The initiation of this level of regional cooperation on training has been made possible by the collective revenue received from the multilateral agreement with the USA, and is now being funded by various donor organizations.

Many developing island and coastal states do not benefit from value added activities associated with their DWF fleets because they do not have fishing ports. Even though international treaties discourage transhipment at sea, most of these host countries do not have facilities to dock fishing vessels such as purse seines. The fish caught in their EEZ's is therefore transhipped at sea, or landed in other regional countries, where it generates value added revenue. Without a port, a country cannot benefit from value added activities such as repair of ships, net making and repair, and other stevedoring services. When fishing vessels dock, the crew needs to make purchases of personal items, and the ships refuel and take fresh water supplies. In order to provide these support services, some host countries have invested in companies around the ports to produce required goods, and also in oil refining. On average, value added activities generate more revenue than direct fees paid for access in several host countries such as Seychelles. Some host countries (such as Mauritius) are therefore now aggressively marketing themselves to attract fishing fleets which operate not only in their waters, but in the region as well, to base or frequent at their ports.

Perhaps one of the ultimate objectives of host countries is to have all fish caught in their waters processed locally up to consumer unit levels. This would create employment, and shift significant value of the fish to the local level. There is even an increase in desire by the DWF fleets to process locally, considering the huge savings they would make by using cheaper local labour, and avoiding high costs of transporting whole fish as opposed to finished products. One of the main drawbacks is the lack of sufficiently skilled personnel in processing in some countries, and infrastructure (laboratories, factories for processing supplies such as packaging material or processing aids, back-up service for factory equipment, and cost-effective electricity and water supply systems). A further challenge is the competition such facilities would face from factories (especially canneries) already established in the Distant Waters Fishing Nations (DWFN), considering that the jobs they create in these developed countries are perhaps the *raison d'être* for the political support of the DWFN to fishing access agreements.

The subsidy issue

Fisheries access agreements do not fit the conventional dynamics of international trade. The goods (access to fish) are usually negotiated (traded) at the government level and distributed to the private sector (DWF fleet) at a subsidised cost. On average, the EU Commission and USA State Department pay for 70% of cost of access for their private (DWF) fleet, which pays the remaining 30%. There have been strong suggestions that these payments are subsidies and should be discouraged. This may threaten incomes in developing coastal and island countries who derive up to 45% of their GDP, such as Mauritania, Kiribati and Tuvalu) from fisheries access payments. Their argument is that payment from a central government is more convenient and sure, compared to having to establish expensive monitoring, control and surveillance to collect revenue from foreign fleets. The central tenet of this argument is that as far as the developing countries are concerned, they are selling a resource to another government for a fee (which is not even as high as it should be), and therefore this is not a subsidy. It is however a subsidy to the DWF fleet when their countries extend it to their private sector, as is currently the case.

The capacity by developing countries fishers to access fish is usually also subsidized. This is done via subsidies in sectors such as shipbuilding, financial and repair services. The classical case of Korea vs. EU in the WTO in 2004 serves to illustrate this fact. The EU accused Korea of subsidizing shipbuilding, via a series of financial subsidies that resulted in Korean made vessels being 40% cheaper than others. Ruling in favour of the EU, the WTO dispute settlement panel report issued in March 2005 said that the EXIM (export import mechanism) financing amounted to prohibited export subsidies under the SCM Agreement. South Korea highlighted what it said was the core issue, in which the panel ruled in its favor by rejecting the EC claim that it had suffered "serious prejudice" from corporate restructuring of loans to three South Korean shipbuilders. In reality however, the EU has been subsidizing several aspects of their DWFN in the past, and still continue to do so especially through programs such as fleet modernization, and therefore South Korea may have been playing catch-up in this matter. In 2006, the European Commission adopted a draft regulation on *de minimis* aid in the fisheries sector. *De minimis* aid is state aid deemed not to distort competition. The current regulation exempts national aid of up to EUR 3 000 per fisherman, over a period of three years, from prior notification to the Commission. Under the new regulation, the ceiling would be set at EUR 30 000 per three-year period, per beneficiary, on condition that the total amount of such aid represents less than 2.5% of the annual national fisheries output. None of this aid may be used to purchase or construct new vessels or to enhance existing fleet capacity. In the fishing industry, the catching sector remains the dominant one and is likely to be the main recipient of *de-minimis* aid. If the ACP could provide their fishers with this level of aid (EUR 10 000 per year), there would certainly be sufficient capacity to domestically exploit EEZ stocks.

WTO Members reached agreement in the Doha Ministerial Declaration (DMD) in 2001 to undertake negotiations on fisheries subsidies. In the context of these negotiations, WTO Members agreed to clarify and improve WTO disciplines on fisheries subsidies, taking into account the importance of this sector to developing countries (Paragraph 28 of the DMD). Since the Doha Ministerial Conference, negotiations have aimed at identifying specific fisheries subsidies that would be targeted for 'disciplining'.

According to the World Bank (1996), global fisheries subsidies vary between USD 14 and USD 20 billion, which is approximately 20 to 25% of the turnover in this sector. These figures have been contested by OECD mainly on grounds of the accuracy of their calculations, especially considering a big proportion of these subsidies are in management and surveillance aspects. Several of these subsidies are mainly in shipbuilding, access fees, processing facilities, partnership joint ventures and financing. It is however difficult to accurately compute global subsidy loads in the fisheries sector, given that subsidies in other sectors such as financial services often have impacts in fisheries also. Proposals in the current WTO negotiations on fisheries subsidies centre around the categorization of subsidy types into the 'red' (prohibited), 'amber' (actionable) and 'green' (permitted) boxes and the need for appropriate special and differential treatment measures. Capacity-building is one of the criteria being discussed to determine the categorisation of subsidies. Capacity-enhancing subsidies (such as shipbuilding subsidies) are generally regarded as falling in the prohibited 'red' box. 'Green' subsidies on the other hand could include subsidies for environmentally-friendly technologies such as those aimed at by-catch reduction for endangered species such as turtles.

Discussions on fisheries access agreements and subsidies have tended to concentrate on two aspects: *(i)* conferment of benefits (through access fees) to developing countries and the DWF fleets; and *(ii)* production and supply distortions resulting from subsidies given to DWF fleets by their home countries. These two aspects may be important, but it is necessary to lay emphasis on a third critical aspect – *fisheries access agreements are rarely based simply on trade and their associated subsidies impede developing coastal and island states from exploiting their EEZ's.* Fishing access subsidies, therefore, should not be seen only in terms of the 'access fee subsidies'. Associated subsidies, such as those for shipbuilding and financing, targeted at DWF fishing, are of greater magnitude and arguably are more trade-distorting. Fishing access subsidies are the main reason why developing countries are unable to exploit their own EEZ, because the price of their catch would never compete with 25% subsidized fish. This, coupled with the fact that developing countries may not negotiate for a fair access agreement (at least as individual countries), means that they are being globalised out, and in effect deprived of the chance to use their fish as a tool for economic development, and this calls for urgent international attention.

Value chain of access fish

Fishing under access arrangements could be a beneficial arrangement to both developing and developed countries if it was conducted responsibly, and according to well established trade practices. There would be value for developing countries in that they would get market based resource rent value for amount of catch accessed, employment for their nationals aboard the vessels, significant stevedoring activities income, and even value from services like provision of food and other basic needs for the workers when they call to port. A flow diagram of possible value chain for access fishing is illustrated below:

Figure 18.4. Direct and Indirect Value from Access Fishing in Developing Countries

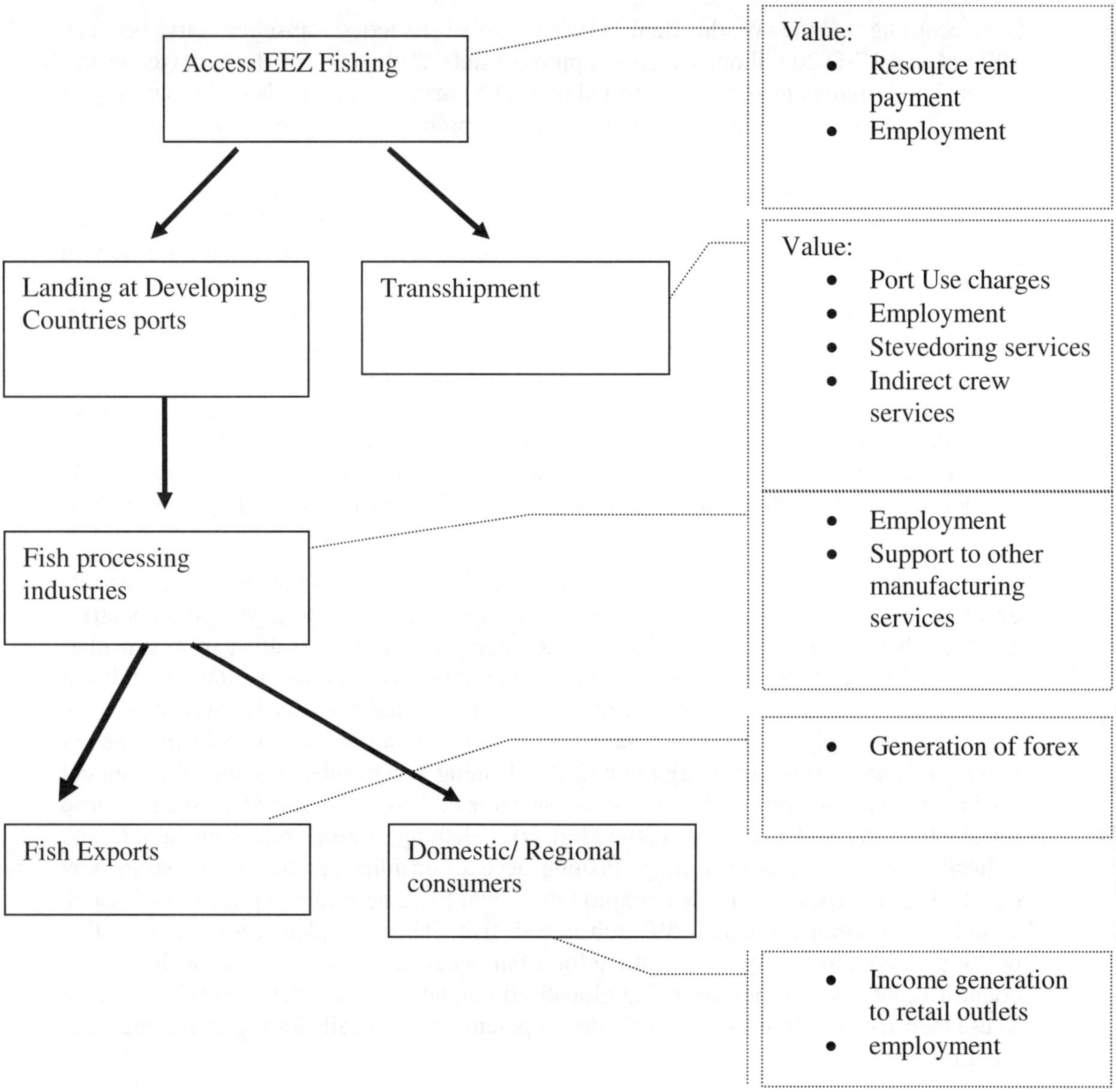

Export Value Addition Activities

Even though export volumes of fishery products from developing countries remains small compared to total production, there is an increasing number of establishments in these countries that are export oriented. Tuna in the world market is still largely supplied though access fishing, but other species such as wild capture shrimps, lobsters, cephalopods and freshwater species such as Nile perch, tilapia and basa are increasingly being supplied by developing countries establishments.

These fisheries are processed into high quality products in establishments in developing countries, and air freighted or shipped to developed countries markets in a timely manner. These establishments have met the stringent sanitary standards imposed by developed countries, as has been attested by numerous inspection teams from destination markets. A summary of products processed is shown in table 18.3 below.

Table 18.3. Value Addition Activities on Fishery Products in Developing Countries

Fish type	Value addition activities and products	Price value added?
Fresh Fin fish: e.g. Tuna hake, Sea bream, Tilapia, nile perch, basa	Filleting, gutting, cleaning. Products: fillets (fresh or frozen), headed and gutted (H&G), whole round	substantial
Frozen fin fish e.g. tuna mackerel	Canned fish products	minimal
Cephalopods Octopus and squid	Cleaning and gutting Exported fresh or frozen	substantial
Crustaceans: Shrimps and lobsters	Lobsters (Cleaning, heading) Shrimps (fresh whole, coking, peeling)	substantial

These products are usually packaged in standardized sizes and packs before being exported to markets in developed countries such as Japan, EU and USA.

Challenges of value addition in developing countries

Fish processing and exporting establishments in developing countries face particular challenges which are either inexistent in developed countries, or whose magnitude are comparably higher. This increases the cost of doing business for traders in developing countries, making them non competitive and hence vulnerable to being globalised-out by traders from developed countries that may be accessible to more trade facilitating privileges (*e.g.* subsidies and affordable credit). These include the following:

Higher costs in sourcing of raw materials

In many developing countries, processors have to bulk economic fish volumes from small scale fishers scattered over a wide area without sufficient road network. There are no auction markets around most fishing areas, and buyers have to rely on a network of agents to collect enough supply. This is expensive when compared to fisheries in developed countries, where processors can purchase sufficient volumes at auction centres even via the internet.

Infrastructure

In developing countries, infrastructure such as roads and landing sites are poorly developed. This not only increases wear and tear of transportation vehicles, but presents a logistical nightmare of having to maintain freshness in this highly perishable product over the long distances travelled over long periods of time. Other services such as ICT (information, communication and technology) are generally poorly developed, and this hinders efficiency in business transactions, raising costs.

Inputs

The average kilowatt electricity cost in developing countries is much higher that in developed countries (often about three times). This is partly because of the technology used in power generation in developing countries (mainly hydroelectricity or diesel fuel) and is too expensive when compared to the cost-effective nuclear technology used in most developed countries. Other inputs such as potable water and processing materials are also comparatively more expensive. Labour is perhaps the only input that developing countries have an advantage over developed countries. Labour costs in developing countries may be up to 10 times cheaper than in developed countries, a factor that is sufficient to encourage a north to south shift in fish processing. Developing countries however lack skilled workforce for processing specialty fishery products such as *surimi* products, *shiokara* and *katsuobushi* consumed in markets such as Japan. Other labour challenges include lack of skilled personnel to maintain sophisticated equipment that may be necessary to improve processing and preservation efficiency.

Affordable credit

A basic fish processing establishment (such as for processing chilled fish fillets) costs a minimum of USD 3 million to construct. This level of investment requires easy access to affordable credit. In developing countries, bank interest rates are about 15-24%, which is well above profit margins of fish export trade. The argument for these high interest rates has been the high risk factor to lending by the banks due to frequent defaults. When compared to offshore and developed countries rates of about 4-6%, developing countries interest rates are one of the main factors limiting investment in this sector.

Increased costs due to SPS Measures

It is expensive, and currently unaffordable by most domestic fisheries entrepreneurs to meet the high costs associated with meeting the stringent SPS (sanitary and phytosanitary) standards imposed by markets in developed countries. At the moment, a fully compliant fish processing (*e.g.* filleting) export facility costs about USD 3 million to establish. This high cost has had the negative effect of forcing closure of most indigenous fish processing establishments, and their replacement by foreign-linked firms. These foreign firms sometimes have vertical integrations with foreign fishing vessels in the countries EEZ, and also linkages with distribution networks in export destinations. Under such circumstances, local firms have found it difficult to compete, which has lead to more marginalization at developing counties level. Some countries such as Namibia have had to come up with programs such as 'Namibianisation' of their fisheries, or 'black empowerment in South Africa' to address this problem, while at the same time upholding the principles of free market economies, and this is a daunting challenge.

Tariffs and rules of origin as fish trade barriers

There has been a tendency by developed countries to impose measures aimed at encouraging exportation of agricultural, fishery, forestry and mining raw materials from developing countries in order to support high value processing in their countries. Fish, both from access arrangements and other export trade is important in supporting processing establishments, port facilities, employment and support service factories in several towns in the EU and far east. Without this raw material fish, the socio economics of these developed country economies would be severely affected. These countries therefore consciously maintain political pressure to ensure that value addition of the target species at source is discouraged, mainly through introduction of rules of origin that reward export of raw materials and punish export of processed products.

The use of tariff escalation is one of the main tools used in discouraging export of value added fish. Fish which is added value to a certain extent (such as canning) is categorized on a different tariff heading when compared with whole gutted fish, or fish fillets. These differentiated products are subjected to different tariff levels, which normally increases with the extent of value addition. The price of the final product is not necessarily the determinant of level of tariffs, but rather the extent of working (processing or other actions), as these are important in creating indirect value added (employment, support industries etc) in destination markets.

ACP countries are fortunate in that they may export fishery products into the EU market duty and quota free. The catch in these arrangement is that the fish must be 'originating', meaning that if harvested from the EEZ, it must have been caught by a vessel owned at least 50% by EU or ACP (including chairman of the Board of Directors), with crew that is at least 50% EU or ACP (including the ships master), and not caught using a chartered vessel outside EU ACP unless the host country had offered to negotiate an access agreement with the EU and the EU refused. In simple terms, fish from ACP EEZ may not be sold into the EU if caught with cooperation from other Distant Waters Fishing Nations. With regard to value addition, an example from Mauritius illustrates the complications this brings. In Mauritius, the canning factory can procure raw material fish from the Asian fleet in its EEZ much cheaper than from the EU fleet. If they do so, they will not be able to export the canned fish (obtained from Mauritius) to the EU duty and quota free. They are therefore obliged to use expensive fish (from the EU fleet) for canning, and this erodes their competitive edge. These rules of origin are unfair in that they not only restrict ACP countries to engage in joint fishing activities, but also hinder competitive value addition of ACP fish.

There is a process underway between the ACP and the EU to revise current rules of origin under the on-going Economic Partnership Agreements (EPA) negotiations. The aim is to achieve the twin objective of ensuring other DWFN in ACP waters do not use ACP countries as a conduit to export fish to the EU (protection from cheap fish) on one hand, and also acknowledging that all fish in ACP EEZ's is originating, and therefore should have a chance to participate in trade with the EU. One of the mechanisms proposed is to replace the 50% rule mentioned above with a value added criterion. The difficulty lies in setting an agreeable level of value addition which previously non originating fish (caught by non EU-ACP vessels) would need to be subjected to in ACP countries in order to qualify.

The argument for use of value added criterion is in line with Kyoto Convention of 1973 which states that:

- Where two or more countries have taken part in the production of the goods, the origin of the goods shall be determined according to the substantial transformation criterion.
- Notes to Provision 3, Annex D1, Kyoto Convention, 1973 state that substantial transformation shall be determined in 3 ways:
  - by a rule requiring a change in tariff heading in a specified nomenclature with lists of exceptions;

– and/or– by a list of manufacturing or processing operations which confer, or do not confer, upon the goods the origin of the country in which the operations were carried out;

– and/or by the ad valorem percentage rule, where either the percentage value of the materials utilized or the percentage of the value added reaches a specified level.

The Kyoto convention therefore does not allow a definition of originating status to be based on crew or vessel ownership nationality, and this has been one of the fundamental incoherencies of the EU-ACP Rules of Origin (RoO) with this international convention. Both the EU and ACP have not disputed that that the review could focus more on value added aspects, but the contention is what level of value addition, or what Change in Tariff Heading (CTH) would be acceptable.

Proponents of CTH argue that it is easy to apply and enforce, as it involves transformation of a product from one form to another (physically discernible), and therefore will not result in increased customs verification costs. The concerns of the EU have however been whether the CTH will be accompanied by 'sufficiently added value' to justify its qualification for duty and quota free access to the EU market. The main thrust of CTH has been ACP countries with canned tuna, who would like to see it qualify because it's a different tariff heading (16) as opposed to tuna loins (heading 3). There have been mentions (during pre negotiations consultations) by the EU of about 30% value addition as being a acceptable threshold, levels which cannot be achieved in canned tuna, and therefore are unacceptable to most ACP countries.

Proponents of value added criterion (with or without a CTH) argue that it is a more sensitive measure of 'sufficient working', and is flexible enough to be used for any fishery product. This arises from an observation that even though some measures such as filleting of fish (which results in yields of 40% on average from non gutted fish) may result in a CTH, others such as peeling of shrimps, flavouring of a loin, portioning and packaging of fish may resulting significant value addition, but not qualify for a change in tariff heading.

This has also introduced discussions on whether to use *ex factory* values or *Net Production Costs* of the various processing steps. A transformation of fish from loins to canned may not always be accompanied by an increase in ex-factory price. In reality, tuna fillets are about USD 4-7/kg on the international market, while canned tuna is about USD 1/kg, meaning that tuna canners make money mainly by ensuring that they trade in huge volumes, given the minimal margins per can. This is almost value-loss from the unit price point of view, but offers traders with a convenient product that is preserved, and therefore can be marketed widely. Other forms of processing such as breading and battering have similar price behaviors, but their value is again more in product diversification and preservation than unit price increase. Fisheries RoO may therefore benefit more from a use of net production costs criteria than *ex-factory price.*

In all, it is acknowledged that there will be need to develop standardised assessment methods for a value added criterion, with agreed limits of level of value added, and this will definitely result in increased cost of customs verification procedures. It is however the only (current) way in which all fisheries products would benefit from new rules of origin that seek to confer originating status to fisheries products that are 'sufficiently worked', and not just a few products. A compromise position would be for the ACP and EU to agree on a level of value addition that both parties see as being sufficient to have fisheries products from ACP states, which may not have qualified on the basis of their area or means of catch, conferred as originating. This value could be set low enough (not more than 15%) to accommodate transformed products such as canned tuna, and this would in turn lead to an almost automatic qualification of other fisheries products once they are processed, regardless of whether there was a CTH or not.

Export Fish Value Chain

As earlier discussed, developing countries are increasingly exporting various fishery products to developed countries. These include fish fillets (fresh and frozen), gutted and cleaned cephalopods, headed (and sometimes peeled crustaceans) and whole round and gutted fish. The products are either air freighted (when fresh) or shipped (frozen) to destinations depending on profit margins after freight costs. A typical value chain of export fin fish (tuna, tilapia or Nile perch) serves to illustrate the value and activities at each level of the chain.

There is at least a USD 10/kg difference between what a consumer in a developed country pays, and what is paid to the developing country exporter. The consumer price of fillet is about 250% more than the export price. Many developing countries assume that this huge difference in price may be explained by value addition activities that take place on the product prior to consumption. In reality however, most fresh fishery products exported from developing countries undergo minimal (if any) value addition at the developed country level. The highest (price) value for fish is in its fresh form; fresh fillets, crustaceans or cephalopods. As soon as any transformation (even freezing) is undertaken, the price falls. In this regard, fresh imports into developed countries markets are transported as soon as possible to retail places (supermarkets and fishmonger shops) and sold on melting ice. Apart from road transportation and some limited warehousing, fresh fish does not undergo any other value addition. Where then does the USD 10 difference go to? It is shared among the 3-5 step distribution chain in the destination market. This level of value capture at this end of the chain is disproportionate to the value captured at developing countries level, considering that the main value addition activities (catching, processing and air freighting) are undertaken there. Some explanation for this may be differences between exporting and importing countries in aspects such as taxation and the cost of doing business, but the margin are still too large to be fully accounted for this way.

Figure 18.5. Export Fish Value Chain

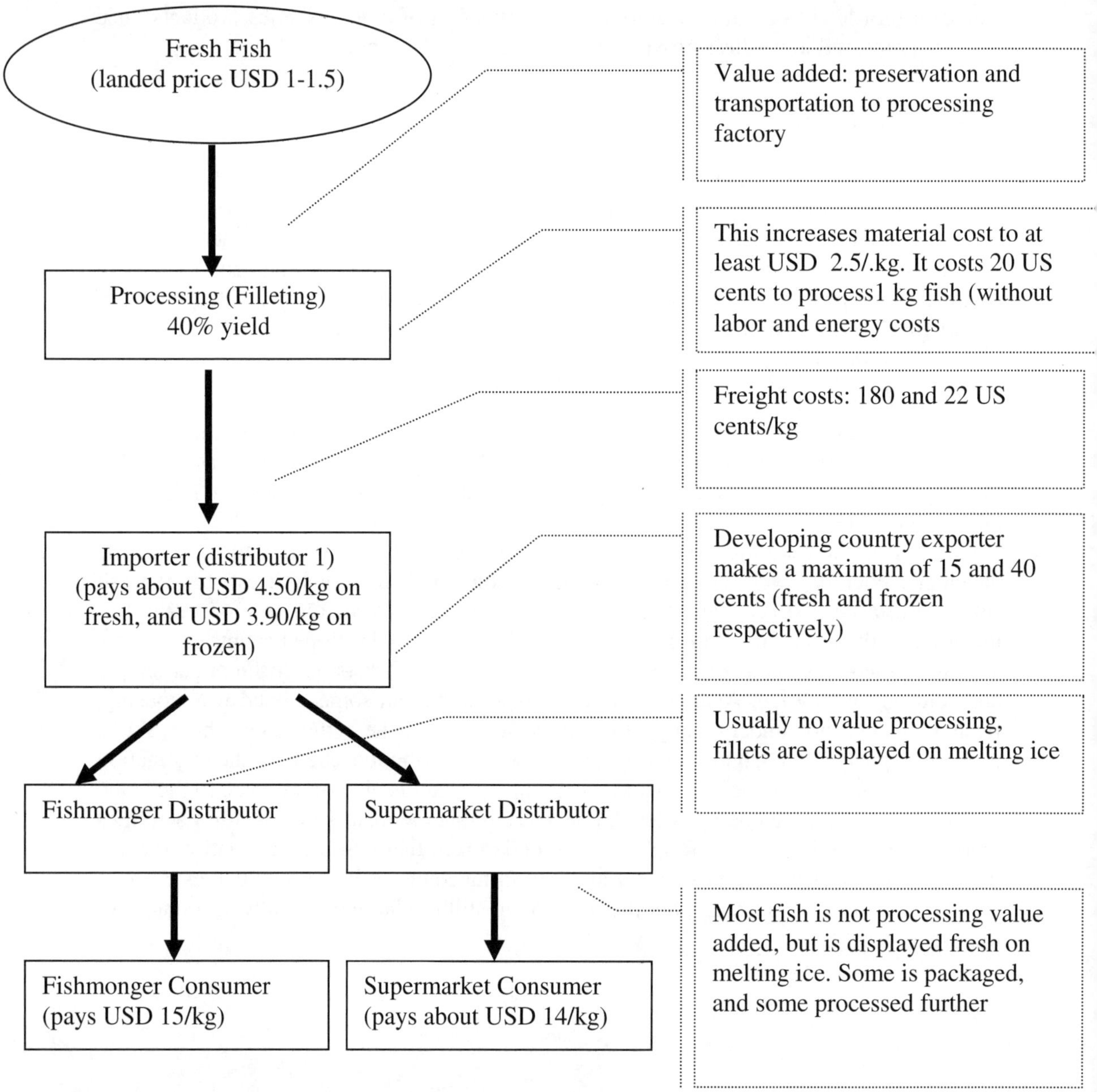

Barriers to Distribution Services in Export Markets

If indeed developing countries feel that their fish value is mainly retained at the market side of the chain, why don't they just set up distribution services and 'sell their own fish' in developed countries, and therefore capture the value themselves. This seems to be the best way of capturing and repatriating value for developing countries fisheries, but it is seriously impeded mainly by restrictions imposed by developed countries on services sector. There are 4 modes of services that are being negotiated at WTO and other multilateral trade negotiations between developing countries and developed countries. These include:

- Mode One (Cross Border) Services refers to cross border trade services where there is no movement of persons, such as financial transactions, consultancies or postal services.
- Mode Two (Consumption) Services refers to those services where the consumer moves to consume the service in a destination country, and returns back to his/her own country (e.g. tourism and medical services)
- Mode Three (Commercial Presence) Services refers to activities which necessitate physical presence of persons in another country to set up businesses. This includes activities associated with processing investment, banking or other Foreign Direct Investment (FDI) activities.
- Mode Four (Temporary Movement of Natural Persons) refers to the whole aspect of free movement of persons.

In order for developing countries to 'sell their own fish', there will be need for developed countries to open up more in mode 3. This should be justified by the fact that if developing countries can be allowed to set up distribution systems for their products such as agrochemicals, banking and pharmaceuticals in developing countries markets, then developing countries should be facilitated to distribute whatever resources they have in developing countries. The ability of Iceland to set up Icelandic fish (Icelandic cod) distribution system in markets such as the USA is usually credited for the positive turnaround in the sector's ability to contribute greatly to the country's economic development. This was because Iceland was able to capture a lot of the margins in the marketplace by selling as close to the consumer as possible.

Value addition of export fish in markets

There is some value addition that takes place on developing countries fish after arrival in market destinations. This is usually targeted at fish that may not be sold in fresh state (either because it is frozen or is received in huge volumes which may not be conveniently sold fresh in the 2-3 days fish is able to stay fresh). These value added activities include:

- Processing of products such as canned fish, fish fingers (breaded and battered) or other preserved fish products (dried, smoked)
- Processing of shredded fish products such as surimi, fish sausages and fish pastes.
- Processing of fish extracts (*e.g.* EPA –*Eicosa pentaneoic acid*, and DHA –*docosa hexaneoic acid* which is used in medical nutrition)

Most of the processing that take place in developed countries can easily be undertaken in developing countries. These include portioning into consumer units, consumer retail packaging and caning. Even the complicated processing of surimi and dried fish technologies in consumer countries such as Japan can be undertaken in developing countries if sufficient capacity building measures are undertaken.

Policy considerations

There is a need for both developed and developing countries to work together in order to ensure that fisheries contribute more to development of economies in developing coastal and island countries. Such considerations may include the following:

1. Policy considerations at developing countries level that seek to minimise wastage of fisheries resources through post harvest spoilage. Such a policy could consider either introducing punitive measures for unhygienic handling and poor preservation of fish, or introducing rights-based fishing where citizens purchase access, therefore have more incentives to preserve their catch (since they paid for it). Post harvest preservation of fish may not be achieved wholly on voluntary basis without legislation back-up, since the price for fish for both those who choose to invest more ice preservation (where ice is available) and those who do not is usually the same at the early periods of a marketing day.

2. Policy considerations at international level aimed at removing the obscurity than now surrounds fisheries access agreements, and ensuring that these arrangements are regulated through international trade instruments implemented through World Trade Organization. This would make fishing access a tradable commodity, with various DWFN bidding for it, hence encouraging developing countries to capture more value. Such measures should also be accompanied by measures to eradicate any current subsidies, and residual impacts of past ones.

3. Policy considerations that aim at eliminating fish trade tariffs to developing countries. Other restrictive measures such as unfavourable rules of origin also need to be eliminated. There is genuine progress in multilateral trade negotiations such as between ACP, MERCOSUR blocks and the EU or USA in addressing these barriers, which should be encouraged.

4. Policies that seek to retain fisheries value at developing countries by discouraging export of raw material fish, and encouraging a certain level of value addition to be retained at country level.

References

Foreign Investment Advisory Service (FIAS) (2000). Capturing Economic Benefits from Tuna Fisheries – A Study of the Pacific Island Countries. International Finance Co-operation and the World Bank. Washington, DC.

IFREMER (1999). French Institute for Research and Exploitation of Fisheries Resources. Evaluation of the Fisheries Agreements Concluded by the European Community. European Union Study, Brussels, Belgium.

Indian Ocean Tuna Commission (IOTC) (2003). Report of the Fifth Session of the IOTC Working Party on Tropical Tunas. FAO-IOTC, Victoria, Seychelles.

Mwikya, M.S. (2006) *Fisheries Access Agreements: Trade and Development Issues,* ICTSD Natural Resources, International Trade and Sustainable Development Series Issue Paper No.2, International Centre for Trade and Sustainable Development, Geneva, Switzerland.

World Bank (1996). Pacific Island Economies: Building a Resilient Economic Base for the 21st Century, World Bank Country Study, Washington, D.C.

Websites

http://ec.europa.eu/comm/fisheries/news_corner/press/inf06_30_en.htm. 2006-Commission release on EFF.

http://ec.europa.eu/comm/fisheries/news_corner/press/inf06_35_en.htm. 2006-CEC release on *de-minimis* aid to fishers

Chapter 19

Perspectives from Canada's Pacific Coast

Christina Burridge, Executive Director, British Columbia Seafood Alliance, Canada

Christina Burridge presented the effects of globalisation on the fishing industry of Canada's Pacific coast. Originally, Canada's Pacific coast was a high cost, low volume producer of fish focussing primarily on salmon and herring fisheries. These fisheries accounted for 85% of total landings. However, the fleet was both inefficient and seasonal. In addition, an inefficient industry structure and protective government regulation left the industry vulnerable, particularly during the mid-1990s, when farmed salmon production increased sharply.

For Christina Burridge the rapidly changing competitive environment, the collapse of the Japanese economy, the strengthening of the Canadian dollar, falling tariffs, rising fuel prices, consolidation of seafood distribution, increased outsourcing of processing and changing consumer preferences, all contributed to the collapse of the salmon and herring industries. By contrast, the groundfish and shellfish sector flourished as these were able to exploit the new competitive environment. While salmon and herring accounted for 85% of the landed value in the 1970's this had fallen to 25% by 2000; similar figures for the share of landed value made up of groundfish and shellfish are 15% and 75%.

The key lessons that can be drawn from these changes in industry structure in the era of globalisation are that an overregulated, inefficient industry structure created vulnerability to external changes. By contrast, fisheries characterised by market driven fisheries management approaches are thriving and can take advantage of the new opportunities created by globalisation.

Introduction

The Organisation for Economic Cooperation and Development (OECD) Fisheries Committee is tapping into the knowledge and experience of fishing industry participants around the world to gain insights on the effects of fisheries globalisation. This paper, offering a perspective from Canada's Pacific coast, is prepared by Stuart Nelson of Nelson Brothers Fisheries Ltd on behalf of Fisheries and Oceans Canada for presentation at the OECD/FAO Workshop on the challenges and Opportunities of Globalisation in the fisheries sector which was held in April, 2007.

To shed light on how British Columbia's fisheries are variously faring under globalisation, three capture fishery case studies are explored: pink salmon, hake (Pacific whiting), and groundfish (trawl). Examination of these fisheries provides sharp contrasts in terms of the nature of underlying fishery resources, fisheries management regimes, and industry responses and financial results under an increasingly globalised business environment.

About Fisheries Globalisation

Before commencing an exploration of the case study on Pacific fisheries, a brief encapsulation of fisheries globalisation, pertinent to the ensuing analysis, is provided.

Globalisation in fisheries is a centuries-old occurrence, a product of the fact that "fisheries resources are not necessarily located in the vicinity of major consumption centres.[1]" Though a flow of fish from regions of capture to sites of consumption is historical, "the market has undergone substantial change at a very fast pace in the last two decades. This is caused by numerous factors including the growing importance of the aquaculture sector, cheaper and faster modes of transport, improved marketing, lower market access barriers, more competition, constantly increasing consumer demand for fish and fish products in OECD countries, combined with declining fish stocks, the emergence of new players (especially China) in world fisheries markets as well as technological improvements.[2]"

The OECD has taken an interest in fisheries globalisation, recognizing that the process represents both exciting opportunities, for seafood businesses and seafood consumers, and serious threats, to economically disadvantaged participants and to vulnerable marine resources.

Fisheries globalisation creates a market "force" that is drawing fish and fish products to their highest value/lowest cost destinations. "Consequently, the fishing industries in OECD countries today are very internationally focussed, sourcing from all oceans, processing raw material in many countries and trading fish products all over the world."

Taken to its logical conclusion, fisheries globalisation will see "trans-border production networks emerge where different elements of the value chain, to varying degrees, take part in the globalisation process by using the most profitable location for their activities.[3]" "In theory, a fully globalised and integrated fisheries economy is characterized by fully transferable (national and international) access rights to fish, no tariffs and trade barriers, no restrictions on foreign direct investments and fisheries services, and no nationality requirements on flags of fishing vessels and fisheries.[4]"

---

1 "Globalisation and the Implications for the OECD Fisheries and Aquaculture Sectors." Aug 30, 2005.

2 *Ibid.*

3 *Ibid.*

4 *Ibid.*

The "force" of globalisation is (partially) frustrated by domestic regulations intended to encourage local utilization of resources. In some jurisdictions, "domestic operators and fisheries management authorities see the fisheries resources as a national endowment that should only be exploited by their own nationals. This is, however, contrary to the process of "globalisation" where market interdependence and liberalization suggests that the best, most cost effective producer with a comparative advantage undertakes the operation.[5]"

The spectre that globalisation-related demand pressures may negatively impact fish stocks is (hopefully) negated by sound fishery management practices that ensure conservation of seafood resources. However, "during the last decade over exploitation of fish resources has added a further dimension to the globalisation process. While consumers in OECD markets have been told that eating fish is healthy, their domestic resources have been dwindling, inter alia, due to poor management. Through necessity, fishing vessels and fish processing companies have been forced to look for raw material further afield.[6]"

There are three globalisation themes introduced in the above passages, with relevance to the Pacific experience described subsequently:

1. Fisheries globalisation, an accelerating phenomenon driven by increasing integration of markets, the growth of aquaculture, and technological advances (among other factors), brings tremendous economic opportunities and pressures to seafood businesses.

2. While globalisation may be seen as an unstoppable force, some jurisdictions use legislation, regulation, policy, or domestic practices to encourage local resource access and utilisation. In other words, seafood resources are "protected" from global exploitation in order to serve national socio-economic objectives.

3. Fisheries globalisation may result in unsustainable fishing pressure on stocks that are not properly managed.

The OECD, through work with fishery sectors and governments in member nations, is seeking to encourage the positives of globalisation – economic efficiency and enhanced profitability – while mitigating the negatives, such as undue displacement of local participation and over-exploitation of seafood resources.

The balance of this paper presents perspectives on fisheries globalisation from Canada's Pacific coast. So that the subject fisheries – pink salmon, hake, and groundfish – can be considered in their proper context, an overview of the British Columbia seafood industry is first provided.

Context - A Recent History the BC Fishing Industry (1970s – mid 1990s)

From the early 1970's through the mid 1990's, the salmon and roe herring sector substantially *comprised* the BC seafood industry, accounting for 85% of landed value in the 1970s. Salmon provided the bulk of harvest volume and revenues, while roe herring

5 *Ibid.*

6 *Ibid.*

was generally a more lucrative fishery, bringing higher margins for fishers and processors alike.

Key elements of salmon/herring industry

Highly seasonal

Salmon are a migratory species, entering BC inshore waters each summer upon their return from the Pacific Ocean to spawn in the stream of their birth. The salmon migration affords a relatively brief harvesting window – the bulk of the fishery occurs between early July and early September.

The roe herring fishery is even shorter in duration than salmon, substantially occurring in as few as five fishing days during the month of March. Because herring are harvested for their roe, fisheries are mounted proximal to spawning grounds immediately preceding (or during) a spawn to maximise roe content.

Thus, BC's "anchor" fisheries were highly seasonal. This necessitated substantial fleet and processing capacities to handle large bursts of production in a relatively short time. It also implied considerable down-time when assets were substantially un-utilised.

Fishing fleet

Commercial salmon fishing is prosecuted by three gear types: seine, gillnet, and troll; roe herring is harvested by seines and gillnets. There is considerable overlap in the assets and individuals involved in the two fisheries.

As with inshore fleets in many fishery jurisdictions, the BC salmon and herring fleet was characterised by substantial numbers (about 4500 vessels until the late 1990's), and duelling objectives: economic returns (to investment and labour), and social benefits such as employment and access to Employment Insurance. Salmon and herring fishing was an important socio-economic activity for harvesters living in BC's many remote coastal communities, including First Nations fishers, who comprised about 1/3 of the fleet.

Fleet size and configuration evolved from historical participation levels. Dividing the available catch amongst the sizeable BC fleet meant marginal financial returns on average, though periodic "boom" years raised expectations that an annual income could be earned from a few weeks' work. It was often observed that "too many boats were chasing too few fish," to the possible detriment of both the resource and harvester incomes, but initiatives to bring economic efficiency to the fleet typically faced strong resistance. The tug between social and economic objectives remains unresolved today.

Processing sector

The processing sector structured around the need to handle the perishable production of thousands of vessels in a short time. The relationship between processors and harvesters has been described as patriarchal, with processors providing much of the infrastructure (fuel, stores, net-lofts, marine repairs, and revolving lines of credit) associated with small vessel operations in remote areas.

The "cannery in every cove" distribution of processing locations gave way to centralisation in Vancouver and Prince Rupert (the urban centres at the south and north borders of BC's coast) by the 1980s. Tender vessels ferried the catch from the fishing grounds to the plants. The processing sector was highly concentrated, with a very few firms accounting for the bulk of fish purchases and wholesale receipts in the industry. Processing sector representatives lamented that BC comprised a high cost salmon/herring production environment: both fish prices paid to fishers and wages paid to shoreworkers were among the highest in the world.

The salmon/herring business is aptly described in the following quote from a 1979 OECD Report on Fisheries: "at the best of times uncertainty is an inescapable feature of the fishing industry from the opening action of dipping in to the sea to harvest its raw material until exposing the product for sale to an often incalculable demand on an uncontrolled market." This speaks of a speculative business, one with high reliance on factors beyond the control of producers, such as commodity market prices, currency exchange rates, and interest rates. To play the salmon/herring game was to roll the dice and trust that circumstances would align favourably. Sometimes they did, and sometimes they didn't. On average, returns to the BC processing sector were less than those earned in more conventional industries. Long term participation required patient, well-capitalized shareholders willing and able to endure a number of lean years in order to reap occasional windfalls.

Markets

The local (Greater Vancouver) market is minuscule, and Canadians, though voracious consumers of canned salmon, consume less than half of BC's canned salmon production. Japan is the only outlet for herring roe. Thus, the BC seafood industry has always exported the bulk of its seafood production.

Canned and frozen salmon (headed and gutted) were the primary product forms. The UK and ANZA were the primary export canned markets, with Europe and Japan the main frozen markets.

Throughout much of the period, frozen sockeye and herring roe were the most lucrative items in the BC seafood product mix; a high level of dependency on the Japanese economy and market had developed.

Vertical integration/Value chain

The BC seafood industry featured a high degree of vertical integration. Its two largest firms, accounting for about 60% of the BC salmon catch, were owned by parent companies whose primary business was retailing food. Canned salmon was thus kept "in-house" from capture until placement on retail shelves[7].

Fishers, typically, had little participation in the value chain other than delivering their catch to a processing company, and trusting that a fair and competitive price would be paid. They had little access to, or interest in, market information. Competition amongst

7 In practice, the fishing companies and retail outlets were autonomous businesses operating under an umbrella corporate structure.

fish buyers was the mechanism that fishers believed ensured them a fair portion of the market value of end-products.

Globalisation during the period

As an exporting industry, the BC seafood industry has always been deeply affected by events occurring outside the Canadian border. A relatively few variables exerted a large impact on financial results:

- Exchange rates – fluctuating rates strongly influenced marketability of product and prices realized in Canadian dollar terms.
- Interest rates – with the salmon pack put-up in a couple of months but sold throughout the year, financing requirements were substantial. Interest rate levels were determined by global economic conditions, and had a major impact on annual financing costs (and profitability).
- Alaskan salmon production – as the major supplier in the world salmon market, the level of Alaskan salmon harvests had a direct effect on export price levels – low production meant firmer prices, high production meant softer prices.
- Strength of national economies – the varying fortunes of economies in key markets had a substantial influence on the attractiveness (pricing, sales volume) of Canadian seafood products.

Collectively, these four factors – all entirely beyond the control of the BC industry - had a huge bearing on the success or failure of BC fishing enterprises. Recognition of the role of luck caused at least one BC participant to humbly observe that: "our good fortune is usually a result of someone else's bad fortune;" and "when we succeed… it's often in spite of ourselves!"

Regulatory Environment

Fishery Management

Fisheries and Oceans Canada (DFO) is responsible for management of marine fisheries in BC. By and large, DFO has managed BC's commercial fisheries conservatively, preserving generally healthy stock conditions. Of course, there is significant variability in fish abundance from year to year related to ocean conditions, natural cyclicity, habitat, and fishing pressure. Salmon populations are particularly volatile.

In managing BC's fisheries, DFO has always sought a balance between ensuring conservation of stocks, while allowing for reasonable economic opportunities. Over time, advances in fishing technology led to enhanced catching power and efficiency of the fleet. The traditional fishery management toolkit allowed few alternatives other than: shorter openings, smaller fishing areas, and limits on effort (gear specifications, size of vessels, number of vessels). Actions of this nature by DFO were inevitably greeted by further innovations by the fleet to ensure that as much fish as ever (if not more) could be caught under the new, restrictive, conditions. Restrictions had the unintended effect of *fuelling* fishing pressure rather than *curbing* it.

By the mid-1990s it was evident that in the salmon/herring sector, the very core of the industry, neither conservation nor economic objectives were being met, and that reform was needed.

Provincial Regulation

The provincial government regulates the buying and processing of the seafood harvest. In the late 1970's two key developments caused the Province to review its policy role with respect to the fishery. First, the extension of the EEZ to 200 miles in 1977 meant that BC fleets would have virtually unimpeded access to the Pacific coast's marine resources. Second, Japanese seafood firms, in order to secure supplies of raw material (mostly sockeye salmon and roe herring) began to invest in the BC seafood processing sector.

This combination of an opportunity to "Canadianise" the fishery, along with the perceived threat that foreign ownership in the processing sector could pose to a Canadianised industry, influenced the Provincial government to re-enforce its policy direction that:

- BC marine resources should to the greatest extent be harvested by the BC fishing fleet.
- The BC marine harvest should to the greatest extent be processed by the BC-based processing sector.
- That the harvest would be largely processed ashore, with processing at sea extremely limited.
- That export of un-processed fish (from fishing vessels or tenders) be tightly restricted.

The Pacific coast marine fisheries were thus viewed by the Provinceas supporting employment and economic activity within BC.

The Province's stance to encourage a "made-in-BC" fish buying and processing sector comprised a competitive shield of sorts for the Canadian shore-based industry, allowing it to develop somewhat sheltered from undue foreign or at-sea competition for resources.

Fully/Over Subscribed Resources

Both the salmon and herring resources in BC have long been "fully subscribed" with the total demands of users and would-be users greater than the available abundance of fish.

These key fish stocks, but particularly salmon, face strong demand from both First Nations (for commercial and food/ceremonial purposes), and from the recreational sector (both sport-fishers and charter/lodge operators). First Nations and the recreational sector have received escalating allocations – and heightened allocation priority – over time, at the expense of the commercial sector.

The public-at-large is also a major stakeholder in salmon – the salmon resource is viewed as an endowment to be protected and preserved. Salmon are embedded in the culture and values of British Columbians (or so the media reports). The public takes solace when salmon stocks are healthy, and express outrage when they are in distress. They expect that salmon can support healthy aboriginal, recreational, and commercial fisheries, but are intolerant if high catch levels are perceived to compromise conservation levels. Given the incredible complexity of salmon – the high number of natural and man-made variables influencing their abundance – the public can never really grasp the overall health of salmon stocks at a point in time.

Thus, the demands placed on salmon, and the governments charged with managing them, are impossible to meet: every user demanding more salmon, concurrent with heightened conservation and protection measures. Over-subscription is yet another reason why considerations other than economics apply to commercial fishery management.

Secondary fisheries

During the 1970s and 1980s, BC's fisheries excluding salmon and roe herring were small scale and slowly developing. These fisheries tended to be conducted by the salmon/herring sector as "sidelines" – as ways of stretching the fishing season – or by very small fishing operations and processing firms comprising the "fringe" of the industry.

The groundfish trawl fishery was probably the most common "diversification" fishery practiced by the major players in the seafood industry. As a year-round, relatively high volume fishery, groundfish was attractive in terms of asset utilization – deploying vessels, plants, and employees that were otherwise idle for much of the year – and a means of absorbing a portion of annual overheads. Major seafood processors were ambivalent about the groundfish trawl fishery. A common sentiment: "absorbing overheads is a polite way of saying we're losing money." One company leader (whose firm is now a major presence in the renewed groundfish trawl industry) said, "I spent a large part of my career trying to get out of the groundfish business."

Summary of the salmon/herring era

This encapsulation describes the state of the BC seafood industry preceding an onset of globalisation factors in the early-mid 1990s. The BC seafood industry was dominated by salmon and herring, both highly seasonal fisheries. Capacities in the industry were geared to peak production, meaning full utilization was achieved only a few days per year, and over (or idle) capacity was the norm.

The fishing industry was a major employer in BC's coastal communities and amongst First Nations. Both DFO and the Province considered social objectives – provision of resource access to competing users and employment opportunities to BC residents - when managing and regulating the fishery. Economic efficiency was rarely an explicit objective of policy. The industry, faced with high wages and raw material costs, was a chronic "high cost producer."

Despite taking a rather insular view of resource utilisation, BC was an exporter of fish products, reliant on access to foreign commodity markets for the bulk of sales.

Under a traditional business environment, and using a speculative business model, the BC seafood industry generated financial results that were volatile, but generally acceptable (as long as the measurement goalposts weren't too high). Under a traditional business environment, the speculative game played by the BC industry made sense, but the industry, lacking internal efficiency, was vulnerable.

Indeed, the business environment was poised for fundamental change, and the prevailing structure of the BC seafood industry was soon to be over-turned.

The "Onslaught" of Globalisation

Beginning in 1990, when farmed salmon first made inroads into wild salmon markets, a succession of events, many of them falling under the globalisation category, conspired to undermine the fragile salmon/herring sector that had dominated the BC seafood industry.

Globalisation – more variables

Over the course of the 1990s and early 2000s, the litany of external variables impacting the BC seafood industry along the value chain grew to include (but was not limited to):

Global economic conditions

- The bursting of the Japanese bubble economy, meaning downward pressure on prices for luxury fishery products like sockeye salmon and herring roe.
- Strengthening of the Canadian dollar against the US dollar (from CAD 0.62 in 2003 to CAD 0.90 in 2006) means lower effective prices for BC seafood exporters in most global markets.
- Rising fuel prices impacting harvesting costs.
- Reductions in tariffs of seafood products.
- A Japanese scandal concerning corporate gift-giving (bribery), which ravaged the market for high-end roe herring gift-packs (the mainstay of the BC industry).
- Consolidation of seafood distribution – fewer wholesalers and retailers, each with greater clout.

Global fish production

- The dramatic growth of farmed salmon (in Norway, Scotland, Chile, and Canada), from alternative to wild, to the pre-eminent product form. Farmed salmon first crimped wild salmon prices, then displaced wild salmon from traditional markets like Europe and Japan. Farmed salmon also stimulated new demand for fresh salmon in the USA market, and conditioned consumers to expect consistency and quality from salmon products.
- Development of other finfish aquaculture species, such as basa and tilapia, increasing the array of high quality seafood alternatives for consumers.
- Record wild salmon production levels in Alaska, Russia, and Japan, even as BC production levels slumped.

- Collapse of cod fisheries in the Atlantic.
- Collapse of groundfish fisheries in the west coast of the United States.
- Development of the Alaskan pollock fishery into the largest volume fishery in the world.

Global political events and rise of developing countries

- With the fall of the Soviet Union, Russian wild salmon was no longer restricted to domestic consumption, but was offered on global markets.
- The end of apartheid in South Africa, allowing that country access to foreign markets (*e.g.* hake).
- The emergence of China, Russia, Thailand, and other developing countries as low-cost re-processing centres, especially for value-added products.

Shifting consumer preferences

- Increasing per capita consumption of seafood in OECD countries.
- A burgeoning consumer concern about the sustainability of wild fish harvests, and awareness of supporting only sustainable fisheries. The growth of "eco-labelling." The establishment of the Marine Stewardship Council as one body providing a "seal of sustainability."
- Consumer awareness – preference in some cases – about "wild" vs. "farmed" seafood… providing the wild fishery is sustainable.
- Consumer acceptance of twice-frozen fish products.
- Traceability – tracking the catch from "boat to throat" as a stringent health and safety requirement.

Globalisation – A new "game"

For the BC seafood industry, the process of globalisation meant more business variables, more complexity, and more competition. It meant that the advantages brought by access to local stocks were diminished; seafood players could access raw material, purchase processing services, and source finished product from anywhere in the world. Proximity to resources and fixed assets were no longer prerequisite to participation in the seafood business; a mere "cell phone" equipped a knowledgeable seafood participant to "play the game".

Globalisation placed greater emphasis on efficiency and adaptability. It meant daunting new challenges and exciting opportunities for the seafood industry.

At the same time, other factors were combining with globalisation to shape the BC seafood sector:

- Anomalous ocean conditions (for example frequent El Niño events) were having negative impacts on salmon ocean survival and recruitment of herring and groundfish stocks.

- DFO fishery management was becoming increasingly precautionary, with conflicts between conservation and economic objectives resolved increasingly in favour of conservation.

Impact on BC fisheries

Globalisation, combined with shifting ocean conditions and evolving fishery management philosophy, has created challenges effectively crippling the salmon/herring sector. Yet, over the same period, this very business and resource environment has provided opportunities allowing other Pacific fisheries – specifically groundfish and shellfish – to grow and flourish.

Groundfish comprises four distinct fisheries[8]: groundfish trawl, sablefish (also known as blackcod), halibut, and hook and line (rockfish, lingcod, and dogfish). The groundfish trawl fishery includes two of the subject fisheries for this study, hake and bottomfish. Shellfish includes shrimp, prawn, crab, sea urchins, geoduck, and sea cucumber.

Figure 19.1 shows the dramatic reversal of fortunes that has occurred in the last 15 years. The "core" and "secondary" sectors of the BC fishing industry have virtually reversed themselves in terms of relative stature.

Figure 19.1. Share of Public Fisheries Landed Value

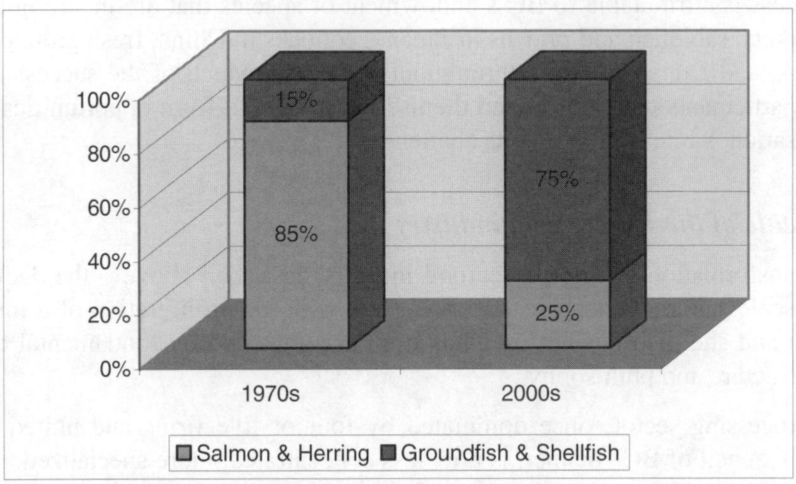

The values underlying these percentages confirm the extent of the diminution of the salmon industry and the startling pace of growth in the groundfish and shellfish sectors[9].

8 The fisheries are distinct in terms of licensing structure and fishery management; in reality, there is significant overlap in terms of species encountered and harvested.

9 Values are not adjusted for inflation.

Table 19.1. Average Annual Landed Value (CAD millions)

Period	Salmon	Herring	Groundfish	Shellfish	Total
1970s	99	35	19	5	158
1980s	199	53	50	21	322
1990s	146	55	112	86	398
2000s	47	36	132	116	331

The salmon sector, unable to cope with the pace and scope of change facing it, has withered. A substantial fleet rationalization program (CAD 285 million government funded licence buyback) reduced the eligible licences by 50% over the period 1996 to 2000, but failed to restore viability. Acrimonious relations between stakeholders – including the commercial and recreational sectors, First Nations, and DFO – preclude meaningful reform. The commercial salmon industry has lost a great deal of its former infrastructure, as exit and cost-cutting became necessary responses to tumbling revenues.

The groundfish and shellfish sectors, by contrast, have featured many successes. Part of the success is attributable to BC's endowment of species that are in strong demand in global markets: sablefish and prawns in Japan, geoducks in China, fresh groundfish fillets in the USA, and Dungeness crab throughout the world. Much of the success is because these BC participants have positioned themselves to benefit from opportunities presented by globalisation, while weathering its challenges.

Current state of the BC seafood industry

The transformation in the BC seafood industry described above – the decline of the industrial-scale salmon/herring sector concurrent with the proliferation of a multitude of groundfish and shellfish businesses – has been accompanied by fundamental changes in business structure and philosophy.

The processing sector, once dominated by four or five firms and united under the "Fisheries Council of BC" banner, is now less concentrated, more specialized, and highly fractured, with each firm pursuing its path independently. Companies avoid the undue risk and high capital requirements of a speculative business model, opting instead for a strategy that promises positive cash flows. The (painful) lessons of the salmon demise have been well-learned: overhead, debt, inventories, and surplus assets are to be avoided. On the positive side, firms are far more responsive to changing market and resource circumstances. On the negative side, access to capital is limited, as is the appetite to invest heavily in the business. This means that the bulk of production remains in commodity form, with value-adding initiatives in the minority.

Industry leadership has migrated from the processing sector to the harvesting sector. Most fisheries are organized into harvesting associations, and most harvesting associations work together in the "BC Seafood Alliance" (salmon is the notable exception). The BC Seafood Alliance strives toward forging effective partnerships with government and improving resource access security for harvesters. Fishers, through harvesting associations, are now immersed in broad industry issues, and more knowledgeable about markets.

Where large processing companies once possessed the majority of fixed assets and expert personnel in the industry, these assets are now dispersed amongst more, smaller processors and a host of service-providers. Custom processors, un-loaders, truckers, and sales brokers abound. The preference for performing all functions "in-house" has been replaced by the practicality of extensive outsourcing. This allows participants to "play the game" without owning a surfeit of assets. A value-chain approach is becoming commonplace, whereby harvesters, custom processors, and brokers coordinate the handling and disposition of the catch, dividing the proceeds according to an agreed-upon formula. A small but increasing number of fishers market their own harvest, either through public sales floats or direct sales to retailers/restaurants.

Whether the transformation seen in the BC seafood industry in recent years is an improvement or a digression, in terms of ability to cope with a rapidly changing competitive environment, is moot. What is clear is that the impacts of globalisation on the BC seafood sector, given historical industry structure, the regulatory regime, and the resource situation, have been *enormous*.

The balance of this paper looks at current performance and future prospects for the three subject fisheries: pink salmon, hake, and bottomfish.

Profile of Subject Fisheries

Following are profiles of the three subject fisheries, providing indications of the scope of each fishery, a summation of its economic stature (past and present) in the BC fishery, an evaluation of the prevailing fishery management program, and an encapsulation of the resource situation.

Pink salmon

Catch levels and values

Pink salmon is the most abundant and least valuable of the Pacific salmon species. Although pinks are BC's most numerous species, BC is a very minor player in the global scheme, averaging only 3% of world supply.

Figure 19.2. Share of Global Pink Salmon Catch by Nation

Average 1993-2004

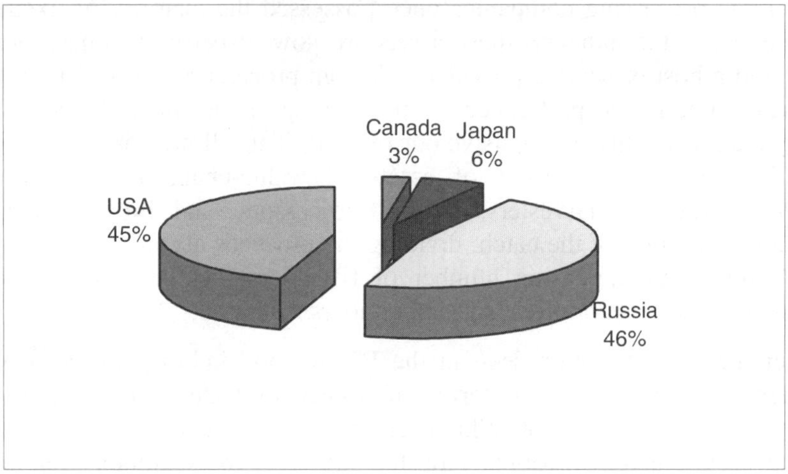

Figure 19.3. British Columbia and Alaska Pink Harvest

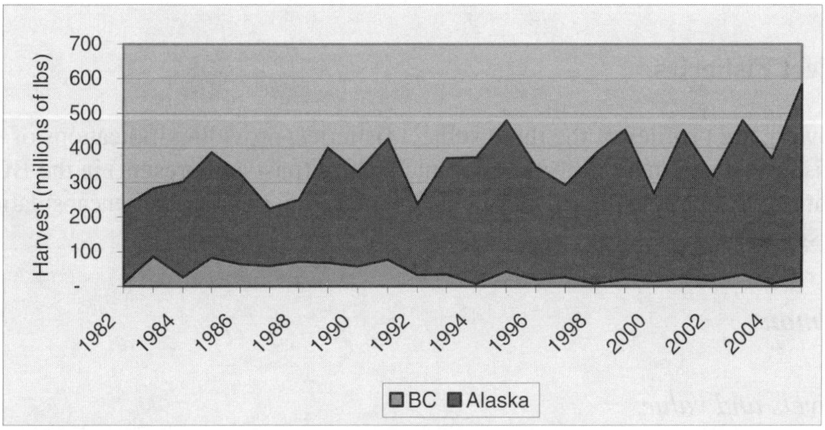

As with many fisheries, BC's direct competitor is Alaska, though in the case of pink salmon, BC poses little true competition to its neighbour. Alaskan catches have been consistently rising, while BC catch levels are in decline (indeed, they appear to have "flat-lined" in Figure 19.3). Reductions in BC pink harvests are attributable to three factors:

- Reduced resource abundance, particularly in north and central coast stocks. Extreme variability is a naturally occurring pink salmon attribute.
- Precautionary fishery management, meaning fewer harvesting opportunities.
- Industry not fully exploiting available pink harvest because of economic constraints.

Figure 19.4 demonstrates the magnitude of the reduction in catch volume and value of the BC pink harvest over time. The decline in landed value is steeper than the drop in landings, indicating that both volumes *and* prices have fallen.

Figure 19.4. BC Pink Landings and Landed Values

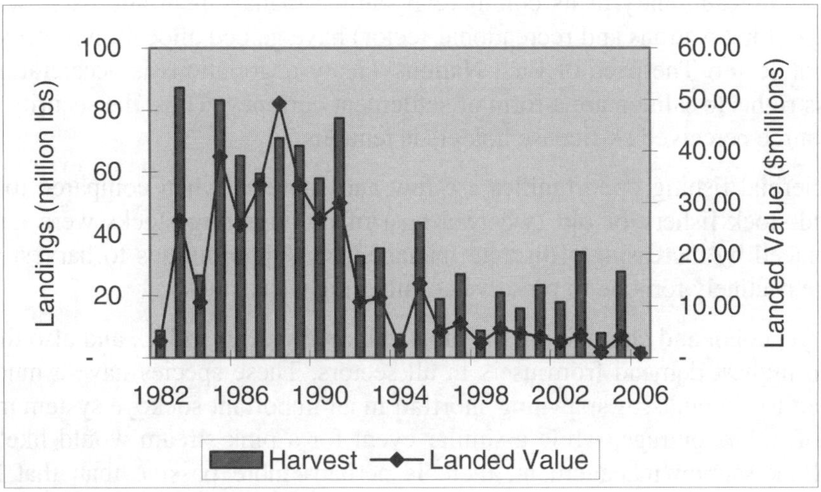

Economic overview

Historically, pinks accounted for about 30% of the BC salmon harvest (by volume) and 10% of the landed value. Virtually all of the catch was canned, with ½ pound cans on export markets the primary product form/market. In general, pinks made a more important contribution to processors bottom-lines that to harvesters – modest raw material prices and generally firm markets meant attractive margins. Pink salmon were a very quiet but important contributor to BC processing firms' financial health.

BC processors have always drawn from the generous supply of nearby Alaskan pink salmon to bolster domestic supply when needed[10]. With the current level of BC landings at less than 1/3 of historic levels, Alaska is now the primary pink salmon source for BC processors, who cannot count on the local supply to serve their markets.

The stature of BC-caught pinks has digressed from that of an important contributor (a backbone, even) to the processing sector, to a purely opportunistic activity around which salmon participants cannot even plan. Annual wholesale receipts of BC canned pink salmon averaged about CAD 100 million in the 1980s, compared to just a few million dollars currently[11]. The decline of the pink salmon business played a large role in the demise of the BC salmon industry.

Fishery management

Pinks are managed along with other salmon species under a "derby" system. Despite programs to reduce the fleet and mitigate fishing pressure, such as a government funded license buyback and area licensing, the capacity of the fleet is still excessive when

10 BC can import round pinks from Alaskan tender vessels so long as Alaskan plants are operating at capacity, which is usually the case. BC processors are generally comfortable with their level of access to Alaskan pink salmon.

11 Estimates.

compared to the magnitude of fishing opportunities. Over time, DFO has become increasingly precautionary in its commercial salmon management approach, while other user groups (First Nations and recreational sector) have gained allocation priority over the commercial sector. The pace of First Nations Treaty negotiations is accelerating in BC, and access rights to salmon are a form of settlement currency. Thus, the security of access to the resource perceived by license holders is tenuous.

Commercial fishing opportunities are few and sporadic when compared to the past. The mixed-stock fishery of old (where a host of co-migrating stocks were targeted) is being replaced by increasingly discrete management. Opportunities to harvest abundant species are routinely forgone to preserve co-mingling weak stocks.

Sockeye, coho, and chinook are the most valuable species in BC, and also the species that are in highest demand from users in all sectors. These species have a much higher profile than lowly pinks. A spawning shortfall in an important sockeye system may spark a frenzy of public outrage, while a similar event for a pink stream would likely go unnoticed. Pink salmon management, then, is perhaps more passive than that for other stocks, and pink harvesting opportunities are at times subordinated to those for more prized species.

The state of fishery management in the salmon fishery is a source of friction within industry and between industry and DFO. The fishery management system results in a pink salmon fishery that:

- Is virtually unpredictable, with openings often announced on short notice.
- Results in few, brief openings, with the bulk of a season's catch landed in a very few days.
- Motivates participants to focus on volume rather than quality.
- Precludes the fleet from accessing all available surpluses.

Initiatives to reform the fishery, to restore a more orderly and quality-driven harvest, have thus far been fruitless. The BC salmon sector has been unable to devise a system that addresses conservation concerns, economic realities, and social objectives. Most salmon industry participants agree that they system is broken, but so far, they have been unable to fix it.

Resource status and outlook

Shifting ocean conditions and other, poorly understood, environmental conditions have caused overall pink salmon abundance to decline in BC, though the state of pink stocks is not as dire as the reduction in catch levels indicates. It is not known whether reductions are transitory in nature – part of a long-term productivity cycle – or permanent. BC, especially the south coast, is perilously close to the southern end of the range of pink salmon in the North Pacific Ocean.

Even at current abundance levels, BC pink salmon stocks have the potential to support higher catch levels and more meaningful business activity.

Groundfish trawl

Catch levels and values

The BC groundfish trawl is a year-round fishery harvesting a suite of bottom-dwelling species. This includes a "core" group of species dominated by rockfish and soles (a variety of species) and including lingcod and Pacific cod; and a group of "developing" species, for which utilisation is steadily growing, including turbot (arrowtooth flounder) pollock, and dogfish.

This is a year-round fishery that, when compared to pink salmon and hake, offers stable harvesting and market-service opportunities.

The groundfish trawl fishery has operated under an Individual Vessel Quota (IVQ) management program since 1997. Under the program, the catch levels for core species have declined (because catch quotas are no longer exceeded) and catch values (both in per-pound and total terms) have risen. Greater value is being derived from less fish (see Figure 19.5).

Figure 19.5. Groundfish Landings and Landed Values (est

Core Quota Species, Annual Averages for Period

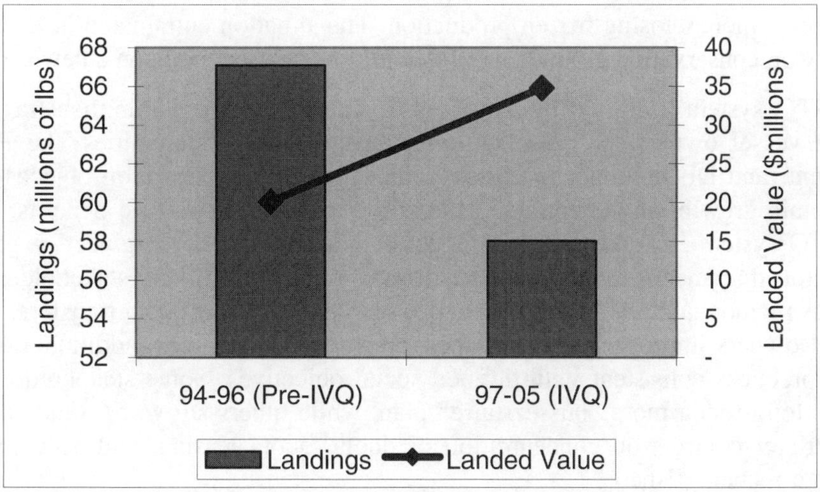

Economic overview

Groundfish trawl was a high-volume, low value fishery for many years. The fishery appealed to salmon processors and "big-boat" fishers as an opportunity to extend asset utilization and amortize fixed costs over a greater production base. The fishery produced frozen fillet blocks at a time when Atlantic cod and lower-48 Pacific groundfish catches were still prolific. BC enjoyed few competitive advantages, and results were marginal.

As groundfish stocks in other jurisdictions were fished to decline, market opportunities improved for BC groundfish. Hard times in the salmon fishery meant greater interest in groundfish. Pressure on groundfish stocks escalated, and a passive fishery management approach was no longer appropriate. A series of increasingly restrictive rules, including periodic trip and species limits, were met with even greater fishing pressure, misreporting of catches, and at-sea discarding. The ranks of participants, both vessels and processors, were swelling. The impacts of fishing on the plethora of groundfish stocks were unknown. Thus, DFO took the bold step of closing the fishery. It opened, months later, under an entirely new and unique management system.

The BC groundfish trawl sector is now the preferred supplier of fresh fillets to western USA markets (also known as the "I-5 corridor," for the name of the interstate highway linking Vancouver and California's urban centres).

Given that it is operating comfortably under DFO's precautionary management mandate, it is clear that the groundfish trawl fishery is meeting conservation objectives. Together with the hake fishery, groundfish trawl is not only BC's highest volume fishery, but it is also the highest landed value fishery in BC.

The bottomfish fishery has progressed admirably from its former "marginal" status.

Fishery management

In the groundfish trawl fishery an individual transferable quota system was implemented in 1997. Prior to that year, groundfish trawl was a derby fishery that combined over-exploitation of the resource and un-even economic results: too many vessels racing for their share of the catch, gluts of fish followed by droughts, and a high percentage of money-losing frozen production. The situation culminated in DFO closing the fishery on conservation grounds in 1996 until industry agreed upon a new system.

The ITQ system[12], crafted by a variety of interests in the groundfish trawl fishery including vessel owners, processors, union reps, coastal communities, the Provincial government, and DFO, sought to ensure achievement of conservation objectives, allow for economic growth, and encourage a fair distribution of economic benefits. This is a unique ITQ system embedded with safeguards in an attempt to "engineer" a groundfish trawl sector that meets the broad and diverse objectives of all stakeholders. These safeguards include quota holdings caps, species caps, curbs on quota transfers, sanctions for vessel owners improperly treating their crews, and incentives (additional quota) for business practices consistent with defined social objectives. Some stakeholders lobbied (and now long for) a more "laissez-faire" plan, while others strove for (and continue to desire) stricter controls on concentration of quota, quota leasing, and accumulation of quota by "armchair" fishers.

While no one admits to being completely happy, it is a compromise plan that all interests can live with. Though it contains curbs on free enterprise, the Plan paved the way for an orderly harvest, meaningful business planning, and improved responsiveness to markets.

The IVQ/GDA plan is extremely complex, with each vessel endowed with about 50 different quotas (combining about 20 species and five management areas). Because the

12 The management system is formally known as the Individual Vessel Quota/Groundfish Development Authority Plan (IVQ/GDA).

plan allows vessels to carry over un-caught quota to the following season[13], few if any of the 50 TACs are reached each year.

Fishers are individually accountable for adhering to each of the 50-odd quotas. Some quotas are generous (target species) while others are small, amounting to a bycatch allowance. If a fisherman exceeds his quota for any species in any area, he is prohibited from bottom fishing in that area until the quota shortfall has been remedied (through quota transfer from another vessel). To ensure compliance, the fishery features 100% at-sea observer coverage, and 100% offload validation. All catch – whether retained or released – is accounted for.

Groundfish trawl licence holders contribute key advice to fishery managers, and contribute financially to (and co-manage) the research and stock assessment program. Given the stringent conservation measures in the plan, and the high costs borne by the fleet (at-sea observers, offload monitors, research and stock assessment, co-management) the BC groundfish trawl fishery proudly claims to be the best-managed trawl fishery in the world.

Given the complexity of the plan, the socio-economic compromises that in contains, the high degree of monitoring it requires, and the high cost to harvesters it brings, the plan was met, upon implementation, with scepticism and doubt. Participants surprised themselves with their adaptability.

Fishing under the new plan has been transformed. To harvest effectively under a multi-quota plan with full individual accountability requires avoiding non-target species for which quotas are slight and metering out the catch over the season. Selectivity and avoidance are keywords in the fishery. The change in fishing culture is startling. Draggers once boasted of how they loaded the boat in record time, of how they out-produced their comrades. Skippers were chosen for their ability to *produce*. Now, trawl fishers brag about how "clean" they can fish. Captains are favoured for their knack of *avoiding* certain species of fish. Fishers cooperate, rather than compete, on the fishing grounds. "Don't set your gear there, Bob… there's a big school of shortraker rockfish!"

As a counterpoint to the fishery's accomplishments, conservation groups persistently note that the BC groundfish trawl fishery still has room for improvement, particularly with respect to bottom impacts and catch utilization.

Resource status and outlook

Many of the groundfish stocks targeted in the bottomfish fishery, such as rockfish and lingcod, are long-lived and relatively sedentary. Fluctuations in population tend to occur over a long period of time. Providing fishing pressure is appropriate, and ocean conditions allow for reasonable survival and periodic strong recruitment, stocks remain quite stable. Indeed, for most species, stock levels and catches have been very stable over the past 10 years.

Other species, most notably Pacific cod, are shorter lived, and abundance is highly variable. For such species, DFO applies its precautionary approach when setting catch levels to mitigate the impact of fishing on year-to-year abundance.

13 Up to 30% can be carried forward.

The IVQ/GDA management plan provides stock assessment scientists with a full picture of groundfish removals (including catches and releases) and a large number of biological samples. The task of performing regular stock assessments for the full suite of species is daunting, so research is performed according to priorities jointly established between DFO and industry.

The current scope and nature of the fishery strikes a pleasing balance for industry and DFO, aligning catch levels with both ecological and market objectives. Subject to the good graces of ocean conditions, the outlook for groundfish stocks – continued stability - is positive.

Hake

Catch levels and values

Hake (Pacific whiting) is to BC groundfish stocks what pink salmon is to BC salmon - the most abundant but lowest value species. As with pink salmon, BC is a relatively minor player in terms of world hake production, accounting for 5-8% of global supply. BC ranks behind Argentina, South Africa, Chile, and the USA (Washington/Oregon) as a hake producer. Total world production ranges from 1 200-1 500 tonnes annually.

Pacific hake is a trans-boundary stock, straddling Canadian and US waters. The two countries in the past disagreed on catch shares, and over-fished the stock to an extent. They have now forged a Treaty, ensuring cooperation in setting harvest levels. Unlike pink salmon, both catch levels and fish values have risen in recent years.

BC production and landed value is shown in Figure 19.6. Hake is BC's highest volume fishery.

Figure 19.6. Hake Landings and Landed Value

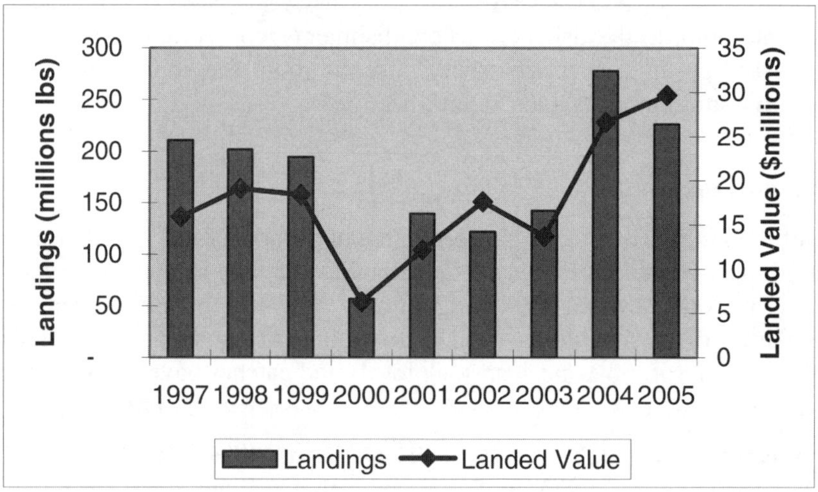

Economic overview

The BC hake fishery is relatively novel, with participation by the BC fleet commencing in 1978. The BC processing industry showed little interest in handling hake until the early 1990s. The Pacific hake species is plagued by an enzyme that degrades the flesh quality unless stringent handling protocols are observed (*i.e.* chilling, prompt processing). And, the BC processing sector was largely focused on more lucrative species, particularly salmon and herring. As an un-utilized resource, the Province made hake an exception to its prohibition on foreign/at-sea processing, allowing foreign motherships to operate offshore, purchasing, processing, and transporting finished hake products overseas. Through the 1980s and early 1990s the so-called "Joint Venture" fishery grew in scale and efficiency.

BC trawlers found the Joint Venture fishery to their liking. A local organization called the Hake Consortium of BC oversaw contracting of the factory ships and established delivery timetables and criteria for participating trawlers. A vessel simply towed for hake using its midwater trawl, and pulled alongside the factory ship, which lifted the cod-end of fish aboard for immediate processing. Never a fish scale onboard, and no need to run the fish to town! Although fish prices were low (less than 10 cents), the volumes and efficiency allowed profitable fishing operations.

By the early 1990s, advances in fish handling practices, processing technology, and seafood markets made the utilization of hake by BC shore plants feasible. The downturn in the salmon business provided motivation to find alternative activities. A few shore based plants located in small communities on the west coast of Vancouver Island invested in processing equipment, mostly surimi capacity. Local governments provided infrastructure to support this much-needed injection of industry. Senior governments confirmed that shore-based hake processors would have priority access to the hake resource, and so the Joint Venture became a "safety valve" fishery, handling only quantities of fish not required by the shore sector. Acrimony over domestic hake allocation – shore vs. Joint Venture – developed. A strong appeal of shore-based processing was the high level of employment associated with hake – large volumes over a 5-6 month season.

Over virtually the same period as the pink salmon business has fallen, the hake business has developed from scratch to become an important contributor to the groundfish sector and BC seafood industry. It has gone from being an entirely un-utilized resource, to the subject of a spirited annual allocation debate.

Fishery management

As noted previously, hake is managed under the same IVQ/GDA plan as bottomfish, though in many respects, hake is a distinct fishery, and business, from groundfish.

While the bottomfish fishery digressed into a race for fish with rampant discarding in the years preceding 1997, the hake fishery has taken a different path. This mid-water trawl fishery is virtually "clean" (there is very little bycatch). The BC hake fishery developed under the Joint Venture system, whereby production was effectively scheduled and the quantity of catch per vessel was known ahead of time; the Joint Venture fishery closely resembled a quota system.

With implementation of the IVQ system, an orderly harvest is now facilitated in the shore-based fishery as well. The system supports sound business planning. Implementation of the Canada-US hake treaty will bring joint fishery management, and agreed upon national catch shares, to hake management.

Resource status and outlook

Hake is among the short-lived (7-8 years), migratory groundfish species that features widely varying abundance. The Pacific hake biomass fluctuates with ocean conditions and recruitment levels, and catch-ability is linked both to abundance and varying fish distribution. While annual TACs will follow abundance patterns, joint Canada-USA fishery management under the Treaty will ensure improved long term stock sustainability.

Impacts of Globalisation on Subject Fisheries

The preceding section provided overviews of each subject fishery. How each fishery has fared under the pressures of fisheries globalisation is strongly influenced by its stature, structure, and recent history. In this section, the globalisation forces impacting each fishery are identified, the responses to date are described, and a brief analysis is offered.

Pink salmon

Globalisation challenges and opportunities

A list of key globalisation-related challenges and opportunities in the BC pink salmon industry follows:

Table 19.2. Globalisation Challenges and Opportunities for Pink Salmon

Pink Salmon Globalisation Challenges	Pink Salmon Globalisation Opportunities
Growth of world pink harvests (Alaska and Russia), concurrent with decline of BC harvests, means BC market share and market clout has declined.	Growing demand for seafood generally. Growing demand for wild salmon. Pink positioned as "most affordable" salmon.
The vast majority of pink production has historically been canned.	For canned product, BC must focus on other product forms, such as ¼ lb, skinless boneless, easy-open, and others.
Canned pink salmon markets are chronically over-supplied.	
Canned pink markets are stagnant (not growing).	
Canned pink price levels have weakened substantially in recent years.	
BC canned pink production is no longer economic	

Pink Salmon Globalisation Challenges	Pink Salmon Globalisation Opportunities
(1/2 pound form); BC is not a low cost producer.	
Alaska has out of necessity developed frozen pink markets. Up to ½ of Alaska's production is now frozen, generally in round or butchered form, for re-processing elsewhere in the world.	BC is in a position to benefit from Alaskan initiatives… let Alaska blaze the trails and make developmental mistakes.
A small fresh and value-added business has also developed.	
While Alaska has been developing new products and markets, BC has been on the sidelines.	
There is a growing consumer movement toward sustainable wild fisheries and eco-labelling.	The conservatively managed BC salmon fishery is in the MSC certification process. Certification may offer new market avenues.
The Alaskan salmon fishery is certified by the Marine Stewardship Council (MSC), while BC is not.	
Emergence of low cost processing operations in China, Thailand, and other locations with low wage rates.	Developing markets represent an outlet for round or butchered frozen BC pinks. A new low value but high volume business.
"Developing countries" as major importers of raw material for value-added processing. BC cannot compete with this cost structure.	BC has the advantage of working on fresh fish, can produce once-frozen fillets and other products.
Acceptance by global consumers of "twice-frozen" product.	BC's proximity to pink stocks and US market provides a differentiation avenue (fresh, once-frozen products).
With farmed salmon, there is a huge volume of salmon in the marketplace. Farmed salmon sets consumer's expectations of quality and consistency at a very high level.	A distinct consumer awareness (sometimes preference) for wild salmon has developed.
	Pink salmon has the opportunity to fill a niche below farmed salmon in the product/value spectrum.
Farmed salmon holds a host of competitive advantages over wild salmon.	Pink salmon is cheap (inexpensive) enough that it can be used as an ingredient in value-added products such as burgers.
Widespread availability of farmed salmon relegates pink to the bottom of the salmon value spectrum.	
BC processors are already involved in the Alaskan salmon industry. Investment and resources may be directed to Alaska rather than BC.	Pink salmon is not a fishery/business that appeals to all BC participants. Those choosing to participate should be able to source raw material on favourable terms.

BC industry response

The BC pink salmon sector has to a much greater extent been floored by the challenges of globalisation than it has benefited from its burgeoning opportunities.

The pink business has shrunk dramatically. A marked reduction in wholesale prices in the ½ pound canned market, BC's primary product form, meant dramatically lower ex-vessel fish prices to harvesters. Prices fell below 15 cents per pound, with fish volumes sporadic. The economics made sense for neither harvesters nor processors. Even with limited fishing opportunities, a portion of the available pink harvest has been forgone.

For a large portion of the industry, pink salmon have ceased to be a meaningful business activity. Some simple rationale for dedicating resources elsewhere: a four-pound pink, at 15 cents/pound, is worth 60 cents; a six-pound sockeye, at CAD 2/pound, is worth CAD 12. It takes 20 pinks to equal the value of one sockeye… and there aren't that many pinks available. So where will effort be directed? Sockeye.

Even those with a will to handle pink salmon struggle with the sporadic, uncertain supply. Developing products, building customer goodwill, and supporting marketing programs is frustrating at best, impossible at worst, without a secure and reliable supply of product.

Not all participants have forsaken pink salmon. A few processors still consider pinks to be an important part of their business, and are focusing on strategies to differentiate their product from pure commodity forms. Effecting such a strategy requires substantial investment in equipment, and use of Alaskan pinks as a backstop of production. It also requires courage; said one processor of his pink salmon program, "we've invested millions of dollars we don't have."

Globalisation has provided another intriguing option – being a marketer of product without being a producer. That is, exploiting "long" global supply by buying processed product on the open market for re-sale. Canada's pre-eminent canned salmon brand, "Clover Leaf" was until recently owned by BC's biggest fish processing company, BC Packers. It made sense that BC's largest fleet supplied Canada's biggest label. When BC Packer's parent company opted to exit the BC seafood business, however, the Cloverleaf label was sold to a US company, Bumble Bee, having no footprint in the salmon industry. Cloverleaf, sourcing strictly from the open market, remains Canada's premiere canned salmon brand.

Even companies that market pink salmon but have not shed their fleet connections have altered their business practices. The adage "catch all you can, and can all you catch" has given way to "source the product you need to fill your markets" (not nearly as clever, but a sensible shift nonetheless).

Analysis

BC pink salmon is a case of a business with fragile fundamentals - including an inefficient industry structure, market-indifferent fishery management, and a volatile resource base – buckling under the pressures of globalisation.

Industry participants are quick to identify the obvious solution to the pink salmon problem – develop a market-responsive fishery management system for the salmon fishery. Pointing to fishery management as a problem does not imply that DFO takes the (sole) blame; it is widely recognized that industry shoulders the primary responsibility for devising a management system that addresses DFO's conservation constraints, while meeting the diverse socio-economic objectives of the industry. But BC's salmon participants are surprisingly willing to endure year-after-year of dismal financial performance rather than accept some necessary compromises.

Given a sound fishery management footing, the BC pink salmon industry is positioned to exploit either or both two strategic directions:

- Becoming an effective supplier of raw material (round-frozen or butchered) to developing nations such as China and Thailand. That is, a low cost, high volume supplier of raw material for re-processing.
- Value-added pink salmon production, exploiting BC's advantage of working from fresh product to service fresh markets or "once frozen" frozen markets.

The pink salmon business in BC enjoys substantial upside. Though pinks will never be a shining-star of the BC seafood industry, there is little business reason that they cannot return to "quiet contributor" status.

Groundfish

Globalisation challenges and opportunities

Globalisation-related challenges and opportunities in the BC groundfish trawl fishery include:

Table 19.3. Globalisation Challenges and Opportunities for Groundfish

Groundfish Globalisation Challenges	Groundfish Globalisation Opportunities
Fluctuating foreign exchange rates have worked against the BC groundfish sector.	
The Canadian dollar has strengthened against the US dollar from CAD 0.62 in 2003 to CAD 0.90 in 2006.	
BC sells the majority of its groundfish products in fresh fillet form to the USA.	
Though prices have remained consistent, or even risen, in US dollar terms, effective prices to Canadian producers have plummeted – a CAD 3.00 US fillet worth CAD 4.80 per pound only netted CAD 3.30 in 2006.	
The primary market for BC groundfish is the western USA, particularly California.	Seafood demand is growing rapidly in BC's primary market (western USA). This provides a partial buffer against growing competition.
This market is being flooded with a host of farmed whitefish species that are being embraced by consumers (eg. basa, tilapia, catfish).	Fresh wild groundfish is an established product with consumer appeal.
BC groundfish products are maintaining sales volume, but losing ground as the overall seafood market grows.	The decline of other groundfish fisheries, particularly that off the coast of Washington-Oregon-California, has provided an opportunity for BC to service fresh markets.
BC must work hard to maintain its primary market, given the array of farmed seafood alternatives.	
The market has a strong preference for fresh product over frozen (especially for wild product).	BC is ideally situated to serve the fresh whitefish fillet market in California.
Frozen BC groundfish fillets are dramatically reduced in value vs. fresh, since there is an enormous world supply of low-cost frozen wild whitefish. A high-quality BC rockfish fillet fits in the same (low) value realm as frozen pollock.	
When the BC groundfish fishery experiences periodic landing "spikes" a portion of production is inevitably frozen. Prices for BC frozen groundfish fillets have declined.	

Groundfish Globalisation Challenges	Groundfish Globalisation Opportunities
Emergence of low cost processing operations in developing countries with low wage rates.	Developing nations represent a new market outlet for round or headed and gutted low-value (developing) groundfish species such as turbot.
"Developing countries" as major importers of raw material for value-added processing. BC cannot compete with this cost structure.	
Acceptance by global consumers of "twice-frozen" product.	
There is a poor public perception of bottom trawling.	The BC groundfish trawl fishery can position itself as a world leader in responsible trawling practices.
Despite the strong conservation record of the BC trawl fishery, constant vigilance is required to defend and justify the fishery.	
Well-intentioned environmental legislation (Species at Risk Act) can prove problematic when applied to multi-species fisheries.	
The increase in global oil prices has caused marine diesel fuel costs to spiral. Rising fuel costs are a major cost factor in the BC groundfish trawl fishery.	Technology is available from around the world to encourage more fuel-efficient operations: fishing gear, flow metres, high efficiency engines.

BC industry response

The BC bottomfish sector has benefited from a combination of good management and good fortune. Concurrent with the BC industry implementing a new fishery management plan to address conservation and socio-economic shortcomings, globalisation events *favouring* BC developed. The collapse of competing groundfish fisheries positioned BC as the preferred supplier of fresh fillets to the western USA (it didn't hurt that American consumers failed to realise that the "local" groundfish species they were enjoying no longer came from their local waters). A weak Canadian dollar relative to the US currency allowed sound values for American customers and strong prices for Canadian exporters. Growing seafood demand in the US helped preserve BC's market position.

This window of agreeable competitive conditions allowed the BC groundfish sector to gain familiarity with a complex new fishery plan, develop its "fishing-to-market" acumen, and enjoy profitable operations. Participants who had initially resisted the new management plan now lauded themselves as business visionaries. During this period, the strategy for improving financial returns was clear – purchase (or rent) more quota; more quota meant improved returns.

Of course, a competitive climate-change was inevitable, and a host of globalisation factors began to heap challenges on the industry. A rapidly strengthening Canadian dollar meant rapid devaluation of product values. An array of farmed white fish products quickly gained consumer acceptance. The fresh market was increasingly intolerant of any spot surge in BC fresh fillet production. Prices for frozen products declined. Fuel costs (a major cost-factor in trawling operations) spiked, tracking the price of crude oil in world commodity markets. In the course of a couple of years, the competitive environment turned decidedly difficult, placing pressure on financial returns.

The groundfish trawl sector was bruised, rather than battered, by these developments. Participants recognised that more-precise execution of harvesting-processing functions was necessary. With lower returns from core operations, they turned greater attention to deriving value from heretofore "marginal" species such as turbot (arrowtooth flounder), dogfish, and pollock. Much of this production was exported in frozen h&g form to developing countries for re-processing. Figure 19.7 shows how utilisation for these developing species has risen. Also, opportunities to hive higher-values species out of the fresh-fillet mix were explored. Ethnic markets for whole red-snapper (rockfish) were exploited.

Figure 19.7. Groundfish Landings

"Core" and "Developing" Species

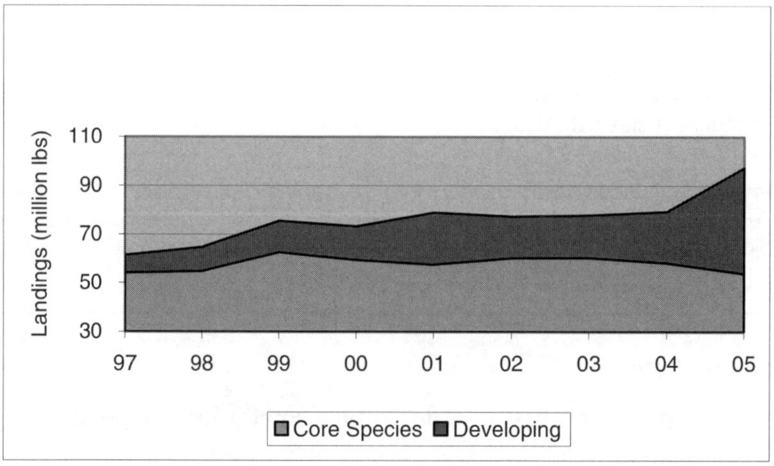

The industry recognised that the "honeymoon" period succeeding implementation of the IVQ/GDA plan was over, and that the future would be characterized by lower margins, the need to further develop products and markets, and the importance of sound execution of business functions.

Analysis

The BC groundfish trawl industry enjoys sound fundamentals – a stable year-round resource, a market-responsive fishery management program, a high degree of industry-government cooperation, and a strong competitive position – and is therefore well-positioned to benefit from the opportunities of fisheries globalisation while enduring its challenges.

An important factor giving the groundfish industry competitive strength and resilience is that it is entirely self-adjusting and largely self-reliant. A high degree of fleet rationalisation has occurred over the past several years, with quotas fished by fewer vessels owned by fewer individuals and companies. There are now 45-50 active trawlers in the fishery, whereas 142 vessels received an initial allocation of quota. There is a high degree on integration between the harvesting and processing sectors. The largest groundfish processors were always significant owners of trawl vessels. Now, successful trawl fishers have entered the processing fray. The security provided by the quota system motivates vessel owners to pursue their own business paths. There is a great deal of individual initiative taken, so product and market development is not the burden of a few processing companies. Virtually every vessel in the groundfish fishery executes a unique business plan, targeting a different mix of fish and different market avenues. Groundfish trawl is a rather close-knit fishery, with generally cooperative and harmonious internal relations, and an effective working rapport with DFO.

The prospects for the groundfish trawl fishery are strong, abetted by a sound management plan that encourages responsible fishing practices and meaningful business planning. The industry requires further business development in order to best deal with future globalisation developments:

- Capital expenditure in the fishery has thus far been dominated by purchases of quota. Investments in vessels and plants have largely been neglected. The industry will need to address physical asset issues in the future to ensure continued competitiveness.
- Over-reliance on the western USA fresh fillet market is recognized, so further product-market diversification measures are required.
- The fishery can continue to improve catch utilization and selectivity in order to pre-empt criticism from the environmental sector, and preclude interruptions under environmental legislation (Species at Risk Act). Further, improved resource stewardship makes sound business sense.

Hake

Globalisation challenges and opportunities

A list of key globalisation-related challenges and opportunities facing the BC hake industry is offered below:

Table 19.4. Globalisation Challenges and Opportunities for Hake

Hake Globalisation Challenges	Hake Globalisation Opportunities
Canada is a minor global hake fishery.	World demand for white fish products is growing.
Larger global hake fisheries such as those in Argentina, Africa, Chile and the USA have a strong bearing on price levels, and marketability, for BC hake.	Hake catch levels in other jurisdictions have been down in recent years, allowing BC hake to make market inroads.
Hake is a generic white fish product. Hake markets are influenced by production of other abundant white fishes around the world.	
There has been rapid development of large scale fisheries in the southern hemisphere for low-value species such as threadfin bream, itoyori, and jack mackerel. Products from these fisheries may go head-to-head with hake (particularly in surimi markets).	
These fisheries provide formidable competition for BC hake from in the surimi business.	
Emergence of low cost processing operations in Russia and China.	Developing markets represent an outlet for round or h&g frozen BC hake. A new low value but high volume business.
"Developing countries" as major importers of raw material for value-added processing. BC cannot compete with this cost structure.	BC has the advantage of working on fresh fish, can produce once-frozen fillets.
Acceptance by global consumers of "twice-frozen" product.	BC's proximity to US market provides a cost and differentiation avenue (once-frozen products).
Much of global white fish harvesting and processing is by factory trawlers or motherships. These operations enjoy greater efficiency and lower costs than shore-based plants.	Joint Venture fishery is used as a "safety net" in the BC hake fishery.
	BC regulations allow a BC catcher-vessel to head, gut, and freeze its own catch (but not fillet).
BC regulations limit the use of factory trawlers and motherships in fisheries, placing BC at a cost disadvantage.	
Fluctuating foreign exchange rates have worked against the BC hake sector.	
The Canadian dollar has strengthened against the US dollar from CAD 0.62 in 2003 to CAD 0.90 in 2006.	

Hake Globalisation Challenges	Hake Globalisation Opportunities
The US is a significant customer of BC frozen hake fillets.	
Many international transactions are conducted in US dollars.	

Industry response

The pace of change in the BC hake industry has been breathtaking. Thirty years ago, hake was completely un-utilized by the BC fishing industry, exploited only by foreign fleets. Fifteen years ago, there was virtually no shore-based processing of hake by the domestic industry. Five years ago, BC hake processing meant surimi processing. Three years ago, surimi processing virtually ceased. The last two years have been, financially, the best ever enjoyed by the hake sector.

The shore hake sector in BC developed around the surimi business. The proximity of prime hake fishing grounds off the west coast of Vancouver Island to the communities of Ucluelet and Port Alberni provided strategic opportunities to accommodate hake processing. Government strongly supported development of a BC shore-based hake sector, according the shore sector priority access to hake stocks (vs. the Joint Venture fishery). By the late 1990s there were three plants operating in the area. As surimi is a pure commodity product, and BC producers were enduring a production learning curve, results were uneven. When market demand was strong, the plants operated to capacity; when demand slackened, the plants reduced throughput. This pattern frustrated fishers, who were accustomed to the greater reliability and higher prices of the Joint Venture fishery. The evolution of the hake fishery, from Joint Venture to shore-based, is charted in Figure 19.8.

Figure 19.8. Pacific Hake Landings

Joint Venture and Shore

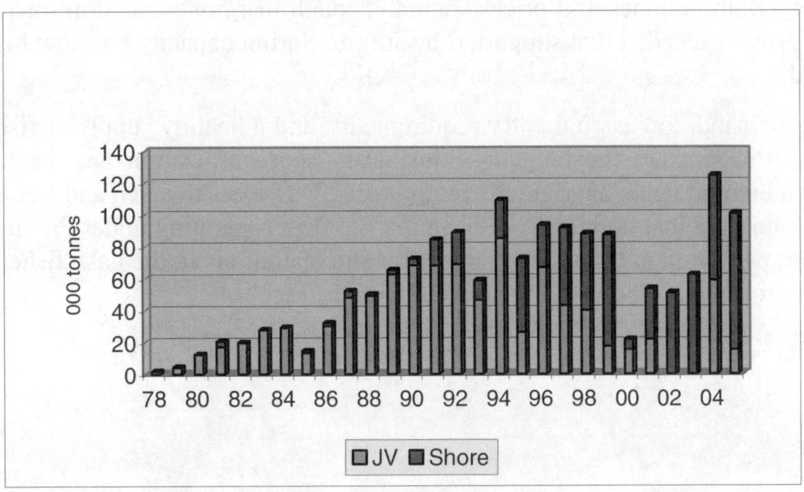

Recognising the frailties of the surimi business, some hake processors began to develop fillet programs.

In both surimi and fillet production, processors faced the challenge of dealing with Pacific hake's high perishability. Quality and recoveries improved over time owing to investments in equipment, improvements in fish-handling practices, and experience in dealing with this finicky species.

BC hake producers developed significant acumen as surimi producers. On a "technology sharing" mission to Chile, a group of BC hake processors magnanimously offered to share some of their surimi insights with their Chilean counterparts. Instead, they found that the Chilean industry had largely progressed beyond surimi, focusing on a host of fillet and value-added products. "We went to see a third world country… and found that we were it," lamented the mission leader.

Indeed, surimi-reliance proved problematic. Markets were chronically over-supplied. "Junk fish" like jack mackerel in Chile, and itoyori in the Indian Ocean, were used to supply an acceptable quality of surimi, at a price that BC could not match. The pattern of volatile utilization by the BC shore sector continued.

The surimi business in BC hit its nadir in 2003, when sales of BC surimi were stalled owing to, of all things, an outbreak of "mad cow disease." Because of a BSE-infected cow originating at an Alberta cattle farm, Canadian beef was, for a period, banned from Asian and American markets. BC surimi uses, as a minor ingredient, a beef-plasma binding agent. This additive was sourced from a US company. BC surimi therefore contained trifling amounts of (American) beef. In the eyes of the international trading community, however, BC surimi effectively became a Canadian beef export, and was treated as such. Even sales to the USA (where the beef additive originated) were prohibited. Globalisation gone mad! Surimi production was halted during the 2003 fishing season, and never re-commenced. Growing fillet production ensured some utilization of hake stocks in 2004, and a substantial Joint Venture fishery was allowed that year.

At the same time, BC hake participants saw sudden, strong interest from Russian and Chinese firms in the purchase of headed and gutted frozen hake. H&g production is very simple, requiring dressing machinery commonplace in BC (much of it idle salmon equipment). Both volumes and prices soared. Though margins were slim for processors, ex-vessel prices exceeded that supported by surimi. Surimi capacity has now been largely dismantled.

Strong demand, low capital entry requirements, and a healthy supply of fish attracted new BC participants to the h&g hake business. Shore utilization has reached record levels. Two entrepreneurs have fitted "factory-scale" trawlers to catch and freeze hake on a scale greater than that previously seen in the BC fleet (operating under Provincial rules, they are not permitted to fillet). The largest fishing operations in the hake fishery are also the largest processors. The sector is now enjoying a "high".

Analysis

The story of BC hake takes many twists because hake is a pure commodity product, highly subject to the shifting forces of globalisation. Globalisation has presented opportunities, snuffed them out, and offered entirely new prospects. Further challenges are clearly on the horizon.

While the BC hake sector is currently enjoying strong results, the industry is clearly vulnerable to shifting global supply and demand conditions. At least half of production is put up in h&g form. The h&g business is highly sensitive to small changes in price, and demand shifts are likely as global fisheries production varies by jurisdiction. A primary appeal of hake for h&g processors is that they can enter and exit with little notice or risk. This suggests an opportunistic activity, not one that is sure to endure.

The largest firms in the hake sector have vertically integrated in vessels, quotas, plants equipment, and market development. They have invested heavily, though they are by no means "heavyweights" in the global seafood business. Fillets and other value added products make up an increasing portion of their product mix. They are attracted by the solid fundamentals of the fishery, including the management plan, the relatively long fishing seasons (May-October or longer), and the growing white fish market segment. This approach, accounting for less than half of the current hake catch, is the one that offers enduring business security.

The hake sector is still in a maturation process – after all the industry is brand new by historical standards – and needs to address outstanding issues:

- Recognising that the h&g sector of the industry is fragile. It emerged from nowhere, and could return there if globalisation circumstances conspire against it. Complacency would be a mistake… full utilisation in the future is not a given.
- The ongoing role, if any, of the Joint Venture fishery. There are those that feel it is anachronistic, and others who believe it is a necessary safeguard for the industry until industry has further matured. Either way, the annual acrimony of the "shore vs. JV" debate is an unproductive disruption.

Pacific Fisheries in the Global Environment

At the outset of this paper three themes of fisheries globalisation, drawn from OECD publications, were identified. Subsequently, globalisation in the BC seafood industry in general, and for three subject fisheries specifically, was examined. To conclude this paper, the BC experience is considered in light of globalisation "theory".

1) Fisheries globalisation brings tremendous economic opportunities and pressures to seafood businesses.

The BC experience fully supports this notion. As an exporter of seafood products, the BC seafood industry has always been subjected to global economic forces. In simpler days, a relatively few factors, such as exchange rates, Alaskan fishery volumes, interest rates, and general economic conditions, influenced results. Though entirely beyond the influence of seafood participants, these factors collectively comprised a tolerable level of risk.

The acceleration of fisheries globalisation, beginning with salmon aquaculture but including a lengthy list of factors, produced a far more complex and risky business environment prompting a fundamental re-structuring of the BC industry. The industry of old was largely dismantled. Capital intensive and speculative strategies were replaced by low risk, cash-flow-positive ones. The industry became by necessity nimbler and more responsive to markets. It also became more fractured, and less willing (and able) to summon capital.

In the various BC fisheries, impacts of fisheries globalisation are closely related to the applicable fisheries management framework. A fishery providing secure and reliable access to the resource, that supports business planning and market service, positions a fishery for success in confronting a complex and challenging environment. A fishery lacking these characteristics sets up a fisheries sector for failure.

In the pink salmon fishery, where fishery management is judged by participants to be a failure, fisheries globalisation has virtually grounded the business. Pink salmon, always a quiet contributor to salmon financial results, has become an economic non-entity. In the groundfish trawl fisheries – bottomfish and hake – an innovative fishery management plan has allowed industry to exploit the opportunities afforded by globalisation and to withstand the challenges. Though still vulnerable to shifts in the competitive landscape, the groundfish fisheries are only ten years into a unique management plan, and will continue to evolve.

2) Some jurisdictions retain legislation, regulation, policy, or domestic practices to encourage use of local resources to serve national socio-economic objectives.

British Columbia is also a prime example of this statement. In BC, the idea that Pacific fish stocks could be handled by a network of global service providers, with the functions of the value chain (harvesting, processing, marketing, distribution) carried out by the most efficient player, has received little support. The federal and provincial governments, and industry itself, view fish resources as an instrument of social, as well as economic, policy. In BC, the issues of "how fish are shared", and "who benefits from the fish", have received as much attention as the issue of "how to be viable in the fish business." Although globalisation requires seafood businesses – especially businesses that are dwarfed by global competition – to be efficient and flexible, the debate over fish in BC is dominated by non-economic considerations.

BC's industry was for many years insulated from some of the realities of the world marketplace; either sheltered by regulation, or un-concerned because of internal concerns. BC's approach to resource management has clearly placed constraints on its seafood industry's ability to compete in the face of accelerated fisheries globalisation. The demise of the salmon/herring sector, and the emergence of a host of fisheries like groundfish (bottomfish and hake) is in part a product of this approach. Today's industry is in some respects successful, but also highly vulnerable.

This is not to say that BC's approach is right or wrong. "Social" fish policy has preserved a mostly "made in BC" seafood industry. The industry may not be a world leader in technology or product development, and it is certainly not the most prosperous in the world (a glimpse at the BC Exhibit in the European Seafood Exposition confirms this), but it is largely owned and controlled by BC interests.

3) Fisheries globalisation can result in unsustainable fishing pressure.

With DFO adopting a precautionary fishery management approach in the late 1990s, and with the advent of environmental legislation such as the Species at Risk Act, BC fisheries are relatively safe from the pressures to over-harvest that may accompany fisheries globalisation.

A lack of responsiveness to markets, in this respect, is a virtue. Virtually all of BC's commercially exploited fish resources are in reasonably healthy condition. However, it is clear from BC's groundfish fishery that designing fisheries that mutually meet conservation and economic objectives is possible, and that doing so better positions businesses to meet the challenges, and exploit the opportunities, of globalisation.

Outlook for Pacific Fisheries

The fundamentals of the BC seafood business – healthy resources, growing global demand for seafood, and a strategic location for servicing Pacific Rim markets – suggest boundless opportunities for the industry. The tendency of this exporting industry to focus inward, to allow resource access and allocation conflicts to fester, and to struggle over an acceptable balance between social and economic objectives, suggests an industry stifled from realizing its full potential. The recent track record of this industry composed of players that are diminutive on a global scale is, indeed, spotty; there are sectors performing well, such as groundfish and hake, and sectors performing poorly, such as pink salmon.

A common-denominator of now-prosperous fisheries sectors is that they have resolved internal differences, and established effective, cooperative relationships with regulators and members of the value chain. A sustainable, co-managed fishery provides a foundation from which to forge viable businesses. Participants that build on a solid fishery foundation by engaging in meaningful business development initiatives, such as product and market development and best-practices in fish-handling and processing, stand the best chance of flourishing in an ever-changing global competitive environment. Sectors failing to adopt a cohesive, cooperative, business-minded approach are destined to flounder.

ANNEX A

BIOGRAPHIES OF SPEAKERS AND EXPERTS

Richard **BATES** (Administrator, Aquaculture Unit of the Fisheries and Martime Affairs, European Commission) has worked for almost 12 years with the European Commission with aquaculture policy, principally in relation to health, environment and trade. He studied marine biology and later food science at UCD Dublin and previously worked for over 16 year in inshore fisheries and aquaculture development with BIM – the Irish Sea Fisheries Board.

Roger **BING** (Vice President, Protein Purchasing, Darden Restaurants, United States) is Vice President of Protein Purchasing for Darden Restaurants, the world's largest casual dining restaurant company. Before joining Darden, Bing spent a good part of his career with Union International Pty., an integrated global primary protein production and marketing company. With a Bachelor of Science degree in Animal Science & Natural Science, he has held numerous leadership positions with significant industry organizations including the National Fishery Institute, the Global Aquaculture Alliance and the International Groundfish Forum.

Christina **BURRIDGE** (Executive Director, the BC Seafood Alliance). The Alliance is an umbrella association representing fisheries on Canada's West Coast. Its 17 members account for about 90% of the value of commercially harvested seafood production in British Columbia, about CAD 750 million annually. The Alliance advocates for the sustainable, profitable management of marine resources in order to supply domestic and local markets with high quality seafood. Christina is a member of numerous Canadian fishery and seafood advisory committees.

John **CONNELLY** (President, National Fisheries Institute (NFI), United States). The NFI is America's leading trade association advocating for the fish and seafood community, representing the fish and seafood commerce chain from "water to table". NFI lobbies Congress and regulatory agencies, serves as the seafood community's spokesperson with the media, and provides technical advice to its members. He served as Chairman of International Coalition of Fisheries Associations and helped found the International Coalition of Aquaculture Associations. Connelly has recently been named to the Board of the Marine Stewardship Council.

Kristjan **DAVIDSSON** (Managing Director of Corporate Banking, Glitnir Bank, Iceland) is currently Managing Director of Corporate Banking in Glitnir Bank Iceland, a Nordic specialist bank with a leading position within the global seafood industry. He holds a Fishing Captains degree from Nautical College of Iceland and a M.Sc. in Fisheries from University of Tromsoe, Norway. Mr. Davidsson has extensive business experience within international seafood and fishing industries, processing, sales, consulting and management. He has been worker at processing plants, crew member of fishing vessels, sales manager and managing director of seafood export companies in Norway and Iceland and CEO of leading Icelandic seafood company HB Grandi. He has served as Director on the Board of Icelandic and Chilean aquaculture and fishing/processing companies.

Sena S. **DE SILVA** (Director General, Network of Aquaculture Centres) is currently Director General, Network of Aquaculture Centres in the Asia-Pacific region and Adjunct Professor, Deakin University, Victoria, Australia With over 35 years of experience in academia, and aquaculture and inland fisheries management research and development, he has held academic positions in universities in Sri Lanka, Stirling, Scotland, Singapore and Australia. He was responsible for development and delivery of post graduate courses in aquaculture in the "distance mode". He is internationally reputed researcher in fin fish nutrition and reservoir fisheries, with expertise in fish introduction and biodiversity in relation to aquaculture. He serves on the editorial board of the journal, Aquaculture International, Aquaculture Research, Fisheries Management and Ecology. Recipient of awards, including the NAGA Award (ICLARM), Deakin University Vice Chancellor's award for "Best Researcher", Asian Fisheries Society Gold Medal and Honorary Life Member of the World Aquaculture Society. Was a founding member of the Asian Fisheries Society and served in the Council for nine years.

Nancy **GITONGA** (Fisheries Consultant, Fishafrica, Kenya) is a Fisheries Consultant for FishAfrica and member of the Board of Directors of the Kenya Forest Service. She was Kenya's Director of Fisheries from 2000 to 2006 and held senior fisheries management positions in the Kenya Government for over 20 years. She has published in a number of FAO publications, and also in a book entitled "Public Participation in the Governance of International Fresh Water Resources" edited by Curl Bruch *et al.* She has received three recognition awards; International Association for Fish Inspectors (IAFI) "Government Inspector of the Biennium"; The Kenyan Presidential Award "Order of the Grand Warrior"; Lake Victoria Fisheries Organization (LVFO) Appreciation Award for contribution to the sustainable fisheries management of Lake Victoria. She is also Member of MSC Developing World Advisory Group and has obtained a Post Graduate Diploma and M. Phil Degree in Food Science and Technology- Fish Option, BSC Degree in Zoology and Biochemistry.

Einar Kristinn **GUÐFINNSSON** (Minister of Fisheries, Iceland) became Minister of Fisheries for Iceland as of September 2005. He was elected to the Parliament in 1991 and has served in the Parliament's finance committee, the social committee and the agriculture committee and the Parliament's committee for foreign affairs. e has been chairman of the fisheries committee and the economics and trade committee. He has also been party group chairman of the Independence Party. He has participated in the international cooperation of Parliamentarians, IPU and has been chairman of its Icelandic group.

Peter **HAJIPIERIS** (Seafood Procurement Strategy Manager, Tesco, United Kingdom) completed studying food technology in 1980, and with an early career in Restaurant management and then Production Management, he then joined the UK Seafish Industry Authority in 1985 as a Quality Assurance (QA) Regional Officer. He led Total Quality Management and QA standards across onshore sectors, launching processor & retailer quality award schemes. He then became QA controller for England, Wales & Northern Ireland, and Supermarket Account Manager for all UK supermarkets. Mr Hajipieris joined Sainsbury's in 1999 as business Development manager for fish, then in 2001 headed the Fresh Foods Training & operational Support Unit, and in 2003 was the Head of Retail Standards. In 2007 he joined TESCO as Commercial Manager of Fish & Fresh Foods, and then led a Group Level Policy role for seafood across 2 800 stores in Chilled / Frozen / Grocery – Canned.

Abraham **IYAMBO** (Minister of Fisheries and Marine Resources, Namibia) has been Namibian Minister of Fisheries and Marine Resources since 1997. From 1995-1997 he held the post of Deputy Minister of Fisheries and Marine Resources and has been appointed as non-voting Member of the National Assembly by President Nujoma, 1995. He has also done extensive consulting work including for the Ministry of Education and Culture of Namibia and for UN Food and Agricultural Organization (FAO).

Poul Melgaard **JENSEN** (Head of Section, Danish Seafood Association) has been a daily leader of the Secretariat in the Danish Seafood Association since June 2005, which comprises 65 trading and processing members located in Denmark with high global orientation. There, he has been an advisor in trade and food security aspects of fish and fish products since 2001. He is Secretary for the Board of Directors (and member since 2002) in the Seafood Importers and Processors Alliance (SIPA). He has four years experience (1996 - 2000) in Brussels with the Confederation of Danish Industries responsible for Trade Policy *vis a vis* EU institutions.

G. Mohan **KUMAR** (Chairman MPEDA, India) obtained an M.Sc. (Chemistry), MBA (U.K) has worked for nearly 27 years in various Government assignments involving public administration, rural development, industrial promotion, etc. was Director (Marketing) of the Marine Products Export Development Authority (MPEDA) from 1995 until 2000. As National Expert (Investment Promotion) in UNIDO 2000-2001, he then worked as Commissioner for Commercial Taxes in the state of Orissa, India. He has been working as Chairman, The Marine Products Export Development Authority since April 2004.

Michael **LODGE** (Counselor, OECD Ministerial Task Force on IUU Fishing) has a long association with the Law of the Sea and fisheries. His positions have included serving as legal Counsel to the South Pacific Forum Fisheries Agency, legal advisor to the International Seabed Authority and Executive Secretary of the Conference to establish the Western pacific Tuna Commission. He has also written widely on Law of the Sea and oceans related issues. Previously with the Secretariat of the High Seas Task Force, Mr Lodge is presently working for the Roundtable on Sustainable Development at the OECD, and is also the Director of an independent panel set up to develop a model for best practice for Regional Fisheries Management Organisations.

Alastair **MACFARLANE** (General Manager – Trade and Information, New Zealand Seafood Industry Council) has held senior management positions with seafood industry organisations for the last fifteen years. Since 1997 he has headed the Seafood Industry Council's Trade and Information Business Unit, with personal responsibility for all international fisheries and trade policy issues. He has line management responsibility for information service delivery to the industry and public. He is a member of the Seafood Standards Council, a joint committee between the seafood industry and the New Zealand Food Safety Authority that consults on food safety in the seafood sector.

Thomas **MAEMBE** (Executive Secretary, Lake Victoria Fisheries Organisation, Tanzania) with a Bachelor of Science in Chemistry and Zoology, and Diploma in Food Science and Technology, Thomas MAEMBE has held positions at the Ministry of Natural Resources and Tourism, Tanzania, including Chief Fisheries Officer from 1974 until 2001 when he became Director of Fisheries. Mr. Maembe is currently Executive Secretary of the Lake Victoria Fisheries Organisation, which is an intergovernmental institution for coordination management measures for Lake Victoria.

Alda B. **MÖLLER** (Seafood Industry Consultant, Iceland) received a B.Sc. honours degree in Food Science and Biochemistry, 1st. class, 1972 from the University of Leeds, England and a Ph.D. degree in Food Science 1976 from the University of Reading, She was Lecturer and later Assoc. Professor in Food Science at the University of Iceland 1978-1986. She was Manager at the Icelandic Freezing Plants Corporation in Iceland 1986-1999, where her responsibilities included marketing retail products, public relations and R&D management for the corporation, which at the time was the largest seafood exporter from Iceland. She has worked as a seafood industry consultant since 1999. Chief projects in this capacity include the Groundfish Forum, where she has been secretary and editor of Proceedings since 1992. Other projects include management of the website www.fisheries.is. for the Icelandic Ministry of Fisheries, management of projects in co-operation with FAO and other international clients and various projects for the Icelandic seafood industry.

Stephen Mbithi **MWIKYA** (Executive Director, AFIPEK, Kenya) is presently a consultant on fisheries trade and development issues working from Nairobi, Kenya. He was the Chief Executive of the Kenya Fish Processors and Exporters Association for about 6 years from 2001 to 2006. Dr Mbithi has also worked as a consultant with COMESA, a regional trading block, Government of Namibia, and organizations such as UNEP and ICTSD. He has written widely on areas such as Sanitary and Phytosanitary Standards, Fishing Access agreements, and project evaluation.

Peter **REDMOND** (Vice President, Wal-Mart Stores, Inc., United States) has been with Wal-Mart for 16 years during which time the Supercenter group has grown from 20 stores to over 2 200 stores. Starting as an hourly associate in 1991, Peter has also held positions as a buyer in the Service Deli and Dairy departments and spent two years in the International Division serving in both Germany and the UK. Current responsibilities include supervising purchasing, marketing and financial planning for the Service Deli, Pre-packed deli, and frozen meat department, fresh and frozen Seafood as Vice President/Divisional Merchandise Manager Seafood and Deli.

Thiraphong **TANGTHIRASUNAN** (President, Bank for Agriculture and Agricultural Cooperatives (BAAC), Thailand) the objectives of which are to provide financial assistance in order to promote agriculture as an occupation for farmers and farmer institutions, to help farmers and farmer institutions increase their productivity and incomes and to offer deposit services to farmers throughout the country similar to those offered by commercial banks. He holds a Bachelor Degree (B. Sc.) in Marine Science from Kasetsart University, Thailand, and a Masters Degree (M.M.) in Agribusiness from the University of the Philippines at Los Banos.He is Advisor to the House of Representatives Standing Committee of Environment, and also the Acting Managing Director of the Thai Farmers Business Promotion Co.,Ltd.

Makhtar **THIAM** (Secretary General of the Employer's Union of Fish Exporters of Sénégal (UPAMES), Sénégal) since 1993.He is the holder and the General Manager of the society named Tropic Fish created in 1993, based in Dakar and specializing in seafood export from Senegal (West Africa) to Europe. In 1992, with some partners he created a society named Okeanos, based in Rungis (France), specializing in import and distribution of seafood in Europe. He was the buyer for the products from Africa (Mauritania, Senegal, and Guinea).Mr. THIAM studied and graduated in physics, chemistry and pollution of air and water at the Universities of Orsay and Paris 6 (Jussieu) in France, and taught there while living in Paris for 10 years.

Cherif Ould **TOUEILIB** (Independent International Consultant, Mauritania), obtained a Doctorate (3rd cycle) in fisheries economics, Master of Science and has occupied positions such as: Director of Fisheries, Director of Studies and Development, Technical Counsellor for the Ministry of Fisheries and Deputy Director General of Research. At the regional level: President of the steering committee of the SRFC (Sub-Regional Fisheries Commission) and of the Regional Surveillance Project for Fisheries. He is currently an independent international consultant.

ANNEX B

LIST OF PARTICIPANTS[1]

Chair	**Ms. Lori RIDGEWAY** Director General Fisheries and Oceans Canada OTTAWA. Ontario KIA OE6 Canada
Austria	**Ms. Birgit HELL** Counsellor for Agriculture Permanent Delegation of Austria to the OECD
Canada	**Dr. Robert DAY** A/Director, International Policy and integration Fisheries and Oceans Canada
	Ms. Kelly MOORE International Policy Analyst Fisheries and Oceans Canada
Denmark	**Mr. Lars B.F. POULSEN** Head of Section Ministry of Food, Agriculture and Fisheries
	Mr. Carsten PHILIPSEN Minister Counsellor Permanant Delegation of Denmark to the OECD
	Mr. Jonhard ELIASEN Embassy Counsellor Mission of the Faroes to the European Union

1. This list does not include the speakers at the Workshop. However they can be found under Annex A where short biographies are included.

Finland

Mr. Harry KAASINEN
Senior Officer
Department of Fisheries and Game
Ministry of Agriculture and Forestry

Mr. Kimmo NÄRHINEN
Counsellor
Permanent Delegation of Finlnad to the OECD

France

Mme Servane GILLIERS
Chargée de mission
Direction des Pêches maritimes et de l'Aquaculture (DPMA) / Bureau de l'économie des pêches
Ministère de l'Agriculture et de la Pêche

Mme Marie-José LLEDO
Chargée de mission
MAP-DPMA
Ministère de l'Agriculture et de la Pêche
Bureau pisciculture

Mr. Philippe PAQUOTTE
Chef de Division
Office interprofessionnel des produits de la mer et de l'aquaculture

M. Jean-Jacques BENEZIT
Conseiller Economique et agricole
Agriculture, environnement, tourisme, transports
Délégation permanente de la France auprès de l'OCDE

Germany

Dr. Markus BRILL
Deputy Head of Division
Federal Ministry of Food, Agriculture and Consumer Protection (BMELV)

Mr. Bernd KREMER
Head of Unit
Federal Office for Agriculture and Food

Dr. Matthias KELLER
Bundesverband der deutschen Fischindustrie und des Fischgrosshandels e.V.

Iceland	**Mr. David Olafur INGIMARSSON** Economist Ministry of Fisheries
Italy	**Mme Paola COLITTI** Commercial Attaché Permanent Delegation of Italy to the OECD
Japan	**Mr. Yasuhiko KURASHIGE** Counsellor, Agriculture Permanent Delegation of Japan to the OECD
	Mr. Mitsuaki SHINDO First Secretary Agriculture, Forestry and Fisheries Permanent Delegation of Japan to the OECD
	Dr. Masahiko ARIJI Principal Analyst AMITA Institute for Sustainable Economies
	Mr. Atsushi TAKAHARA AMITA Institute for Sustainable Economics
Korea	**Mr. Dohyung KOO** Deputy Director Ministry of Maritime Affairs
	Dr. Jung-Hee CHO Associate Research Fellow Korea Maritime Institute (KMI)
Mexico	**Mr. Gerardo BRACHO CARPIZO** First Secretary ELSAC, DEV Permanent Delegation of Mexico to the OECD
Netherlands	**Mme Lucie WASSINK** Ministère de l'Agriculture de la Nature et de la Quallté des Aliments
	Mr. Ancel VAN ROYEN Agricultural Counsellor Permanent Delegation of the Netherlands to the OECD

New Zealand	**Ms. Jane WILLING** Manager International and Biosecurity Ministry of Fisheries
Norway	**Mr. Jan Frederik DANIELSEN** Deputy Director General Ministry of Fisheries and Coastal Affairs
	Mr. Dag Erling STAI Deputy Director General Ministry of Fisheries and Coastal Affairs
	Ms. Nina DRANGE Adviser Ministry of Fisheries and Coastal Affairs
	Mr. Sverre JOHANSEN Higher Executive Officer Ministry of Fisheries and Coastal Affairs
	Ms. Anita STEINSEIDE Senior Adviser Directorate of Fisheries
Portugal	**Mr. Pedro LIBERATO** Counselor Permanent Delegation of Portugal to the OECD
Spain	**Mrs. Carmen ASENCIO** Sous-Directeur Général Adjoint des Relations Internationales de la Pêche Ministère de l'Agriculture, de la Pèche et de l'Alimentation
	Mr. Alberto LOPEZ GARCIA-ASENJO General Director General Direction of Structures and Fisheries Markets Ministry of Agriculture, Fisheries and Food
	Ms. María Isabel ARAGÓN Chef de Service de Commercialisation de la Pêche Secrétariat Général de la Pêche Maritime Ministère de l'Agriculture, de la Pêche et de l'Alimentation
	Mr. Ramón FRANQUESA ARTES Conseiller Cabinet de l'économie de la mer Université de Barcelone

Spain (cont.)	**Mr. Vicente FLORES REDONDO** Conseiller pour l'Agriculture, la Pêche et l'Alimentation Délégation permanente de l'Espagne auprès de l'OCDE
	Miss Carmen GONZÁLEZ MARTÍN Attachée Bureau de l'Agriculture, de la Pêche et de l'Alimentation Délégation permanente de l'Espagne auprès de l'OCDE
Sweden	**Mr. Rolf AKESSON** Deputy Director Ministry of Agriculture & Fisheries
	Mrs. Malin WILHELMSSON Economist Swedish Board of Fisheries
Turkey	**Mr. Erdinç GUNES** Program Coordinator General Directorate for Agricultural Research
United Kingdom	**Mr. Andy GOODWIN** Economic Adviser -- Fisheries Policy DEFRA
United States	**Mr. Greg SCHNEIDER** Senior International Trade Specialist NOAA Fisheries, IA
	Dr. Rebecca LENT Director, Office of International Affairs NOAA Fisheries, IA
European Commission	**Mr. Antonio FERNANDEZ-AGUIRRE** Policy Desk Officer DG FISH B-1
	Mr. Miguel PENA CASTELLOT DG Fisheries and Maritime Affairs European Commission

OBSERVER ECONOMIES

Argentina	**Mr. Juan IRIARTE VILLANUEVA** Director Special Economic Affairs Ministry of Foreign Affairs, International Trade
	Mrs. Elisa CALVO Coordinator and Fisheries Expert Sous-Secrétariat à la Pêche et à l'Aquaculture
Chinese Taipei	**Mr. Chin-Yaw WANG** Senior Specialist Fisheries Agency Council of Agriculture
	Mr. Tsung-Yi CHEN First Economic Secretary Taiwan Economic Delegation in France – CAPEC
	Ms. Anne Jing-Hwai CHOW First Secretary Taiwan Economic Delegation in France – CAPEC
Russian Federation	**Mr. Àlexander OKHANOV** Head, Aquatic Bioresources Department Federal Agency for Fisheries
Thailand	**Dr. Somying PIUMSOMBUN** Deputy Director General Department of Fisheries
	Mrs. Natthariya KIATPHAIBOOL Economist Department of Fisheries

OBSERVER ORGANISATIONS

European Free Trade Association (EFTA)	**Mrs. Merete FAGERHAUG** Senior Officer Trade Relations Division
ITF, Seafarers, Fisheries & Inland Navigation Sections	**Ms. Katie HIGGINBOTTOM** Senior Section Assistant
U.N. Environment Programme (UNEP)	**Mr. Charles ARDEN CLARKE** Senior Programme Officer, Division of Technology, Industry and Economics
	Ms. Emilie BABUT Researcher Technology, Industry and Economics
World Bank	**Mr. Lidvard GRONNEVET** Senior Fisheries Specialist Environment Department
	Mr. Xavier VINCENT Fisheries Specialist Sustainable Development Department Africa Region

EXPERTS

African Union Commission	**Mr. Edson MPYISI** Head of Food Security and Agriculture Division African Union Commission Addis Ababa
ASEAN Secretariat	**Dr. Somsak PIPOPPINYO** Assistant Director
Bank for agriculture and agricultureal cooperatives	**Mr. Luck WAJANANAWAT** Senior Executive Vice President
	Mr. Nipath KUASAKUL Senior Vice President

Carrefour DMMG	**Mme Stéphanie MATHEY** Quality Manager Social and Environmental Affairs
Coalition for Fair Fisheries Arrangements	**Mme Béatrice GOREZ** Coordinator
Danish Seafood Association	**Mr. Morten Haugaard JACOBSEN** Advisor
Enda Diapol/REPAO	**Mr. Papa Gora NDIAYE** Programme Officer on Fisheries
Eurofish International Organisation	**Ms. Ann-Mari HARAM** Project Director
European Bureau for Conservation & Development (EBCD)	**Mr. Konstantinos KALAMANTIS** Fisheries Policy Officer
Federacion Espanola de Organizaciones Pesqueras	**Mr. Javier GARAT PEREZ** Secretary General
Féderation Nationale de Pêche	**Mr. Mohamed Mahmoud Ould SADEGH** Secretary General
FINDUS	**Mme Anna ROSLUND** Directrice Qualité
GEMBA Seafood Consulting	**Mr. Jens Henrik MOELLER** Partner
GEXPROMER	**Mr. Jaco Chan Kit WAYE** President COPEFRITO S.A.
International Fishmeal and Fish Oil Organisation (IFFO)	**Mr. Jean-François MITTAINE** Director Market and Trade Issues
Marine Stewardship Council (MSC)	**Mr. Rupert HOWES** Chief Executive

Marketing Seafood	**Mme Marie Christine MONFORT** Consultant
Namibia	**Mr. Michael MUTONGA** First Secretary Namibian Embassy
NEAFC	**Mr. Kjartan HOYDAL** Secretary
Norwegian Agency for Development Co-operation (NORAD)	**Ms. Kirsten BJORU** Senior Fisheries Officer
	Mr. Jan ERIKSEN Senior Adviser
Organisation for Environment and Sustainable Development (OPED)	**Mr. Jonas KEMAJOU SYAPZE** Director
Seachill	**Mr. Nigel EDWARDS** Technical Director
Seafood Services Autralia	**Ms. Jayne GALLAGHER** Business Development Manager
The Scottish White Fish Producers Association Ltd.	**Mr. Michael PARK** Executive Chairman
University of Delaware	**Mr. Lee ANDERSON** Policy Analyst, Professor College of Marine and Earth Studies
US Agency for Internaitonal Development (USAID)	**Mr. Richard VOLK** Coastal and Aquatic Resources Advisor

JOURNALISTS

INFOMER	**Mr. Bruno VAUDOUR** Journaliste
Intrafish Media AS	**Mr. Karl-Erik STROMSTA** Journalist

ORGANISERS

UN Food and Agricultural Organization (FAO)	**Mr. Grimur VALDIMARSSON** Director Fishery Industries Division
	Mr. William EMERSON Senior Fishery Industry Officer Fishery Industries Division
	Dr. Audun LEM Fishery Industry Officer Fishery Industries Division
	Mr. Angel GUMY Senior Fishery Planning Officer, Development Planning Service, Fishery Policy and Planning Division
OECD	**Mr. Stefan TANGERMANN** Director Trade and Agriculture Directorate
	Mr. Ken ASH Deputy Director Trade and Agriculture Directorate
	Mr. Carl-Christian SCHMIDT Head, Fisheries Policies Division Trade and Agriculture Directorate
	Mr. Anthony COX Senior Economist Fisheries Policies Division Trade and Agriculture Directorate
	Mr. Sibiri Jean ZOUNDI Head of Unit, Rural Transformation & Sustainable Development Sahel and West Africa Club
	Ms. Courtney Sara MINARD Socio-economist, Medium and Long-Term Development Perspectives Unit Sahel and West Africa Club
	Mr. Doan JEONG Project Manager Fisheries Policies Division Trade and Agriculture Directorate

OECD (Cont.)

Ms. Ingrid KELLING
Fisheries Economist
Fisheries Policies Division
Trade and Agriculture Directorate

Mr. Thomas BINET
Junior Consultant
Fisheries Policies Division
Trade and Agriculture Directorate

Mr. Simon GREGG
Fisheries Policies Division
Trade and Agriculture Directorate

Mrs. Emily ANDREWS-CHOUICHA
Fisheries Policies Division
Trade and Agriculture Directorate

OECD PUBLICATIONS, 2, rue André-Pascal, 75775 PARIS CEDEX 16
PRINTED IN FRANCE
(53 2007 04 1 P) ISBN 978-92-64-03777-9 – No. 55735 2007